MW00845504

Perspectives in Physiology

This fascinating series seeks to place medical science inside a greater historical framework, describing the main pathways of development and highlighting the contributions of prominent investigators.

More information about this series at http://www.springer.com/series/11779

John B. West

Essays on the History of Respiratory Physiology

 Springer

John B. West
Professor of Medicine and Physiology
School of Medicine
University of California, San Diego
La Jolla
California
USA

Perspectives in Physiology
ISBN 978-1-4939-2361-8   ISBN 978-1-4939-2362-5 (eBook)
DOI 10.1007/978-1-4939-2362-5

Library of Congress Control Number: 2014959856

Springer New York Heidelberg Dordrecht London

© American Physiological Society 2015
This work is subject to copyright. All rights are reserved, whether the whole or part of the material is
concerned, specifically the rights of translation, reprinting, reuse of illustrations, recitation, broadcast-
ing, reproduction on microfilms or in any other physical way, and transmission or information storage
and retrieval, electronic adaptation, computer software, or by similar or dissimilar methodology now
known or hereafter developed.
The use of general descriptive names, registered names, trademarks, service marks, etc. in this publication
does not imply, even in the absence of a specific statement, that such names are exempt from the relevant
protective laws and regulations and therefore free for general use.
The publisher, the authors and the editors are safe to assume that the advice and information in this book
are believed to be true and accurate at the date of publication. Neither the publisher nor the authors or the
editors give a warranty, express or implied, with respect to the material contained herein or for any errors
or omissions that may have been made.

Printed on acid-free paper

Springer New York is part of Springer Science+Business Media (www.springer.com)

Preface

This book consists of 23 essays on interesting people or events in the history of respiratory physiology. It does not attempt to cover the history of this topic in a comprehensive way but consists of events that I have found particularly fascinating. Many of the essays have to do with the history of pulmonary gas exchange, but some are in the related field of high altitude physiology, and one is on space physiology. All of the articles have previously been published in journals of the American or British Physiological Society. However in a few places some changes have been made to lessen duplication or to correct minor errors. The articles can stand alone but they have been arranged in approximate chronological order.

One of the reasons for writing the book is to try to interest medical and graduate students in history. Many are sadly ignorant. Of course the curriculums of both medical school and graduate school are very demanding, and many students cannot find time for anything else. However the hope is that some students will find these essays stimulating.

Several years ago I taught an elective course for medical and graduate students at the University of California San Diego Medical School using some of the topics covered here. The course lasted about 10 weeks (one quarter) and we met once a week for an hour or so. The format was that one of the students gave a presentation on an interesting person or topic, and this was followed by a class discussion.

One of the problems of this course was finding an appropriate text, and my hope is that this book would be suitable. A student could choose any of the topics that particularly interested him or her, present it to the class in 30 to 40 min, and this could be followed by a 20-min class discussion. This would be a painless way of exposing students to some aspect of the humanities in addition to medical science.

Although each essay is centered on some aspect of physiology, other broader topics are covered to some extent. For example, the first chapter on Galen and the beginnings of Western physiology deals, of course, with Galen's teachings and their links with ancient Greece, but goes on to discuss the medieval period and the challenges that scientists had in the Renaissance because of dogma held by the state and church. Another example is Chap. 2 on Ibn al-Nafis and the Islamic Golden Age. This introduces early views on the pulmonary circulation, but at the same time emphasizes the role of the Arab world in preserving and extending the Greco-Roman science during the dark ages of the medieval period in Western Europe. Another example is Chap. 11 on Joseph Priestley and the discovery of oxygen. Of course this is centered on his research, but it also explores Priestley's response to the Enlightenment which eventually resulted in him being hounded out of England and taking refuge in Pennsylvania. On a more modern note, Chap. 21 deals with the remarkable discoveries made by Fenn, Rahn and Otis at the University of Rochester during World War II, but it makes the point that this trio were very poorly prepared to work in respiratory physiology. However because they were well-trained in science, they responded superbly to the exigencies of war.

There is one feature of these essays that deserves some apology. They were written separately over a time span totaling 30 years. For example Chap. 8 on Stephen Hales dates from 1983 whereas some other essays were written this year. Initially there was no thought of bringing them together and, as a result, the reader will find some overlap. However each essay was written to stand alone, and this is perhaps an advantage if it is to the topic of a seminar as suggested above.

The essays fall into three groups. Numbers 1 through 14 generally follow the development of respiratory physiology up to the end of the eighteenth century. Then numbers 15 through 20 are on aspects of high altitude physiology in approximately chronological order. Finally the last three essays are about more modern topics.

It is worth pointing out that a few of the essays involved much more research than the others. Of course the famous scientists such as Boyle, Hooke, Priestley and Lavoisier have been the subject of many articles, and the essays here mainly assemble some of the most salient facts relative to their role in the history of respiratory physiology. By contrast, several essays such as those on Jourdanet and Finch, and in particular the two on Kellas and Ravenhill, are the result of extensive original research. In fact when the article on Kellas was written, it was the only substantial account in English about this important man, and there still is nothing elsewhere about Ravenhill.

I hope readers of the book will obtain some of the pleasure that I derived from writing the essays. So often a topic that appeared to be of minor interest at the beginning developed into an absolutely fascinating story. It is hoped that some of our medical and graduate students can share this.

So many people have contributed in various ways to these essays that it would be invidious to name any. However I would like to acknowledge the enormous help that I have received from the staff of the Biomedical Geisel Library at UCSD. In addition, my assistant, Keith Lander, helped greatly in preparing the material for publication. Some of the earlier essays were prepared with Amy Clay. I also received much help from the Publications Department at the American Physiological Society and Springer Press.

Information on Where These Essays were Originally Published

All the essays originally appeared in journals of the American or British Physiological Society. The references are as follows.

West JB. Galen and the beginnings of Western physiology *Am J Physiol Lung Cell Mol Physiol.* 307:L121-L128, 2014.

West JB. Ibn al-Nafis, the pulmonary circulation, and the Islamic Golden Age. *J. Appl. Physiol.* 105: 1877–1880, 2008.

West JB. Torricelli and the ocean of air: the first measurement of barometric pressure. *Physiology (Bethesda)* 28:66–73, 2013.

West JB. Robert Boyle's landmark book of 1660 with the first experiments on rarified air. *J. Appl. Physiol.* 98: 31–39, 2005.

West JB. The original presentation of Boyle's Law. *J. Appl. Physiol.* 87: 1543–1545, 1999.

West JB. Robert Hooke: Early respiratory physiologist, polymath, and mechanical genius. *Physiology (Bethesda)*, 29:222–233, 2014.

West JB. Marcello Malpighi and the discovery of the pulmonary capillaries and alveoli. *Am J Physiol Lung Cell Mol Physiol.* 304:L383–90, 2013.

West JB. Stephen Hales: neglected respiratory physiologist. *J. Appl. Physiol.* 57: 635–639, 1984.

West JB. Joseph Black, carbon dioxide, latent heat, and the beginnings of the discovery of the respiratory gases. *Am J Physiol Lung Cell Mol Physiol.* 306:L1057–1063, 2014.

West JB. Carl Wilhelm Scheele, the discoverer of oxygen, and a very productive chemist. *Am J Physiol Lung Cell Mol Physiol,* 307:L811–816, 2014.

West JB. Joseph Priestley, oxygen, and the Enlightenment. *Am J Physiol Lung Cell Mol Physiol.* 306: L111–119, 2014.

West JB. The collaboration of Antoine and Marie-Anne Lavoisier and the first measurements of human oxygen consumption. *Am J Physiol Lung Cell Mol Physiol.* 305: L775–785, 2013.

West JB. Henry Cavendish, hydrogen, water, and the weight of the earth. *Am J Physiol Lung Cell Mol Physiol.* 307:L1–6, 2014.

West JB. Humphry Davy, nitrous oxide, the Pneumatic Institution, and the Royal Institution. *Am J Physiol Lung Cell Mol Physiol* 2014 Aug 29 [307:L661–667, 2014].

West JB and Richalet J-P. Denis Jourdanet (1815–1892) and the early recognition of the role of hypoxia at high altitude. *Am J Physiol Lung Cell Mol Physiol.* 305: L333-L340, 2013.

West JB. Centenary of the Anglo-American High Altitude Expedition to Pikes Peak. *Exp Physiol.* 97:1–9, 2012.

West JB. Alexander M. Kellas and the physiological challenge of Mount Everest. *J. Appl. Physiol.* 63, 3–11, 1987.

West JB. Ravenhill and his contributions to mountain sickness. *J. Appl. Physiol.* 80: 715–724, 1996.

West JB. George I. Finch and his pioneering use of oxygen for climbing at extreme altitudes. *J. Appl. Physiol.* 94: 1702–1713, 2003.

West JB. Joseph Barcroft's studies of high altitude physiology. *Am J Physiol Lung Cell Mol Physiol.* 305: L523–529, 2013.

West JB. The physiological legacy of the Fenn, Rahn and Otis school. *Am J Physiol Lung Cell Mol Physiol.* 303: L845–51, 2012.

West JB. The physiological challenges of the 1952 Copenhagen poliomyelitis epidemic and a renaissance in clinical respiratory physiology. *J. Appl. Physiol.* 99: 424–432, 2005.

West JB. Historical aspects of the early Soviet/Russian manned space program. *J. Appl. Physiol.* 91: 1501–1511, 2001.

Contents

1 Galen and the Beginnings of Western Physiology 1
  1.1 Introduction 1
  1.2 Brief Biography 3
  1.3 Physiology in Ancient Greece 3
  1.4 Physiology of the Galenical School 6
  1.5 Galen's Legacy 9
  1.6 Andreas Vesalius and the Rebirth of Anatomy and Physiology 10
  1.7 Michael Servetus and His Assertion of the Pulmonary
     Transit of Blood 13
  1.8 William Harvey and the Beginnings of Modern Physiology 14
  References 15

2 Ibn Al-nafis, the Pulmonary Circulation,
  and the Islamic Golden Age 17
  2.1 Introduction 17
  2.2 Islamic Science in the Eighth to Sixteenth Centuries 18
  2.3 Ibn Al-Nafis 19
  2.4 Pulmonary Circulation 20
  2.5 Note on Sources 23
  References 24

3 Torricelli and the Ocean of Air: The First Measurement
  of Barometric Pressure 25
  3.1 Torricelli's Great Insight: The Ocean of Air 25
  3.2 Galileo's View on the Force of a Vacuum 28
  3.3 Gasparo Berti's Experiment with a Long Lead Tube 29
  3.4 Weighing the Air 31
  3.5 The Decrease of Barometric Pressure with Altitude 32
  3.6 Demonstration of the Enormous Force that can be
     Developed by the Barometric Pressure 33
  3.7 Subsequent Studies of the Effects of Reducing
     the Barometric Pressure 35
  References 35

4 Robert Boyle's Landmark Book of 1660 with the First
   Experiments on Rarified Air 37
   4.1   The Setting 38
   4.2   The Man 39
   4.3   The Book 40
   4.4   The Pump 42
   4.5   The Experiments 46
   4.6   Conclusion 52
   References 53

5 The Original Presentation of Boyle's Law 55
   Appendix 59
   References 60

6 Robert Hooke: Early Respiratory Physiologist, Polymath,
   and Mechanical Genius 61
   6.1   Introduction 61
   6.2   Brief Biography 62
   6.3   Air Pump and the First Experiments on Rarified Air 64
   6.4   Artificial Ventilation 64
   6.5   Human Decompression Chamber 65
         6.5.1   Micrographia 67
   6.6   Mechanical Inventions 70
   6.7   Architecture 73
   6.8   Diary 74
   References 75

7 Marcello Malpighi and the Discovery of the Pulmonary
   Capillaries and Alveoli 77
   7.1   The Man 77
   7.2   Discovery of the Pulmonary Capillaries 79
   7.3   Discovery of the Alveoli 82
   7.4   Insect Respiration 83
   7.5   Embryological Studies 85
   7.6   Botanical Studies 87
   7.7   Other Studies 88
   7.8   Malpighi's Difficulties 88
   7.9   Note on Sources 90
   References 90

8 Stephen Hales: Neglected Respiratory Physiologist 91
   References 97

9 Joseph Black, Carbon Dioxide, Latent Heat, and the
   Beginnings of the Discovery of the Respiratory Gases 99
   9.1   Introduction 99

9.2   Brief Biography.. 100
9.3   The Chemistry of Alkalis and Carbon Dioxide............................... 102
9.4   Latent Heat.. 107
9.5   Joseph Black and James Watt.. 109
References 111

10 Carl Wilhelm Scheele, the Discoverer of Oxygen, and a Very
Productive Chemist... 113
10.1   Introduction... 113
10.2   Brief Biography... 115
10.3   The Discovery of Oxygen... 116
10.4   Scheele's Other Discoveries... 121
10.5   Scheele's Death... 122
References 123

11 Joseph Priestley, Oxygen, and the Enlightenment 125
11.1   Introduction... 125
11.2   Brief Biography... 127
11.3   First Production of Oxygen... 129
11.4   Oxygen is Produced by Green Plants.. 133
11.5   Who Discovered Oxygen?... 134
11.6   Other Gases Discovered by Priestley.. 135
11.7   Two Revolutionaries: Lavoisier and Priestley................................ 136
11.8   Priestley's Contributions to Electricity 137
11.9   Priestley and the Enlightenment... 138
References 139

12 The Collaboration of Antoine and Marie-Anne Lavoisier
and the First Measurements of Human Oxygen Consumption 141
12.1   Introduction... 141
12.2   Antoine Lavoisier's Contributions to Respiratory Physiology 143
12.3   Contributions of Marie-anne Lavoisier... 146
12.4   Personal Backgrounds of Antoine and Marie-anne Lavoisier......... 156
References 158

13 Henry Cavendish (1731–1810): Hydrogen, Carbon Dioxide,
Water, and Weighing the World 161
13.1   Introduction... 161
13.2   Hydrogen... 163
13.3   Carbon Dioxide... 166
13.4   Composition of Atmospheric Air.. 166
13.5   Composition of Water.. 167
13.6   Electricity and Heat... 167
13.7   Density of the Earth 168
References 170

14 Humphry Davy, Nitrous Oxide, the Pneumatic Institution, and the Royal Institution 173
   14.1 Introduction 173
   14.2 Early Years 174
   14.3 The Pneumatic Institution 175
   14.4 The Royal Institution 179
   14.5 Further Chemical Researches 182
   14.6 Davy Safety Lamp 183
   14.7 Later Years and Michael Faraday 183
   References 185

15 Denis Jourdanet (1815–1892) and the Early Recognition of the Role of Hypoxia at High Altitude 187
   15.1 Introduction 187
   15.2 Brief Biography 190
   15.3 Jourdanet's High Altitude Studies 192
   15.4 Aerotherapy 196
   15.5 Criticism of the Claim by Jourdanet and Bert that Hypoxia is the Critical Factor in the Physiological Responses to High Altitude 198
   15.6 Relations between Jourdanet and Bert in Their Later Years 199
   References 200

16 Centenary of the Anglo-American High Altitude Expedition to Pikes Peak 203
   16.1 Acute Mountain Sickness 208
   16.2 Alveolar Gases During Acclimatization 208
   16.3 Arterial PO_2 and Comparisons with the Alveolar Values 209
   16.4 Respiratory Gas Exchange During Rest and Exercise 210
   16.5 Periodic Breathing 210
   16.6 Estimates of Blood Circulation Rate 210
   16.7 Blood Studies 211
   16.8 Principal Factors in the Process of Acclimatization 211
   16.9 Contributions of Mabel Purefoy Fitzgerald (1872–1973) 212
   16.10 Conclusion 215
   References 216

17 Alexander M. Kellas and the Physiological Challenge of Mt. Everest 219
   References 233

18 T. H. Ravenhill and His Contributions to Mountain Sickness 235
   18.1 Family and Early Years 236
   18.2 High-Altitude Studies 238

18.3   War Experiences... 245
18.4   Post World War I and Archeology.. 246
18.5   Painting 248
18.6   Conclusion... 249
References 251

**19 George I. Finch and His Pioneering Use of Oxygen
     for Climbing at Extreme Altitudes** 253
19.1   Use of Oxygen at High Altitude Before 1921............................... 254
19.2   Early History of G. I. Finch 255
19.3   Preparations for the Expedition of 1921 257
19.4   Expedition of 1922... 263
19.5   1924 and After.. 266
Appendix 1 268
Appendix 2 269
Appendix 3 271
References 272

20 Joseph Barcroft's Studies of High Altitude Physiology 273
20.1   Introduction.. 273
20.2   Glass Chamber Experiment of 1920 274
20.3   Toxicity of Hydrocyanide Gas 279
20.4   International High Altitude Expedition to Cerro De Pasco, Peru ... 280
20.5   Exercise at Extreme Altitude While Breathing 100% Oxygen....... 285
References 288

21 The Physiological Legacy of the Fenn, Rahn and Otis School 289
21.1   Introduction.. 289
21.2   Unlikely Beginnings... 290
21.3   The Initial Research Topic 291
21.4   Pulmonary Gas Exchange 296
21.5   Pulmonary Mechanics.. 298
References 302

**22 The Physiological Challenges of the 1952 Copenhagen
     Poliomyelitis Epidemic and a Renaissance in Clinical
     Respiratory Physiology**... 305
22.1   The Poliomyelitis Epidemic... 307
22.2   The Renaissance in Clinical Respiratory Physiology
       During the 1950s... 314
Appendix 319
Postscript 321
References 322

23 Historical Aspects of the Early Soviet/Russian Manned
   Space Program 325
   23.1   Tsiolkovsky, An Early Russian Space Visionary............................ 326
   23.2   Wernher von Braun and Peenemünde... 328
   23.3   Sergei Korolev, the Principal Architect of the Soviet/
          Russian Manned Space Program.. 330
   23.4   Flight of the Dog Laika, the First Living Creature in Space........... 334
   23.5   Flight of Yuri Gagarin, the First Human Being in Space................ 336
   23.6   Alexei Leonov Performs the First Extravehicular Activity............. 339
   23.7   Salyut and Mir, the First Permanently Manned Space Station 340
   Appendix... 341
   References... 341

Chapter 1
Galen and the Beginnings of Western Physiology

Abstract Galen (129-c. 216 AD) was a key figure in the early development of Western physiology. His teachings incorporated much of the ancient Greek traditions including the work of Hippocrates and Aristotle. Galen himself was a well-educated Greco-Roman physician and physiologist who at one time was a physician to the gladiators in Pergamon. Later he moved to Rome where he was associated with the Roman emperors Marcus Aurelius and Lucius Verus. The Galenical school was responsible for voluminous writings many of which are still extant. One emphasis was on the humors of the body which were believed to be important in disease. Another was the cardiopulmonary system including the belief that part of the blood from the right ventricle could enter the left through the interventricular septum. An extraordinary feature of these teachings is that they dominated thinking for some 1300 years and became accepted as dogma by both the State and Church. One of the first anatomists to challenge the Galenical teachings was Andreas Vesalius who produced a magnificent atlas of human anatomy in 1543. At about the same time Michael Servetus described the pulmonary transit of blood but he was burned at the stake for heresy. Finally with William Harvey and others in the first part of the seventeenth century, the beginnings of modern physiology emerged with an emphasis on hypotheses and experimental data. Nevertheless vestiges of Galen's teaching survived into the nineteenth century.

1.1 Introduction

Claudius Galenus (129-c 216 AD) (Fig. 1.1) who is universally known as Galen of Pergamon was a famous Greco-Roman physician and physiologist. He is an appropriate subject for this essay on the beginnings of Western physiology for several reasons. First he and his school had an extraordinary influence on medical science including physiology for about 1300 years. It is not easy to think of a comparable situation in any other area of science where one school dominated thinking for so long. In fact some medical students were still studying Galen's writings in the nineteenth century and some of the practices that he advocated, for example bloodletting, were still being used at that time.

© American Physiological Society 2015

J. B. West, *Essays on the History of Respiratory Physiology,*
Perspectives in Physiology, DOI 10.1007/978-1-4939-2362-5_1

Fig. 1.1 Galen of Pergamon
(Claude Galien in French).
Lithograph by Pierre Roche
Vigneron, Paris ca. 1865

CLAUDE GALIEN

Another reason for choosing Galen as an introduction to Western physiology is that he was heavily influenced by the teaching of the ancient Greeks. He therefore allows us an opportunity to summarize this important body of work. Admittedly many of these mainly theoretical musings of over two millennia ago do not resonate with present-day physiologists but some of the influences of this group can still be seen.

Finally the writings of Galen and his school were extremely voluminous and much of the material still survives. For example Karl Gottlob Kühn collected no less than 22 volumes [11]. Other large collections of Galen's writings also exist. Therefore there is a wealth of information about the man, his school, and his teachings.

The main purpose of this essay, as with the others in this book, is to introduce medical and graduate students to his work, and show how his teachings lasted up to the Renaissance. Then with the advent of Vesalius, Harvey, Boyle and many of their contemporaries, a sea change in attitudes occurred, and the beginnings of modern physiology are clearly seen. For graduate and medical students who would like an introduction to Galen, the short books by Singer [19; 20] are recommended. More recent extensive studies have been carried out by Nutton [12]. An article by Boylan [2] has a useful list of primary and secondary sources.

1.2 Brief Biography

Galen was born in Pergamon (modern day Bergama, Turkey) which at the time was a very lively intellectual center. The city boasted a fine library which had been greatly expanded by King Eumenes II, and it was only bettered by the famous library in Alexandria. Galen's father was a well-educated and affluent man who had high hopes that his son would continue in his own philosophical traditions. However a remarkable event narrated by Galen was that when he was about 16, he had a dream in which the god Asclepius urged his father to have his son study medicine. His father agreed and Galen was initially a student in Pergamon which was a famous medical center and attracted many sick people who could afford to get the best treatment. Three years later his father died leaving him wealthy, and he was able to travel widely and visit the most important medical centers including the outstanding medical school in Alexandria.

At the age of 28 Galen returned to Pergamon where he became a physician to the gladiators. This institution was run by the High Priest of Asia who was enormously influential. Galen spent 4 years treating the gladiators for their wounds and also emphasizing their training, fitness and hygiene. It is said that during his period, there were only five deaths among the gladiators while he was in charge, and this was an enormous improvement over the previous period when many gladiators died of their wounds. There is an interesting recent article on head injuries of gladiators whose bodies were exhumed from a cemetery in Ephesus, Turkey [9]. A feature was the large number of extensive fractures of the skull in spite of the fact that most gladiators are believed to have worn helmets. Perhaps the blows were delivered after the victims had received other serious injuries in order to put them out of their misery.

When he was 33 Galen went to Rome, then the center of the civilized Western world. However he recounted that he fell out with some of the prominent physicians there and, fearing that he might be harmed, he moved away from the city. A few years later he was recalled by the Roman emperors Marcus Aurelius and Lucius Verus, who ruled together, to serve in the army. A great plague broke out in Rome at that time and large numbers of patients developed severe skin lesions and many died. It is now thought that the disease was smallpox. Galen remained in Rome for the rest of his life and there has been much discussion on when he died. However many historians now believe that his death occurred in about 216 when his age was 87. This was an exceptionally long life in those times.

1.3 Physiology in Ancient Greece

As indicated above, Galen and his school were much influenced by earlier Greek thinking. Students of today often find it difficult to see the relevance of many of these developing ideas and there is some reluctance to grapple with them. However many vestiges of early Greek thinking remained in the work of the Galenical school, for example the notion of how four humors determine the medical status of an individual, and it is interesting to review how such concepts developed.

Historians often chose to start the beginnings of early Greek physiology with the work of Anaximenes (ca. 570 BC). He argued that "pneuma" (πνεύμα, Greek for breath or spirit) was essential for life. This is hardly surprising because death is often signaled by a cessation of respiration. However Anaximenes expanded the idea of pneuma which was seen as an all-pervading property that was essential for life everywhere. For example he stated "As our soul, being air, sustains us, so pneuma and air pervade the whole world" [20].

About a hundred years later, Empedocles (490–430 BC) wrote about the movement of blood, and he developed the idea that this ebbed and flowed from the heart in a reciprocating manner. A related notion was "innate heat" which was seen as a life-giving principle and was distributed by the blood throughout the body. Empedocles was also one of the first philosophers to suggest that all things are made up of four essential elements: earth, air, fire and water. This notion evolved into the philosophy of the four humors which persisted in different forms right up to the European Renaissance. Earth, air, fire, and water represented the concepts of solidity, volatility, energy, and liquidity. This idea was taken up a hundred years later by Aristotle (384–322 BC) and was still part of physiological dogma 2000 years later.

Hippocrates (c. 460–360 BC) was one of the giants of the ancient Greek period. His school produced an enormous volume of work known as the Hippocratic Corpus which was studied extensively until the European Renaissance. The emphasis here was on the practice of medicine rather than its physiological principles. For example this was the origin of the Hippocratic Oath which sets out ethical principles for physicians and is still often used in one form or another for graduating medical students.

Many clinical signs that are still taught to medical students can be found in the Hippocratic Corpus. For example there is a description of the succussion splash, that is the sound that can be heard if a patient with air and fluid in the pleural cavity or an abdominal viscus such as the stomach is moved from side to side. Another sign included in the Corpus is the pleural friction rub. This is the sound heard through a stethoscope, or the ear applied directly to the chest, when there is disease of the pleural membranes, and they move over each other during breathing with a rasping sound like sandpaper. Hippocrates also described some of the clinical features of pulmonary tuberculosis which was rife at the time. For example he stated that the disease was associated with fever, the coughing up of blood, and that it was usually fatal [8].

The Hippocratic Corpus also continued the belief, earlier enunciated by Empedocles, that the heart is the origin of innate heat and that the primary purpose of respiration is to cool this fiery process. Plato (428–348 BC) expanded on these views in his book *Timaeus* where he stated "As the heart might be easily raised to too high a temperature by hurtful irritation, the genii placed the lungs in its neighbourhood, which adhere to it and fill the cavity of the thorax, in order that their air vessels might moderate the great heat of that organ, and reduce the vessels to an exact obedience" [15].

Aristotle was not only the most eminent biologist of Greek antiquity but many would say that he deserved this accolade up to the time of the European Renaissance. He was a pupil of Plato, and incidentally also tutored Alexander the Great (356–323 BC). Aristotle had an inexhaustible curiosity and his writings on various animals give pleasure even today. One of his great strengths was in the classification of animals. It is said that he described 540 different species. Aristotle's colorful text *De Partibus Animalium* (*On the Parts of Animals*) [1] still makes enjoyable reading. For example here is his description of the elephant trunk. "Just then as divers are sometimes provided with instruments for respiration, through which they can draw air from above the water, and thus may remain for a long time under the sea, so also have elephants been furnished by nature with their lengthened nostril; and, whenever they have to traverse the water, they lift this up above the surface and breathing [sic] through it" [1]. It could be argued that Aristotle's contributions to systematizing biology were not equaled until the time of Carl Linnaeus (1707–1778). Aristotle's three great books on biology were *History of Animals*, *Parts of Animals*, and *The Generation of Animals*. His contributions include a diagram of the "ladder of nature" (scala naturae) shown in Fig. 1.2.

However although Aristotle had such remarkable insights into the diversity of nature, his footing in physiology was not always secure. For example although it had previously been concluded by others that the brain was the seat of intelligence, Aristotle made a backward step and elevated the heart to this status. As we saw, Empedocles had initiated this idea many years before. Also strangely, Aristotle believed that the arteries normally contained air. This error resulted from the fact that

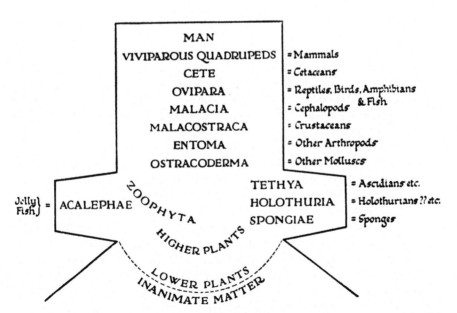

Fig. 1.2 The Ladder of Nature (scala naturae) of Aristotle demonstrating the great breadth of his interests in the whole animal kingdom, plants and inanimate materials. (From [19] by permission)

in preparing some animals for dissection by strangulation, the arteries were left virtually empty. Another interesting feature of Aristotle's beliefs was that living creatures were fundamentally different from inanimate objects because there was a special principle essential for life. This notion, known as vitalism, has recurred many times in the history of physiology, and it could be argued that it survived until Claude Bernard (1813–1878) finally put it to rest. Having said that, even the great British physiologist J.S. Haldane (1860–1936) believed that the lung secreted oxygen and his arguments supporting this were associated with a nod towards vitalism [5].

Erasistratus (c. 304–250 BC) was one of the last, and some would say the greatest Greek physiologist prior to Galen. He is credited with promulgating the pneumatic theory of respiration. This recognized the critical importance of inspired air but was curious in that it taught that air from the lungs passed by way of the pulmonary circulation to the left ventricle where it was endowed with "vital spirit". This was distributed by air-filled arteries to the various tisssues. Some of the vital spirit found its way to the brain where it was changed to "animal spirit" and then distributed via the hollow nerves to the muscles. Venous blood was believed to contain products of food, and this was modified by the liver and delivered to the right heart. As we shall see, a variant of this scheme was adopted by the Galenical school and dominated cardiorespiratory physiology for some 1300 years.

1.4 Physiology of the Galenical School

The teachings of the Galenical school were based on those of the ancient Greeks especially Hippocrates for whom Galen had enormous admiration, but also Aristotle, Plato and Erasistratus. Two major areas of teaching stand out. The first was the dominating effects of the four humors emanating from the four bodily fluids: blood, yellow bile, black bile and phlegm. This tradition closely followed the work of Hippocrates. Health was seen as a situation where the humors were equally balanced. An imbalance resulted in a particular type of temperament or disease. For example an overemphasis of blood led to a sanguine personality, too much black bile made the subject melancholic, an excess of yellow bile resulted in a choleric temperament, and too much phlegm caused the subject to become phlegmatic. We can easily see how the present use of these terms reflects the supposed pathological basis. People with a sanguine temperament were happy, optimistic, extraverted, and generally good company. We could wish we were all like that. People with a superabundance of yellow bile had excessive energy and were likely to have short tempers. By contrast, an excess of black bile resulted in melancholy, a subdued temperament, and perhaps a bipolar personality with periods of depression. Finally the phlegmatic personality was on a more even keel but perhaps with a tendency to occasional depression. On the other hand these people were affectionate.

We shall see later that this theory of the four humors had an enormous influence on medicine up to the Renaissance. For example the popularity of bloodletting was in part due to the belief that if one of the humors was dominating the patient, removal of some blood could reduce its influence. Even today we frequently characterize people based on their temperament and this is not so different from invoking one of the humors.

The second major area of teaching that influenced medical thinking right through to the Renaissance was Galen's scheme for the cardiopulmonary system. This is shown in Fig. 1.3 and clearly derives from the work of Erasistratus. In this concept food that was absorbed by the gut underwent "concoction" and then was transported by the blood to the liver where it was imbued with "natural spirit". The blood then entered the right ventricle of the heart and most of it flowed through the pulmonary artery to nourish the lungs. However some passed through "invisible pores" in the interventricular septum to the left ventricle where the blood was mixed with pneuma from the inspired air and thus endowed with "vital spirit". This air reached the left ventricle from the lungs via the pulmonary vein. "Fuliginous (sooty) wastes" traveled back from the heart to the lungs along the same blood vessel. From the left ventricle the blood with its "vital spirit" was distributed throughout the body in the arteries. Blood that arrived in the brain formed "animal spirit" and was distributed to the various organs of the body through the hollow nerves. As we have seen, much of this scheme was originally suggested by Erasistratus. Also there are clear parallels with modern views on pulmonary gas exchange. For example the addition of pneuma to the blood to provide vital spirit has similarities with the process of oxygenation, and the removal of fuliginous wastes via the lung reminds us of the elimination of carbon dioxide which is a product of metabolism.

Galen's scheme may seem strange to us today but it included some basic physiological principles having to do with the movement of substances through tubes. Admittedly some features of the scheme were difficult to understand. For example his notion that air entered the left ventricle from the lung via the pulmonary vein, but in addition, "fuliginous wastes" traveled back from the ventricle to the lungs along the same route puzzled Harvey 1400 years later [3]. He could not understand how a tube could carry flow in both directions simultaneously. Having said this, the great Harvey was confused about some aspects of the pulmonary circulation. For example he wondered why the lung needed such an enormous blood flow stating "it is altogether incongruous to suppose that the lungs need for their nourishment so large a supply of blood" [7].

In addition to his teachings on the importance of the four humors, and his elaborate scheme for cardiopulmonary function, Galen made many other contributions to physiology especially in the area of respiration. As has already been pointed out, Galen was the physician to the gladiators in Pergamon and as such he must have seen many serious injuries. One of the most interesting observations he made had to do with the effects of dislocations of the neck. He recognized that some of the injured men died immediately but others who were paralyzed below the neck were able to continue to breathe because the diaphragm was still active. Remarkably Galen was able to make sense of these observations using experiments on pigs. He found that when the spinal cord was cut halfway through affecting only one side, the muscles below this were paralyzed. Furthermore he was able to demonstrate that if

Fig. 1.3 Galen's cardio-pulmonary system which held sway for 1300 years. During inspiration, pneuma entered the lung through the trachea and reached the left ventricle via the pulmonary vein. Blood was formed in the liver, and imbued there with natural spirit, and entered the right ventricle. Most then entered the lung but a portion passed through minute channels in the interventricular septum to the left ventricle. Here vital spirit was added and this was distributed through arteries to the rest of the body. The blood that reached the brain was charged with animal spirit which was distributed through the hollow nerves. (From [19] by permission)

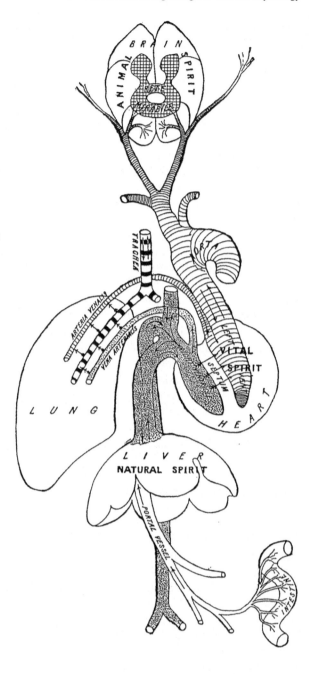

the spinal cord was completely cut at the level of the third cervical vertebra, the animal stopped breathing. However when the spinal cord was severed between the seventh and eighth cervical vertebrae, he showed that the animal continued to breathe with the diaphragm. Related to these studies, he reported that the diaphragm was controlled by the phrenic nerve that had its origin from cervical levels three, four and five. A good source for these observations of Galen can be found in Fulton [4].

The Galenical school made other contributions to the physiology of respiration. The diaphragm was seen as not only a partition between the thorax and the abdomen, but it was also recognized to be an important muscle of respiration. It was also stated that the abdominal muscles were used for forced expiration and also for the production of voice [16]. Galen described the larynx in some detail and remarkably reported that cutting the recurrent laryngeal nerve prevented squealing in pigs. He also understood how the lung expands when he stated "When the whole thorax expands in inspiration… [this] causes the entire lung to expand to fill the space left vacant". He reported that on inspiration, the airways increase both their caliber and length, and he recognized the importance of the upper airways in modifying the inspired air. He wrote that one of the functions of the nose and nasopharynx was to warm the inspired gas and filter out dust particles. He went on to say that the walls of the airways were lined with sticky substances that retain the dust that fell on them.

In addition to these very perceptive insights on physiology, the Galenical school made important advances in anatomy. Here they were at a disadvantage because dissection of human cadavers had been forbidden in Rome since about 150 BC. As a result Galen turned to the Barbary macaque ape, *Macaca inuus*, which resembles humans in many respects but of course is not identical. Other studies were carried out on pigs. The anatomical work was reported in 16 books. For description of bones, Galen had access to human skeletons and the long bones and spine were described in some detail. Pioneering work was done on the anatomy of muscles although necessarily most of this was on animals. Galen's description of the cranial nerves was the basis of teaching until the Renaissance. He stated that there were seven pairs of cranial nerves including the optic, oculomotor, trigeminal, facial, glossopharyngeal and hypoglossal nerves. In his work on the anatomy of arteries he corrected the error made by Erasistratus that the arteries contain air, showing by applying ligatures proximally and distally that they only contain blood.

Galen was a skilled surgeon and operated on many human patients. Some of the procedures that he initiated were not used again for many centuries. Remarkably he attempted to cure blindness caused by cataract by removing the opaque lens using a needle.

1.5 Galen's Legacy

In the years following Galen's death in 216 there was a dramatic decline in science in Western Europe. In 391 the great library in Alexandria where Galen had studied was destroyed by Christian fanatics. Shortly after, in 410, Rome was conquered by the barbarians and theologians such as St. Augustine of Hippo emphasized preparing for the afterlife rather than survive in the dismal conditions that prevailed at the time. In central Europe the centers of learning were mainly limited to the monasteries, and while theology was studied, science withered.

It was very fortunate however that much of the knowledge accumulated in Greek antiquity and expounded at great length by the Galenical school was picked up by the Islamic civilization. In fact the period from the eighth to the fifteenth century is sometimes described as the Islamic Golden Age. This is discussed at length in Chap. 2, particularly the contributions made by Ibn al-Nafis. This important figure built on the work of Galen but also was responsible for one of the first challenges to Galen's scheme of the cardiopulmonary system shown in Fig. 1.3. Ibn al-Nafis stated very vigorously that blood could not pass through the interventricular septum because there are no invisible pores. The statement is important because it shows that although the academics of the Islamic Golden Age are mainly praised for their work in preserving the advances of the Greco-Roman schools, they also challenged some of Galen's writings.

It is not easy at first to understand how Galen's writings developed such an enormous influence which lasted for 1300 years until the end of the sixteenth century. But it is a fact that the prodigious written output of the school became the official canon not only of medicine but of the Church itself. Remarkably his scientific edifice was seen as consistent with Christian dogma and indeed Galen's authority became so great that people who challenged his doctrine could be branded heretical by both Church and State. His books were copied and re-copied by innumerable scribes before the dawn of printing and many manuscripts are still extant. The fact that Galen's views were adopted as official canon of the Church is often not widely appreciated but it helps to explain why the Church reacted so violently to his critics such as Vesalius and Servetus when the European Renaissance began.

1.6   Andreas Vesalius and the Rebirth of Anatomy and Physiology

One of the first big challenges to Galen's teachings came from Andreas Vesalius (1514–1564) who published a splendid atlas of human anatomy in 1543 titled *De humani corporis fabrica* (On the fabric of the human body). This immediately challenged many of Galen's conclusions and resulted in a sea change in thinking about human anatomy. Vesalius founded anatomy as a modern science by reestablishing the experimental method. This inspired people to throw off the encumbrances of theological dogma and start thinking again for themselves.

Vesalius was born in Brussels, Belgium and received his early training in Louvain. He then moved to Paris where he carried out dissections, and when he was only 22 he began to have doubts about some features of the Galenical texts. These attitudes worried his director, Joannes Guinterius, who was an ardent Galenist and other people as well. At the age of 23 Vesalius moved to the University of Padua where one of his responsibilities was to conduct dissections for both the students and also the public. The university had a famous steeply raked lecture room for anatomy demonstrations which was built in 1594 and which remarkably still exists.

Fig. 1.4 Typical woodcut from *De humani corporis fabrica* by Vesalius. (From [21])

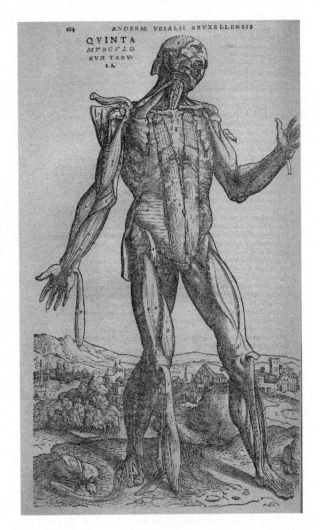

At the age of 28 Vesalius published his masterpiece *De humani corporis fabrica*, a magnificent production of nearly 700 pages. Many reproductions of the famous woodcuts exist. An accessible series is on the Web at https://tinyurl.com/k43mh3n. Another useful source is Saunders and O'Malley [17]. Figure 1.4 shows one of the most famous plates. Note that the artist has decorated the rather stark illustration of anatomy with part of the Italian countryside. The genius of this book was that for the first time the structure of the human body was systematically described and accurately depicted in detailed images. It is interesting that the book was published in the same year as that by Nicolaus Copernicus's *De revolutionibus orbium coelestium* (On the revolutions of the celestial spheres) that removed the earth from the center of the universe and thus helped to overturn the medieval teachings on cosmology.

Fig. 1.5 Michael Servetus
(1511–1553). He was the
first European to describe the
pulmonary transit and was
burned at the stake for heresy
as indicated in the upper
part of the image. (From an
engraving by Christian Friedrich Fritzsch (1719–1774))

Vesalius's masterpiece *De humani corporis fabrica* consists of seven books, the first being on bones and joints. An interesting tidbit is that a skeleton prepared by him is still extant. Vesalius was passing through Basel where his book had been printed several years before and he presented the skeleton to the University where it remains to this day and is probably the oldest anatomical preparation in existence. The seven books of *De humani corporis fabrica* deal with the skeleton, muscles, vascular system, nervous system, abdominal viscera, heart and lungs, and brain respectively. A number of differences from the structures described by the Galenical school are noted. Of particular interest to us are Vesalius' comments on the interventricular septum of the heart. Recall that in the Galenical scheme shown in Fig. 1.3, part of the blood from the right ventricle was thought to enter the left through pores in the septum. However in the second edition of *De humani*, Vesalius wrote "Not long ago I would not have dared to turn aside even a nail's breadth from the opinion of Galen the prince of physicians... But the septum of the heart is as thick, dense, and compact as the rest of the heart. I do not, therefore, know... in what way even the smallest particle can be transferred from the right to the left ventricle through the substance of the septum..." [18].

Vesalius was subjected to a number of attacks because his great book challenged the accepted views of Galen and the Church. At one stage the Emperor Charles V set up an inquiry to determine the religious errors in his work, and although Vesalius was not found guilty at that time, some attacks continued. Many people were unwilling to accept the new anatomy and clung to the teachings of Galen. A remarkable rebuke came from Jacobus Sylvius (1478–1555) with whom Vesalius had worked in Paris and who was an ardent Galenist. Sylvius stated in writing that the anatomy of the human body must have changed since Galen described it!

Fig. 1.6 Title page of William Harvey's book *Exercitatio anatomica de motu cordis et sanguinis in animalibus.* (From [6])

1.7 Michael Servetus and His Assertion of the Pulmonary Transit of Blood

Interestingly Vesalius did not remark on the possibility of the movement of the blood from the right to the left side of the heart through the lungs, although Galen recognized this as did Ibn al-Nafis. However a famous early statement of the pulmonary transit was made by Michael Servetus (1511–1553) who was a physician, physiologist and theologian. He was born in Villeneuve in northern Spain and studied law in Toulouse in southern France at the university there. Later he traveled to Paris to study medicine where his teachers included the anatomist Jacobus Sylvius, known for aqueduct of Sylvius in the brain, and Jean Fernel who apparently was the first person to introduce the term "physiology" into medicine.

Servetus has the distinction of being the first European to state categorically that blood could not pass through the interventricular septum and, in keeping with this, that it moved from the pulmonary artery to the pulmonary vein. This was the first assertion of the pulmonary transit. Curiously, the statement was made in a theological context in his book *Christianismi Restitutio* (The Restoration of Christianity). Servetus was concerned with how the God-given spirit could be spread throughout the human body. He argued that it did this by entering the blood in the lungs and was thus delivered to the left ventricle and from here to the rest of the body. He wrote about the pulmonary blood flow as follows "However, this communication [from the right to the left ventricle] is made not through the middle wall of the heart, as is commonly believed, but by a very ingenious arrangement the refined blood is urged forward from the right ventricle of the heart over a long course through the lungs; it is treated by the lungs, becomes reddish-yellow and is poured from the pulmonary artery into the pulmonary vein" [13]. As we have seen, Ibn al-Nafis made a similar statement 200 years earlier but it was probably not known to Servetus.

The writings of Servetus were very confrontational in many ways and he was accused of heresy by both the Church of Rome and the Protestant Calvinists. The main theological issue was that he believed that the manifestation of God in Jesus occurred at the moment of conception and was not eternal. He therefore referred to Christ as "the Son of the eternal God" rather than "the eternal Son of God". For this heresy he was condemned to be burned at the stake. The great physician William Osler [14] described the terrible event as follows. In the procession to the place of execution Servetus was exhorted by the pastor accompanying him to change his statement about Christ. After being bound to the stake he cried out "Jesus, thou son of the eternal God, have mercy upon me" but the fire was lit and he perished. Strange, is it not, that could he have cried, "Jesu, thou eternal Son of God" even at this last moment he would have been spared. Such were the monstrous attitudes of those times.

This ghastly episode in the chronicle of man's inhumanity to man reminds us of one more that took place about one hundred years later. Galileo Galilei (1564–1642), having observed the moons circling Jupiter, promoted the heliocentric view of the world first suggested by Copernicus in the same year as the publication of *De humani corporis fabrica*. As a result, the Holy Office of the Inquisition ordered Galileo to stand trial. He was found vehemently suspected of heresy, he was shown the instruments of torture, he was required to abjure, curse and detest his views, he was forbidden to publish anymore, and he remained under house arrest for the rest of his life. Little had changed in the hundred years.

1.8   William Harvey and the Beginnings of Modern Physiology

William Harvey (1578–1657) is a convenient figure with which to announce the emergence of modern physiology. The publication of *Exercitatio anatomica de motu cordis et sanguinis in animalibus* (An anatomical exercise on the motion of the heart and blood in living beings) in 1628 describing the circulation of the blood was a turning point in the history of physiology. The book was an example of the vigorous use of scientific method and the result was an abrupt acceleration of knowledge. During the ensuing 50 years a host of new advances in physiology were made and men such as Torricelli, Pascal, Boyle, Hooke, Malpighi, Lower, and Mayow all made important contributions. The attitudes towards hypothesis, experimental data, and other evidence were very different from the situation one hundred years before.

Having said this, it should not be thought that physiologists of the seventeenth century necessarily made a clean break with the past. For example John Aubrey (1626–1697) was a well-known writer, and his selection of biographies known as *Brief Lives* contained colorful accounts of many seventeenth -century luminaries. He met Harvey on several occasions in Oxford and elsewhere and in 1651 when Aubrey was contemplating a trip to Italy he wrote the following. "He [Harvey] was very communicative and willing to instruct any that were modest and respectful to him. And in order to my journey gave me, that is dictated to me, what company to keep, what bookes to read, how to manage my studies; in short, he bid me goe to the fountain head and read Aristotle, Cicero, Avicen [Avicenna], and did call the neoteriques shitt-breeches" [10]. Neoteriques referred to people with the latest ideas.

Finally it should be noted that although there was a great renaissance in physiology in the seventeenth century, vestiges of Galenism lasted well into the eighteenth and even nineteenth centuries. Bloodletting continued to be prescribed although increasingly leeches were used rather than venesection. Indeed there are images of bloodletting by venesection as late as 1860. Even today venesection is occasionally used for diseases such as hemachromatosis and polycythemia although of course the reason is not because of the humoral theory.

In conclusion, Galen was a key figure in the early history of Western physiology. The voluminous writings of his school were based on the advances of the classical Greeks including Hippocrates and Aristotle. An extraordinary feature of the Galenical school was that its influence on medicine and physiology lasted some 1300 years right up to the European Renaissance. One of the first challenges to the teachings of the school came from Andreas Vesalius who produced a magnificent book on human anatomy. Finally in the early seventeenth century William Harvey used modern scientific methods including hypotheses and reliance on experimental findings. This resulted in an enormous acceleration of new knowledge in the mid-seventeenth century. However vestiges of Galen's teachings could still be seen in the nineteenth century. It is not easy to find another example of a school whose teachings had such a long-lived influence as that of Galen.

References

1. Aristotle. De Partibus Animalium. Oxford: Clarendon; 1911. p. 658b.
2. Boylan M. Galen (130–200 C.E.). In: The internet encyclopedia of philosophy. http://www.iep.utm.edu/galen/.
3. Fleming D. Galen on the motions of the blood in the heart and lungs. Isis. 1955;46:14–21.
4. Fulton JF. Michael Servetus, humanist and martyr. New York: Herbert Reichner; 1953.
5. Haldane JS, Priestly JG. Respiration (preface). Oxford: Clarendon Press; 1935.
6. Harvey W. Exercitatio anatomica de motu cordis et sanguinis in animalibus. Frankfurt a. M.: William Fitzer; 1628.
7. Harvey W. The works of William Harvey, translated by Willis R. Philadelphia: University of Pennsylvania Press; 1989.

8. Hippocrates. Aphorisms. Translated by Thomas Coar. London: Valpy; 1822.
9. Kanz F, Grossschmidt K. Head injuries of Roman gladiators. Forensic Sci Int. 2006;160:207–16.
10. Keynes G. The life of William Harvey. Oxford: Clarendon Press; 1966.
11. Kühn KG, editor. Medicorum Graecorum opera quae exstant. Leipzig: C. Cnobloch; 1821–1833.
12. Nutton V. From democedes to Harvey. London: Variorum Reprints; 1988.
13. O'Malley CD. Michael Servetus. Philadelphia: American Physiological Society; 1953.
14. Osler W. Michael Servetus. London: Oxford University Press; 1909.
15. Plato. Timaeus, c. 360 BC. Translation from: Thomson T. Chemistry of animal bodies. Edinburgh: Adam and Charles Black; 1843. p. 604.
16. Proctor DF. A history of breathing physiology. New York: Dekker; 1995.
17. Saunders CM, O'Malley CD. The illustrations from the works of Andreas Vesalius of Brussels; With Annotations and Translations, A discussion of the plates and their background, Authorship and Influence, and A biographical sketch of Vesalius. Cleveland: World Pub. Co.; 1950.
18. Singer C. The discovery of the circulation of the blood. London: Bell; 1922.
19. Singer C. A short history of anatomy and physiology from the Greeks to Harvey. New York: Dover Publications; 1957.
20. Singer C. A short history of scientific ideas to 1900. London: Oxford; 1959.
21. Vesalius A. De humani corporis fabrica. Basel: Joannis Oporini; 1543.

Chapter 2
Ibn Al-nafis, the Pulmonary Circulation, and the Islamic Golden Age

Abstract Ibn al-Nafis (1213–1288) was an Arab physician who made several important contributions to the early knowledge of the pulmonary circulation. He was the first person to challenge the long-held contention of the Galen School that blood could pass through the cardiac interventricular septum, and in keeping with this he believed that all the blood that reached the left ventricle passed through the lung. He also stated that there must be small communications or pores [*manafidh* in Arabic] between the pulmonary artery and vein, a prediction that preceded by 400 years the discovery of the pulmonary capillaries by Marcello Malpighi. Ibn al-Nafis and another eminent physiologist of the period, Avicenna (ca. 980–1037), belong to the long period between the enormously influential school of Galen in the 2nd century, and the European scientific Renaissance in the sixteenth century. This is an epoch often given little attention by physiologists but is known to some historians as the Islamic Golden Age. Its importance is briefly discussed here.

2.1 Introduction

Ibn al-Nafis (1213–1288) was an Arab physician who made significant contributions to the early knowledge of the pulmonary circulation. However little has been written about him in the physiological literature. He forms a link between the early studies of the school of Galen (130–199) in the second century, and the European Renaissance scholars such as Michael Servetus (1511–1553), Realdus Columbus (1516–1559), Andreas Vesalius (1514–1564), and William Harvey (1578–1657). The intervening period of some 1300 years includes a section which is sometimes referred to as the Islamic Golden Age, is often largely ignored, and some of its contributions are emphasized here. Ibn al-Nafis was a remarkable man and deserves to be better known.

© American Physiological Society 2015

J. B. West, *Essays on the History of Respiratory Physiology,*
Perspectives in Physiology, DOI 10.1007/978-1-4939-2362-5_2

2.2 Islamic Science in the Eighth to Sixteenth Centuries

As alluded to above, there is a tendency for people who are interested in the history of physiology to move rapidly over the 1300 years between the flowering of the Greco-Roman school of Galen in the second century to the beginnings of the European Renaissance. One of the reasons for this is the extraordinary influence that Galen's teaching had for upwards of 1400 years. For example, when William Harvey was at Cambridge University in the late 1500s, part of his instruction included Galen's writings [11]. In fact some of Galen's teachings, for example on blood-letting, were still being followed in the eighteenth century. However with the blossoming of the scientific Renaissance in Europe in the fifteenth and sixteenth centuries, Galen's teachings were increasingly questioned by scholars such as Michael Servetus, Realdus Colombus, Juan Valverde (ca.1525–ca.1587), Andreas Vesalius, and finally William Harvey. There is therefore a temptation to ignore the intervening 1300 years or so.

One the other hand, some historians of science refer to the period from the eighth to the sixteenth centuries as the Islamic Golden Age. This terminology is inexact but is shorthand for the scientific activity that took place in a substantial area of Europe and Asia from the Iberian Peninsula and north Africa in the west, to the Indus Valley in the east, and from southern Arabia in the south, to the Caspian Sea in the north. Some scholars prefer the term "Arab science" because most of the documents were written in Arabic which was the lingua franca of the region. However not all the scientists were Arabs; indeed some of the most distinguished such as Avicenna (ca. 980–1037) were Persians. In addition although most of the scholars were Muslim, this was not true of all.

A number of important scholarly institutions developed in this period. Some of the most significant centers were Baghdad, Damascus, and Cairo. The institutions included groups of scholars in schools that were like emerging universities in that they were made up of collections of like-minded academics and teachers. There were also academic hospitals, libraries, and observatories. For example, Damascus where Ibn al-Nafis trained, boasted the Nasiri Hospital in the twelfth century which attracted many academic physicians including Ad-Dakhwar who amassed a large library of medical texts. According to one authority, the University of Al Karaouine in Fes, Morocco, can claim to be the oldest university in the world being founded in 859. Cairo had the Al-Azhar University which began in the tenth century and offered academic degrees.

A feature of these institutions was the emergence of polymaths, that is scholars who worked in a large number of different areas. We are certainly aware of this to some extent in the European Renaissance when people such as Robert Boyle (1637–1691) made important contributions in chemistry, physics, mechanics, and physiology. However as we shall see, Ibn al-Nafis wrote in a bewildering array of fields including physiology, medicine, ophthalmology, embryology, psychology, philosophy, law, and theology.

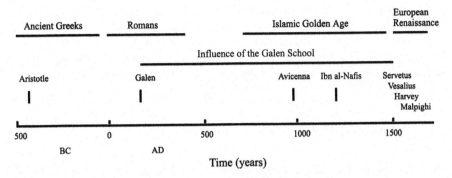

Fig. 2.1 Approximate time-line showing the period of the Islamic Golden Age, and the long influence of the teachings of the Galen School

One of the most important writings of Ibn al-Nafis was his *Commentary on Anatomy in Avicenna's Canon* (Sharh Tashrih al-Qanun Ibn Sina). Avicenna is usually referred to by his Latin name rather than Ibn Sina, and he was one of the most illustrious scholars of the period although he preceded Ibn al-Nafis by some 200 years. Avicenna was born in Persia in Bukhara Province, now part of Uzbekistan, and is sometimes spoken of as the father of modern medicine. His teachings persisted at many Islamic and European universities up to the early nineteenth century. He was particularly interested in clinical pharmacology, experimental physiology, infectious diseases and clinical trials, but also made contributions to physics. His most famous textbooks were *The Canon of Medicine*, and *The Book of Healing*. Because of political problems he was forced to move frequently as an adult but spent most of his life in what is now modern Iran. Avicenna was perhaps the most eminent scholar of the Islamic Golden Age (Fig. 2.1).

2.3   Ibn Al-Nafis

His full name was Ala al-Din Abu al-Hassan Ali Ibn Abi-Hazm al-Qarshi al-Dimashqi, and so not surprisingly he is commonly referred to as Ibn al-Nafis [13–15]. He was born in Damascus (or very nearby) in 1213 and had his medical education there at the Medical College Hospital (Bimaristan al-Noori). At the age of 23 he moved to Cairo where he first worked at the Al-Nassri Hospital, and subsequently at the Al-Mansouri Hospital, where he became physician-in-chief. When he was only 29 he published his most important work, the *Commentary on Anatomy in Avicenna's Canon* which included his ground-breaking views on the pulmonary circulation and heart that are referred to below [1–6, 8]. He also worked on an enormous textbook, *The Comprehensive Book of Medicine*. This was never completed but was the largest medical encyclopedia to be attempted at the time and is still consulted by scholars.

Ibn al-Nafis was an orthodox Sunni Muslim and, as mentioned above, wrote extensively in areas outside of medicine including law, theology, philosophy, sociology, and astronomy. He also authored one of the first Arabic novels translated as *Theologus Autodidactus*. This is a science fiction story about a child brought up on an isolated desert island who eventually comes in contact with the outside world.

2.4 Pulmonary Circulation

The teachings of the Galenical school about the pulmonary circulation were discussed at length in Chap. 1 and depicted there in Fig. 1.3. A feature was that some of the blood in the right ventricle passed into the left ventricle through "invisible pores" in the interventricular septum. The existence of these so-called pores was a puzzle to some anatomists but they were a necessary feature of the Galen scheme because it was not appreciated that a large amount of blood flowed from the lungs to the heart.

In his *Commentary on Anatomy in Avicenna's Canon*, Ibn al-Nafis made three important advances with respect to Galen's scheme:

1. He stated categorically that the interventricular septum between the right and left ventricles was not porous, and could not allow blood to travel through it as was critical in the Galen model (Fig. 1.3 in Chap. 1). Here is the English translation made by Meyerhof [15] of the section of the book by Ibn al-Nafis identified as fol. 46 r:

   but there is no passage between these two cavities [right and left ventricles]; for the substance of the heart is solid in this region and has neither a visible passage, as was thought by some persons, nor an invisible one which could have permitted the transmission of blood, as was alleged by Galen. The pores of the heart there are closed and its substance is thick.

This forceful denial of the permeability of the interventricular septum is also repeated elsewhere in the commentary. For example, in the section identified as fol 65 r and v, Meyerhof's translation is as follows:

   There is no passage at all between these two ventricles; if there were the blood would penetrate to the place of the spirit [left ventricle] and spoil its substance. Anatomy refutes the contentions [of former authors]; on the contrary, the septum between the two ventricles is of thicker substance than other parts in order to prevent the passage of blood or spirits which might be harmful. Therefore the contention of some persons to say that this place is porous, is erroneous; it is based on the preconceived idea that the blood from the right ventricle had to pass through this porosity--and they are wrong!

2. Since there is no communication between the right and left ventricles through the interventricular septum, it follows that the output of the right ventricle can only reach the left ventricle via the pulmonary circulation. In the section of the Commentary identified as fol. 46 r, Meyerhof's translation reads: "the blood after it has been refined in this cavity [right ventricle], must be transmitted to the left cavity where the [vital] spirit is generated." In the section identified as fol.

— ٧ —

قوله : « و فيه ثلاثة ^(٥) بطون » هذا الـكلام لا يصح فإن القلب له بطنان

فقط أحدهما مملوء من الدم وهو الايمن والآخر مملوء من الروح وهو الايسر ، ولا

منفذ بين هذين البطنين ^(٦) البتة و إلا كان الدم ينفذ إلى موضع الروح فيفسد

جوهرها . والتشريح يكذّب ما قالوه والحاجز بين البطنين أشد كثافة من غيره

لئلا ينفذ منه شيء من الدم أو من ^(٧) الروح فيضيع ^(٨) . فلذلك قول من قال ^(٩)

إن ذلك الموضع كثير التخلخل باطل ، والذي أوجب له ذلك ظنه أن الدم الذي

فى البطن الايسر إنما ينفذ اليه من البطن الايمن من هذا التخلخل وذلك باطل ،

فان نفوذ الدم الى البطن الأيسر إنما هو من الرئة بعد تسخنه وتصعده من البطن

الأيمن كا قررناه أوّلا .

Fig. 2.2 Section of the Arab text from the *Commentary on Anatomy in Avicenna's Canon* by Ibn al-Nafis dealing with the pulmonary circulation. This extract states that there is no connection between the two cavities of the heart [right and left ventricles] and that blood cannot pass through the [interventricular] septum. An interesting feature of this extract is that although it was written nearly 800 years ago it can be read by people familiar with Arabic today. (From [15])

65 r and v, the translation reads, "For the penetration of the blood into the left ventricle is from the lung, after it has been heated within the right ventricle and risen from it, as we stated before."

3. In a further short passage, Ibn al-Nafis states that there must be small communications between the pulmonary artery and the pulmonary vein. This was an inspired prediction of the existence of the pulmonary capillaries because these were not seen until 400 years later by Marcello Malpighi (1628–1694). Here is the translation of the relevant section in fol. 46 r, "And for the same reason there exists perceptible passages (or pores, *manafidh*) between the two [blood vessels, namely pulmonary artery and pulmonary vein]". Figure 2.2 reproduces part of the Commentary relating to the pulmonary circulation in the original Arabic text. Figure 2.3 is a statement by Ibn al-Nafis in his own hand that one of his students has satisfactorily understood one of his other writings.

It was not until 300 years later that scholars in Europe came to the same conclusion, that is that the blood had to pass through the pulmonary circulation and could not move directly from the right to the left ventricle. The first person to state this was Michael Servetus (1511–1553) who wrote

"However, this communication is made not through the middle wall of the heart, as is commonly believed, but by a very ingenious arrangement the refined blood is urged forward from the right ventricle of the heart over a long course through the

Fig. 2.3 A statement in his own hand by Ibn al-Nafis that one of his students named Shams al-Dawlah Abu al-Fadl ibn al-shaykh Abi al-Hasan al-Masihi has understood his commentary on a Hippocratic treatise. (From the National Library of Medicine, Bethesda, by permission)

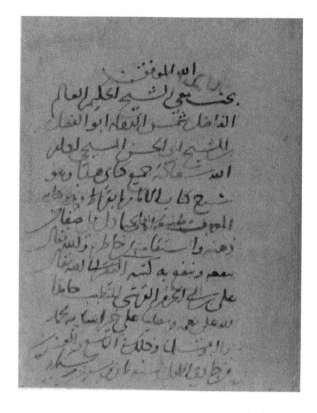

lungs; it is treated by the lungs, becomes reddish-yellow and is poured from the pulmonary artery into the pulmonary vein" (translation modified from O'Malley [16]). The Latin version is as follows. "Fit autem communicatio haec non per parietem cordis medium, ut vulgo creditur, sed magno artificio a dextro cordis ventriculo, longo per pulmones ductu, agitatur sanguis subtilis: a pulmonibus praeparatur, flavus efficitur: et a vena arteriosa, in arteriam venosam transfunditur" (text in Izquierdo [10] and Fulton [7] although the spelling is slightly different between the two).

As described in Chap. 1, the statement was actually made in a theological treatise, "Christianismi Restitutio" ("The Restoration of Christianity") which was considered heretical by both Catholics and Calvinists, and as a result Servetus was burned at the stake in Geneva. Only three copies of the book survive today. Later the same ideas were expressed by Realdus Columbus, his student Juan Valverde, Andreas Vesalius, and William Harvey.

However it is interesting that Harvey was initially puzzled by the physiology of the pulmonary circulation. For example he stated, "It is altogether incongruous to suppose that the lungs need for their nourishment so large a supply of blood, so pulsatorily delivered" [9]. In fact it was only a few years before his death that Harvey reported experiments where he conclusively demonstrated the passage of liquid from the pulmonary artery to the left ventricle. This was done in a letter to a German friend named Slegel. Harvey described how he had ligated various vessels in the lung of the cadaver of a man who had died as a result of strangulation, and showed how water passed freely through the lungs from the pulmonary artery to the left ventricle [11].

A fascinating sidelight on Ibn al-Nafis is how the manuscript *Commentary on Anatomy in Avicenna's Canon* was made known to the Western world only about 80 years ago. A young Egyptian physician, Muhyo Al-Deen el Tatawi, came across the manuscript in the Prussian State Library in Berlin in the course of writing his doctoral thesis for the medical faculty of Albert Ludwig's University of Freiburg im Breisgau, Germany [20]. The young doctor was subsequently employed by the Egyptian Public Health Service and transferred to small provincial towns where he could not carry out further research. Happily, Max Meyerhof, an eminent medical orientalist in Cairo, was made aware of the discovery and wrote a short commentary on Dr. Tatawi's thesis to save it from oblivion. This was fortunate because the thesis was never published and only five type-written copies of it were made. Meyerhof subsequently published German, French, and English translations of the relevant parts of the Commentary of Ibn al-Nafis. One of these contains a reproduction of the Arabic text [15].

A final interesting question is whether Michael Servetus whose book was dated 1553, and later Colombus, Valverde, Vesalius, and Harvey were aware of the work of Ibn al-Nafis on the pulmonary circulation published over 300 years before. The statements on the pulmonary transit of blood by Ibn al-Nafis and Servetus are rather similar and might suggest that the latter was aware of the earlier work although one study points out that there are some inconsistencies between the two statements [21]. One reason for raising the issue is that in 1547 there was a Latin translation of another book by Ibn al-Nafis entitled *Commentary on Compound Drugs*. This dealt with the last part of Avicenna's book *The Canon of Medicine* concerning lists of drugs. The translation was made by Andrea Alpago of Belluno but apparently it did not include any reference to the pulmonary transit of blood. Other translations of Alpago have also been discussed in this context [12, 17]. However, many historians believe that Servetus was not aware of the work of Ibn al-Nafis and if so, both deserve credit for the discovery of the pulmonary transit of blood.

2.5   Note on Sources

Many people have written short pieces about Ibn al-Nafis and so it might be useful to identify the primary sources that I have found helpful. First, there is the doctoral thesis by Tatawi, and the three articles by Meyerhof who was responsible for introducing the thesis to the scholarly community. Tatawi's thesis and Meyerhof's 1935b article in German both include reproductions of the Arab text. In addition the articles by Chehade, and Haddad with Khairallah add interesting information and also reproduce the original text. In fact the latter authors actually owned one of the original manuscripts. Bittar wrote a thesis on Ibn al-Nafis for Yale University and his three articles have much of interest. Sezgin [18] recently collected all these articles with others into a book which is invaluable.

References

1. Bittar EE. A study of Ibn Nafis. Bull Hist Med (Baltimore). 1955a;29:352–68.
2. Bittar EE. A study of Ibn Nafis. Part III: a study of Ibn Nafis's commentary on the anatomy of the Canon of Avicenna. Bull Hist Med (Baltimore). 1955b;29:429–47.
3. Bittar EE. The influence of Ibn Nafis: a linkage in medical history. Med Bull (Ann Arbor). 1956;22:274–8.
4. Chehade AK. Ibn an-Nafis et la decouverte de la circulation pulmonaire. Damascus: Institut Francais de Damas; 1955.
5. Chehade AK. Ibn An-Nafis et la decouverte de la circulation pulmonaire. Maroc Med (Casablanca). 1956;35:1013–6.
6. Coppola ED. The discovery of the pulmonary circulation: a new approach. Bull Hist Med (Baltimore). 1957;31:44–77.
7. Fulton JF. Michael Servetus: humanist and martyr. New York: Herbert Reichner; 1953.
8. Haddad SI, Khairallah AA. A forgotten chapter in the history of the circulation of the blood. Annals Surg (London/Philadelphia). 1936;104:1–8.
9. Harvey W. The works of William Harvey. Trans., R. Willis. Philadelphia: University of Pennsylvania Press; 1989.
10. Izquierdo JJ. A new and more correct version of the views of Servetus on the circulation of the blood. Bull Hist Med. 1937;5:914–32.
11. Keynes G. The life of William Harvey. Oxford: Clarendon Press; 1978.
12. Loukas M, Lam R, Tubbs RS, Shoja MM, Apaydin N. Ibn al-Nafis (1210–1288): the first description of the pulmonary circulation. Am Surg. 2008;74:440–2.
13. Meyerhof M. La decouverte de la circulation pulmonaire par Ibn al-Nafis, medecin arabe du Caire (XIIIe siecle). Bull l'Institut d'Egypte (Cairo). 1934;16:33–46.
14. Meyerhof M. Ibn An-Nafis (XIIIth cent.) and his theory of the lesser circulation. Isis (Philadelphia). 1935a;23:100–20.
15. Meyerhof M. Ibn an-Nafis und seine theorie des lungenkreislaufs. Quel Stud Gesch Naturwissenschaften Med (Berlin). 1935b;4:37–88.
16. O'Malley CD. Michael Servetus. Philadelphia: American Philosophical Society; 1953.
17. O'Malley CD. A Latin translation of Ibn Nafis (1547) related to the problem of the circulation of the blood. J His Med Allied Sci (Minneapolis). 1957;12:248–53.
18. Sezgin F. Ali ibn Abi I-Hazm al-Qarshi ibn al-Nafis (d.687/1288): texts and studies/collected and reprinted. Frankfurt a. M.: Institute for the History of Arabic-Islamic Science at the Johann Wolfgang Goethe University; 1997.
19. Singer C. A short history of anatomy and physiology from the greeks to Harvey. New York: Dover; 1957.
20. el Tatawi MD. Der Lungenkreislauf nach el Koraschi. Wortlich Iibersetzt nach seinem Kommentar zum Teschrih Avicenna. Med. diss., University of Freiburg, Germany; 1924.
21. Temkin O. Was Servetus influenced by Ibn an-Nafis? Bull Hist Med. 1940;8:731–4.

Chapter 3
Torricelli and the Ocean of Air: The First Measurement of Barometric Pressure

Abstract The recognition of barometric pressure was a critical step in the development of environmental physiology. In 1644, Evangelista Torricelli described the first mercury barometer in a remarkable letter which contained the phrase "We live submerged at the bottom of an ocean of the element air, which by unquestioned experiments is known to have weight". This extraordinary insight seems to have come right out of the blue. Less than ten years before, the great Galileo had given an erroneous explanation for the related problem of pumping water from a deep well. Previously Gasparo Berti had filled a very long lead vertical tube with water and showed that a vacuum formed at the top. However Torricelli was the first to make a mercury barometer and understand that the mercury was supported by the pressure of the air. Aristotle stated that the air has weight although this was controversial for some time. Galileo described a method of measuring the weight of the air in detail but for reasons that are not clear his result was in error by a factor of about two. Torricelli surmised that the pressure of the air might be less on mountains but the first demonstration of this was by Blaise Pascal. The first air pump was built by Otto von Guericke and this influenced Robert Boyle to carry out his classical experiments of the physiological effects of reduced barometric pressure. These were turning points in the early history of high altitude physiology.

3.1 Torricelli's Great Insight: The Ocean of Air

On June 11, 1644, Evangelista Torricelli (1608–1647) (Fig. 3.1) wrote a remarkable letter to his friend Michelangelo Ricci who was a mathematician and also a cardinal in Rome. Torricelli himself was a mathematician and physicist, originally from Faenza, but now in Rome. Both men were part of the extraordinarily effervescent scientific activity in Italy in the early and mid-seventeenth century. The letter contained the wonderful statement "We live submerged at the bottom of an ocean of the element air, which by unquestioned experiments is known to have weight" (Fig. 3.2). This has to be one of the most dramatic statements in the early history of atmospheric science, and therefore, by implication, in the early development of high altitude medicine and physiology. One wonders how many medical and graduate students today appreciate the fact that they are living at the bottom of a sea of

© American Physiological Society 2015

J. B. West, *Essays on the History of Respiratory Physiology,*
Perspectives in Physiology, DOI 10.1007/978-1-4939-2362-5_3

Fig. 3.1 Evangelista Tor-
ricelli (1608–1647). (From
(Lezioni d'Evangelista
Torricelli and available at
http://en.wikipedia.org/wiki/
File:Libr0367.jpg))

EN VIRESCIT GALILÆVS ALTER
Anagr.
EVANGELISTA TORRICELLIVS
Sereniſſimi M. Ducis Hetruriæ
Mathem.^{us} & Philos.^{us}
Obijt Anno Dom. MDCXLVII. Ætꞌ XL.

Fig. 3.2 Portion of the text
of Torricelli's letter to Ricci
containing the phrase "We
live submerged at the bottom
of an ocean of the element
air, which by unquestioned
experiments is known to have
weight". (From [5, 8]

Noi viviamo sommersi nel fondo d'un pelago
d'aria elementare, la quale per esperienze
indubitate si sa che pesa, e tanto che questa
grossissima vicino alla superficie terrena, pesa
circa la quattrocentesima parte del peso
dell'acqua.

We live submerged at the bottom of an ocean of
the element air, which by unquestioned
experiments is known to have weight, and so
much, indeed, that near the surface of the earth
where it is most dense, it weighs [volume for
volume] about the four-hundredth part of the
weight of water.

air that bears down upon them and is responsible for the barometric pressure. This remarkable insight apparently came right out of the blue. For example it eluded the great scientist Galileo who, as we shall see, only a few years before gave an erroneous explanation for related phenomena.

Torricelli then went on to describe how he made the first barometer, and how he recognized that it was the weight of the air that supported the column of mercury. He took a glass tube about 2 cubits (about 110–120 cm) long and filled it with mercury (Fig. 3.3). He then placed a finger over the end and inverted the tube in a basin containing mercury. He saw that the mercury fell until its height above the surface in the trough was "a cubit and a quarter and an inch besides". A cubit and a quarter is probably about 73 cm so he reported the height as in the region of 76 cm of mercury.

Torricelli correctly reasoned that the space above the mercury contained nothing and therefore was a vacuum. Previous experimenters using water (see below) had seen a similar behavior in much longer water-filled tubes and it had been argued that the column of liquid was held up by the properties of the vacuum above it. Incidentally this is apparently why Torricelli used two tubes, one with a simple blind end and the other with a small sphere on the end as shown in Fig. 3.3. He argued that if a vacuum was responsible for attracting the mercury, the heights of the columns would be different because the differences in shape of the end of the tube would change the properties of the vacuum. However, as Fig. 3.3 shows, the heights were the same. Torricelli went on to argue that the vacuum was irrelevant to maintaining the height of the mercury column. After remarking that the space above the mercury

Fig. 3.3 Torricelli's drawing of his barometer in his letter to Ricci. (From [5])

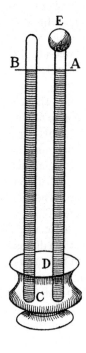

contained nothing and therefore could have no attractive effect, he stated that "On the surface of the liquid which is in the basin, there gravitates a mass of air 50 miles high". In other words he clearly saw that it was the pressure of the "ocean of air" on the mercury in the trough that was responsible for maintaining a column of about 76 cm.

The actual experiment was not done by Torricelli himself but by his colleague Vincenzo Viviani (1622–1703). Viviani was an assistant to Galileo at Arcetri near Florence from the age of 17 to Galileo's death in 1642, and went on to edit the first edition of Galileo's collected works. The fact that it was Viviani who actually carried out Torricelli's experiment emphasizes the close link between Torricelli and Galileo. Torricelli was invited to work with Galileo in Arcetri but arrived just a few months before the latter's death.

3.2 Galileo's View on the Force of a Vacuum

There was much interest at the time in the problem of raising water from a deep well by means of a suction pump. This was extensively discussed before Torricelli's experiment and it was well known that it was not possible to pump water from a well if the pump was more than about 9 m above the surface of the water. A related interest was the behavior of siphons which were used to transport water in a pipe over a small hill. In 1630, Giovanni Galliano (1582–1666) had written to Galileo asking him why a siphon that Galliano had designed to carry water over a hill 21 m high refused to work. If the tube of the siphon was filled with water by means of a pump, and then the pump was stopped, the water separated high in the tube and flowed out at both ends.

Galileo discussed this problem in some detail in his last book "Discourses Concerning Two New Sciences" [4]. This was published in 1638 in Leiden, far from Rome. The reason is that Galileo had been under house arrest at Arcetri since 1633 because he was "vehemently suspected of heresy" by the Holy Office of the Inquisition. This came about because Galileo argued in his previous great book "Dialogues Concerning the Two Chief World Systems" [3] that the earth circled the sun. In fact the original sentence was prison but this was reduced to house arrest. Galileo was precluded from publishing anything after 1633 but since Leiden in the Netherlands was outside the influence of the Church, it was possible to have his book published there.

"Discourses Concerning Two New Sciences" makes excellent reading. The format is a discussion between three people, Salviatti who is a spokesman for Galileo, and two others, Sagredo and Simplicio who continually challenge Galileo on various points. This is the same format that Galileo used in "Dialogues Concerning the Two Chief World Systems".

Galileo takes up the issue of why water cannot be raised more than a certain amount from a well by describing a "thought experiment" illustrated in Fig. 3.4. CABD represent the cross-section of a cylinder either of metal or, preferably of glass, hollow inside and accurately turned. Into this is introduced a perfectly fit-

Fig. 3.4 Galileo's drawing of the "thought experiment" to measure the force of a vacuum. See text for details. (From [4])

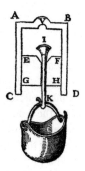

ting cylinder of wood represented in cross-section by EGHF, and capable of up and down motion. A hole is bored through the cylinder to receive an iron wire carrying a hook at the bottom. The conical head at the top of the wire makes a perfect fit with the countersunk wooden cylinder. For the experiment, the cylinder is carefully filled with water so that no air remains and weights are added gradually until the water separates and the weights fall. The weight of the stopper, wire and bucket with its contents then measure the force of the vacuum [forza del vacuo].

Segredo, one of the discussants, then remarks "Up to this time I had been so thoughtless that, although I knew a rope... if sufficiently long would break by its own weight when held at the upper end, it never occurred to me that the same thing would happen only much more easily to a column of water. And really is not that thing which is attracted in the pump [raising water from the well] a column of water attached at the upper end and stretched more and more until finally a point is reached where it breaks, like a rope, on account of its excessive weight". To which Salviati responds "That is precisely the way it works; this fixed elevation of 18 cubits (about 10 m) is true for any quantity of water whatever, be the pump large or small". In other words Galileo was thinking in terms of a force to break the vacuum rather like a force can break a wire by stretching it.

This section is quoted at some length to emphasize how revolutionary was Torricelli's new insight. Only some six or so years before Torricelli's experiment, one of the greatest scientists of all time held an entirely different view.

3.3 Gasparo Berti's Experiment with a Long Lead Tube

Torricelli's experiment was revolutionary, but as so often happens in science, it had been preceded by other somewhat similar activities. One of the most important of these was a remarkable demonstration by Gasparo Berti (c. 1600–1643) who was another Renaissance Italian mathematician and physicist. Unfortunately the details of Berti's experiment are not as clear as those of Torricelli's because the accounts were written several years later. Indeed the actual date of the experiment is uncertain but was probably between 1639 and 1644. The best account is by Emmanuel Maignan (1601–1676) who taught natural philosophy in a convent in Rome [6].

Maignan first refers to Berti as "indeed my greatest friend" and then goes on to describe how Berti erected a long lead tube on the outside wall of the tower of his house (Fig. 3.5). The length was about 11 m and initially Berti filled it with water and sealed it at both ends. He then placed the bottom of the tube in a tank of water and opened the seal. Some of the water flowed out of the tube but much remained so that the height of the column of water was about 10 m. Berti referred to the space above the water as a vacuum.

A number of variations of this arrangement were made. One was to attach a large sphere at the top of the tube which was initially filled with water as well and then became part of the container for the vacuum. The sphere can be seen in Fig. 3.5. The properties of the space above the water created a great deal of controversy. One experiment was to place a bell in the sphere and arrange for this to be struck in some way because it was argued that sound could not travel through a vacuum. However there was no way of supporting the bell so that the sound could not travel along its support. Interestingly although this was a very remarkable experiment, it was not well known at the time, and as indicated above, accounts were written only some years later. There was never any suggestion that the column of water was supported by air pressure.

Fig. 3.5 Illustration of Gasparo Berti's experiment using a very long lead tube containing water. See text for details. (From [7])

3.4   Weighing the Air

Torricelli's letter (Fig. 3.2) extends the passage quoted earlier as follows "We live submerged at the bottom of an ocean of the element air, which by unquestioned experiments is known to have weight, and so much, indeed, that near the surface of the earth where it is most dense, it weighs [volume for volume] about four-hundredth part of the weight of water". This raises the issue of how these early scientists were able to weigh air.

Torricelli is here referring to a section in Galileo's "Discourses Concerning Two New Sciences" where Salviati states "But can you doubt that air has weight when you have the clear testimony of Aristotle affirming that all elements have weight including air, excepting only fire? As evidence of this, he cites the fact that a leather bottle weighs more when inflated than when collapsed". Actually Aristotle's remarks on this are extremely brief. He states in De Caelo Book IV.4, 311b, lines 6–11 "Earth, then, and bodies in which earth preponderates, must needs have weight everywhere, while water is heavy anywhere but in earth, and air is heavy when not in water or earth. In its own place each of these bodies has weight excepting fire, even air. Of this we have evidence in the fact that a bladder when inflated weighs more than when empty" [1].

In fact Aristotle's statement that "a bladder when inflated weighs more than when empty" was the subject of much subsequent controversy. Even Galileo followed up his remarks cited above with the following "I am inclined to believe that the increase of weight observed in the leather bottle or bladder arises, not from the gravity of the air, but from the many thick vapors mingled with it in these lower regions. To this I would attribute the increase in weight in the leather bottle".

It is interesting that today, teachers of elementary physics often use an experiment with an inflated balloon to make the point that air has weight. There are several examples of these on the Internet. In a typical demonstration, two inflated toy balloons are suspended at the ends of a long stick such as a meter rule, and this is supported in the middle so that it is balanced. One balloon is then burst by putting a match under it, and the other balloon tilts the balance downward.

In fact this is a misleading demonstration because the reason why the intact balloon falls is that it contains air under pressure. If the two balloons are inflated with air at normal atmospheric pressure, deflating one balloon will not change the balance. The reason is that a balloon at normal pressure, such as a thin plastic bag that has been partly inflated, receives support by buoyancy from the air around it which cancels the weight of the air inside it.

Galileo presumably understood this although he does not appear to state it. What he does do is describe in detail a method for measuring the weight of the air. This is included in "Discourses Concerning Two New Sciences" [4]. Basically he takes an empty bottle containing air at normal pressure and weighs it. He then blows air into it so that the pressure is increased and weighs it again. A second bottle filled with water is then connected to the first bottle so that the air escaping from the first bottle displaces water from the second, and the volume of the displaced water is measured.

Now he has an accurate measurement of the weight of the air introduced into the first bottle, and its volume measured from the displaced water. This allows him to determine the weight per unit volume, or as he actually calls it, the specific gravity of the air. At the end of the description he states that this allows him to determine "definitely how many times heavier water is than air, and we shall find, contrary to the opinion of Aristotle, that this is not ten times, but as our experiment shows more nearly four hundred times".

The last part of this statement is surprising because the actual value is about 800 times at sea level at a temperature of 20 °C. In other words Galileo's figure is too low by a factor of about two. The procedure is described in such detail that it is difficult to understand how such a large error was incurred. One wonders whether this too was in part a "thought experiment" rather than something actually carried out as described.

3.5   The Decrease of Barometric Pressure with Altitude

In his letter, Torricelli makes an oblique reference to the fact that barometric pressure may decrease with altitude. First he states "Those who have written about twilight, moreover, have observed that the vaporous invisible air rises above us about 50 or 54 miles". This statement implies that above this altitude, there is no air, so that as we approach the altitude the pressure will fall. Torricelli goes on to say that "I do not, however, believe its height is as great as this, since if it were, I could show that the vacuum would have to offer much greater resistance than it does". Here Torricelli is presumably arguing that if the air extended as high as 50–54 miles, the height of the column of mercury in his barometer would be greater. He then continues "The weight referred to by Galileo applies to air in very low places, where men and animals live, whereas that on the tops of high mountains begins to be distinctly rare and of much less weight than four-hundredth part of the weight of water". In other words Torricelli is certainly suggesting that barometric pressure declines with altitude.

The definite proof of the fall in barometric pressure with altitude is usually ascribed to Blaise Pascal (1623–1662). He was an infant prodigy, particularly in the area of mathematics, but he also did extensive work on pressure in fluids. Students are taught Pascal's Law which states that the pressure at any point in a liquid is equally transmitted in all directions. Indeed up to now we have been concentrating on the intellectual ferment in Italy in the early and mid-seventeenth century. However French scientists such as Pascal and René Descartes (1596–1650) were also extremely active.

Pascal had the idea of taking a barometer up the Puy-de-Dôme near Clermont in central France where he was born. Rather than do this himself, he asked his brother-in-law, Florin Perier who lived in Clermont, to carry out the experiment. The results were subsequently sent to Pascal in a delightful letter by Perier dated September 22, 1648. Note that this is only four years after Torricelli's letter to Ricci and this short period is further evidence of the lively intellectual activity of the time.

Perier described the project in considerable detail. A group of people met early in the morning in a garden in Clermont and filled several mercury barometers which gave a pressure of about 710 mm Hg. One barometer was left in the garden where it was observed all day by the Reverent Father Chastin who reported that there were no changes in the height of the mercury. However on the summit of Puy-de-Dôme, the altitude of which was described as 500 fathoms above the garden, the mercury had a height of only about 625 mm Hg. This meant that there was a fall in barometric pressure of about 12 %. The result was considered so remarkable that the experiment was repeated a number of times. In fact Perier subsequently ascended to the summit again and found the same result.

Perier was so impressed by the results of this experiment that he then took the barometer up the tower of the cathedral in Clermont to see whether there would be a measureable change of pressure as a result of this much smaller ascent. Indeed there was a fall of about 5 mm Hg which gave him great satisfaction. Perier's letter of 1648 to Pascal is so beautifully written that it is embarrassing to state that there is some doubt about its authenticity [6]. Some historians have contended that Pascal's request to Perier to carry out the experiments could not have been as early as November 1647 as he claimed. Other commentators have argued that the whole idea of the experiment came from Descartes rather than Pascal.

Torricelli's experiment had an enormous effect on the scientific community, not only in Italy but throughout Europe. This was in contrast to the experiment of Gasparo Berti described earlier which resulted in little interest and in fact was only reported several years later. However Torricelli's discovery had rapid repercussions.

3.6   Demonstration of the Enormous Force that can be Developed by the Barometric Pressure

One of the most colorful and best known subsequent experiments was carried out in 1654 by Otto von Guericke (1602–1686) who was the mayor of the city of Magdeburg in central Germany (Fig. 3.6). He has the distinction of making the first air pump. He did this by modifying a water pump that previously had been used for fighting fires. Von Guericke then constructed two copper hemispheres which fitted together so accurately that they were airtight when they were evacuated. When he pumped the air out of the hemispheres which had diameters of about 50 cm, the force developed by the pressure of the air was so great that two teams of horses were unable to pull the hemispheres apart. As can be expected, this dramatic demonstration provoked widespread interest.

It is interesting to put some numbers on von Guericke's experiment. First why did he need two teams of horses? One team would have given the same result although the demonstration would perhaps have been less arresting. In fact one team of horses would have resulted in the same tensile force on the hemispheres if they had been fastened to a solid structure such as a large wall. The second team simply provided a counterforce which otherwise would have been provided by a wall.

Another interesting point is how close did the horses come to separating the two hemispheres? Probably not close at all. If we assume as a first approximation that von Guericke's pump was capable of removing all of the air, the force holding the two hemispheres together was equal to the barometric pressure times the area of a circle of 50 cm diameter. Using English units and taking the radius as 10 in., the area of the circle is πr^2 or 3.14×100, that is 314 square inches. The barometric pressure is 14.7 lbs per square inch giving a total force of about 46,000 pounds weight. In SI units, the area of the circle is 0.0625 m², and the total force is about 10,300 N.

Next, how much force can be developed by a horse? There is a competition in some rural areas known as "horse pulling" in which the maximum force developed by a horse is measured by a dynamometer. A medium sized horse can pull about 1500 pounds weight. Therefore it would take 46,000/1500 horses, that is about 30 horses in one team to separate the hemispheres. Von Guericke's team of 8 horses on one side (Fig. 3.6) was therefore far less than required. Of course if the air in the hemispheres could only be pumped down to half the normal pressure, the force required to separate them would be reduced to one half.

Fig. 3.6   Demonstration by Otto von Guericke of the attempt to separate two evacuated copper hemispheres by two teams of horses. (From http://commons.wikimedia.org/wiki/File:Magdeburg. jpg)

3.7   Subsequent Studies of the Effects of Reducing the Barometric Pressure

One person who was greatly influenced by the news of von Guericke's demonstration was Robert Boyle (1627–1691) in England. He read about the new pump in a book by Schott [7] and realized the potential of making scientific studies in an experimentally produced low-pressure environment. He then persuaded his brilliant colleague Robert Hooke (1635–1703) to make an air pump that could evacuate a glass sphere into which small animals and other objects could be introduced. This was impossible with the metal hemispheres used by von Guericke. Boyle's experiments were initially described in his influential book "New Experiments Physico-Mechanicall, Touching the Spring of the Air and its Effects" [2]. This publication marked the beginning of a new era of what we now know as high altitude physiology [10].

References

1. Aristotle. De Caelo Book IV, 311b, lines 6–11. Translated by Stocks JL and Wallis HB. Oxford: The Clarendon Press; 1922. http://archive.org/details/decaeloleofric00arisuoft.
2. Boyle R. 1660. New experiments Physico-mechanicall, touching the spring of the air, and its effects. Oxford: H. Hall for T. Robinson; 1660.
3. Galileo G. Dialogo dei due massimi sistemi del mondo. Florence: Landini; 1632 ([Dialogues concerning the two chief world systems], translated by Stillman Drake. Berkeley: University of California Press; 1967).
4. Galileo G. Discorsi e dimostrazioni matematiche, intorno à due nuove scienze. Leiden: Elzevir; 1638 ([Discourses and mathematical demonstrations relating to two new sciences], translated by Henry Crew and Alfonso di Salvio. New York: Macmillan; 1914. pp. 74–5). http://oll.libertyfund.org/index.php?option=com_staticxtstaticfile=show.php%3Ftitle=753&Itemid=99999999.
5. Loria G, Vassura G. Opere di Evangelista Torricelli [Works of Evangelista Torricelli]. Vol. III. Faenza: G. Montanari; 1919. p. 186.
6. Middleton WEK. The history of the barometer. Baltimore: Johns Hopkins Press; 1964.
7. Schott GP. Gasparis Schotti Mechanica Hydraulico-Pneumatica. Francofurti ad Moenum: Sumptu Heredum Joannis Godefridi Schönwetteri Bibliopol. Francofurtens, excudebat Henricus Pigrin, Typographus Herbipoli; 1657.
8. Spiers IHB, Spiers AGH. The physical treatises of Pascal; the equilibrium of liquids and the weight of the mass of the air (1663). Trans. New York: Columbia University Press; 1937.
9. von Guericke O. Experimenta nova (ut vocantur) Magdeburgica de vacuo spatio. Amsterdam: J. Jansson-Waesberg, 1672. Translated as The new (So-Called) Magdeburg experiments of Otto von Guericke. Dordrecht: Kluwer Academic Publishers; 1994.
10. West JB. Robert Boyle's landmark book of 1660 with the first experiments on rarified air. J Appl Physiol. 2005;98:31–9.

Chapter 4
Robert Boyle's Landmark Book of 1660 with the First Experiments on Rarified Air

Abstract In 1660, Robert Boyle (1627–1691) published his landmark book *New Experiments Physico-Mechanicall, Touching the Spring of the Air, and its Effects.* in which he described the first controlled experiments of the effects of reducing the pressure of the air. Critical to this work was the development of an air pump by Boyle with Robert Hooke (1635–1703). For the first time, it was possible to observe physical and physiological processes at both normal and reduced barometric pressures. The air pump was described in detail, although the exact design of the critical piston is unclear. Boyle reported 43 separate experiments, which can conveniently be divided into 7 groups. The first experiments were on the "spring of the air," that is the pressure developed by the air when its volume was changed. Several experiments described the behavior of the barometer invented by Torricelli just 16 years before when it was introduced into the low-pressure chamber. The behavior of burning candles was discussed, although this emphasized early misunderstandings of the nature of combustion. There were some physiological observations, although these were later extended by Boyle and Hooke. The effects of the low pressure on such diverse physical phenomena as magnetism, sound propagation, behavior of a pendulum, evolution of gases from liquids, and the behavior of smoke were described. This classic book is brimming with enthusiasm and fresh ideas even for today and deserves to be better known.

In the history of the physiology of high altitude, Paul Bert's book *La Pression Barométrique* [2] stands out as a watershed contribution. This contained the first definitive demonstration that the deleterious effects of high altitude were due to the low partial pressure of oxygen, whether this was caused by a reduction of barometric pressure or a reduced oxygen concentration at normal pressure. As a result, Bert is often rightly referred to as the father of high-altitude physiology.

However, some 200 years before the publication of *La Pression Barométrique*, there was another landmark book in high-altitude studies. The author was Robert Boyle (1627–1691) and the full title was *New Experiments Physico-Mechanicall, Touching the Spring of the Air, and its Effects (Made, for the Most Part, in a New Pneumatical Engine) Written by Way of Letter to the Right Honorable Charles Lord Vicount of Dungarvan, Eldest Son to the Earl of Corke* [3]. The importance of this book was that it described the first controlled experiments of the effects of reducing

© American Physiological Society 2015

J. B. West, *Essays on the History of Respiratory Physiology*,
Perspectives in Physiology, DOI 10.1007/978-1-4939-2362-5_4

the pressure of the air. The book was not limited to physiology, and in fact this topic occupied only a small portion of the text. But the book was so innovative and stimulating, and such a joy to read, that it deserves to be better known by physiologists.

4.1   The Setting

It is interesting to trace the events leading up to the work described in Boyle's book partly because of the great rapidity of progress over a period of less than 20 years. The crucial breakthrough was made by Evangelista Torricelli (1608–1647), who in 1643 invented the mercury barometer. He took a long glass tube, which was closed at one end, filled this with mercury, put his thumb over the open end, and inverted the tube in a dish of mercury. The level of the mercury in the tube fell to ~76 cm, and the nature of the space above it, which we now know as a vacuum, was a topic of great controversy. Torricelli recognized that the column of mercury was supported by the pressure of the atmosphere acting on the surface of the mercury in the dish, and in his letter to Michelangelo Ricci in 1644 he made the dramatic statement that "We live submerged at the bottom of an ocean of the element air, that by unquestioned experiments is known to have weight" [19]. This concept was a major breakthrough in an area that had caused much confusion in the past, and, for example, even the great Galileo only some 6 years before was confused about the reason why a suction pump could only raise water ~9 m. The explanation he gave was related to the presumed tensile strength of water [14].

Torricelli's simple but dramatic experiment resulted in a flurry of activity over the next 16 years. In 1648, Blaise Pascal (1623–1662) persuaded his brother-in-law Florin Perier to take a Torricellian barometer up the Puy de Dôme outside Clermont in the center of France, and, as expected, the level of the mercury fell because, on the top of the hill, the pressure of the atmosphere was less. A major advance was then made in 1654 by Otto von Guericke (1602–1686) when he constructed an air pump and demonstrated to the Diet of Regensburg that it was possible to reduce the amount of air in a glass vessel. However, his most famous experiment took place in 1657 when two metal hemispheres were constructed so that they fitted accurately with an airtight seal. When the air was pumped out of these, two teams of eight horses each were unable to break the seal. This was a very colorful demonstration of the enormous force that could be developed by atmospheric pressure.

Boyle learned of Guericke's experiments from a book written by Gaspar Schott in 1657 [17], but he soon realized that Guericke's apparatus had two serious deficiencies. The most important was that the hollow vessel that was evacuated could not easily be opened, and therefore it was impossible to put anything inside it. Because Boyle intended to compare the behavior of various processes both in rarified and normal air, this design was inadequate. Another problem was that Guericke's pump was so inefficient that it required "the continual labour of two strong men for divers hours." Boyle obviously wanted something more convenient than this.

As indicated above, the book by Schott was published in 1657, but by early 1659 Boyle had a much-improved air pump and was ready for experiments. The pump was designed and constructed by Robert Hooke (1635–1703), who was a mechanical genius. He made important contributions to an extremely wide field, including microscopy, horology, mechanics, and architecture. Boyle hardly mentioned Hooke in the 1660 book, but later he acknowledged the great contributions of his assistant. In fact, it seems likely that a number of the experiments described in the 1660 book owed their origin to Hooke's interests. An example was *experiment 26* on the effect of rarified air on the swing of a pendulum, since the motion of a pendulum was one of Hooke's many interests. Boyle finished writing the book on December 20, 1659, and it was published in 1660. Therefore, it was only 16 years between Torricelli's seminal discovery of the barometer and Boyle's completion of this remarkable book.

4.2   The Man

The Honourable Robert Boyle (Fig. 4.1) was one of the most important figures in the history of science in the middle of the seventeenth century and not surprisingly has been the subject of extensive studies. This is therefore only a brief summary of his career. Robert Boyle was the 14th son of the first Earl of Cork, a man with extensive landholdings in Ireland. As a consequence, Robert was wealthy and lived

Fig. 4.1 Portrait of Robert Boyle (1627–1691). (Reproduced from Ref. [18])

the life of a seventeenth-century gentleman. He was initially educated at home by tutors and then at Eton College. After spending some time on the Continent, during which he met Galileo in Florence in 1641–1642, he returned to England and lived in London. He was one of a circle of friends who discussed contemporary scientific issues, and Boyle referred to the group as the "invisible college." This later became the Royal Society in 1662. Boyle moved to Oxford in 1654, taking a house next to University College, where a plaque on the wall can still be found.

Boyle was a prolific writer, and even now some of his material remains unpublished [16]. His 1660 book discussed here is arguably his best, but there were several sequels to this, and in addition Boyle wrote extensively on theology, ethics, philosophy, and chemistry.

The prodigious literary output of Boyle occurred despite tremendous political and social upheaval in England. There was a civil war from 1642 until 1646 with continuing unrest after that until King Charles I was beheaded in 1649. The Commonwealth that followed under Oliver Cromwell was also an unsettling period, and when Cromwell died in 1658 and was succeeded by his son Richard, there was even more unrest. The monarchy was restored with Charles II in 1660. Boyle refers to these difficulties in his preface, where he states "I need not perhaps represent to the equitable Reader, how much these strange Confusions of this unhappy Nation, in the midst of which I have made and written these Experiments, are apt to disturb that calmness of Minde, and undistractedness of Thoughts, that are wont to be requisite to Happy Speculations."

4.3   The Book

The title page is shown in Fig. 4.2. This is followed by an 11-page preface in which Boyle gave some of his reasons for the style of the book. An interesting feature is the extent to which he emphasized that he did all the experiments himself (albeit with the help of Hooke). The experiments were described in considerable detail so that readers could reproduce them if they wished. For example, on the second page of the preface, which is headed "To the Reader," Boyle stated, "On my being somewhat prolix in many of my Experiments, I have these Reasons to render, That some of them being altogether new, seem to need the being circumstantially related, to keep the Reader from distrusting them." It might seem strange to the present-day reader that Boyle put so much emphasis on the fact that all the experiments were actually carried out. However, a number of books before this contained a mixture of actual and "thought" or imaginary experiments. As an example, Galileo in his great book *Dialogues Concerning Two New Sciences* [14] discussed how the force or resistance of a vacuum could be measured. He described how a piston could be made to fit perfectly in an inverted cylinder filled with water with all the air excluded. When weights were added to the piston until it fell, the force of the vacuum could be derived. However, this is not an experiment in the sense we now use the term, but an imaginary situation that Galileo developed to explain a concept. Boyle was at

Fig. 4.2 Title page of the 1660 book

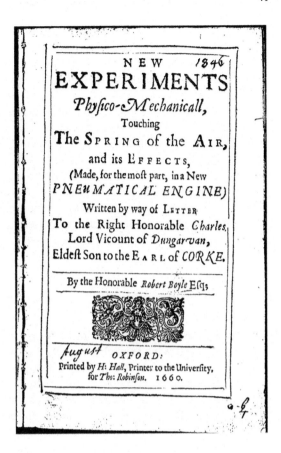

great pains to emphasize that his experiments were actually done, and in fact some of the detail is rather tedious.

Another issue briefly referred to in the preface is why the book was written in the form of a letter to Boyle's nephew. In fact, throughout the book, there are numerous allusions to the recipient of the letter, which allowed Boyle to emphasize important points. For example, on *page 17*, after describing the construction of the pump, Boyle writes "Your Lordship will, perhaps, think that I have been unnecessarily prolix in this first part of my Discourse: But if you had seen how many unexpected difficulties we found to keep out the externall Air, even for a little while, when some considerable part of the internal had been suckt out; You would peradventure allow, that I might have set down more circumstances than I have." This little conceit of imagining that he is talking directly to his nephew is somewhat like the format Galileo used in his book referred to above [14] in which the whole of the scientific thesis was given in the form of a discussion between three people.

After Boyle's preface titled "To the Reader," there is another brief introduction headed "Friendly Reader," but this was written by the editor of the book, Robert Sharrock (1630–1684), who took Boyle's presumably handwritten manuscript and prepared it for publication. At the same time, he prepared a Latin edition because, as he said, "Since the Mountain cannot come to Mohamet, Mohamet will go to the Mountain." By this, he meant that many scientists outside England wanted to read the book but could not understand English. Latin was still the lingua franca of the intelligentsia of the civilized world.

The next section of the book is an expanded table of contents under the heading "A Summary of the chief Matters treated of in this Epistolical Discourse." Boyle explained that the book begins with a "Praemium," an old term for an introduction, that is devoted to a description of the pump. This is followed by brief descriptions of the 43 experiments in the book and a short conclusion at the end.[1]

It should be added here that this 1660 edition does not include what we now know as Boyle's Law, that is the inverse relationship between the volume of a gas and its pressure [20]. This was added in the second edition of 1662, as described later.

4.4   The Pump

The first 19 pages of the book describe the air pump in considerable detail. The description in the text is made much clearer by the fine engraving of the pump shown in Fig. 4.3. A modern reconstruction of the pump is shown in Fig. 4.4.

Boyle divided the description into two parts. The first is the glass "receiver" on the top in which the partial vacuum was developed and in which the experiments were carried out. The most striking feature of this is its great size. Boyle stated that it contained ~30 wine quarts, each of them containing "near 2 pound of water." In modern units, this is a volume of ~28 l, which, if the receiver were spherical, would mean a diameter of ~38 cm. This is the size of the receiver in two modern reconstructions.[2] In fact, Boyle wanted a larger receiver, stating "We should have been better pleas'd with a more capacious Vessel, but the Glass-men professed themselves unable to blow a larger, of such a thickness and shape as was requisite to our purpose." Boyle apparently recognized the enormous compressing force that would be developed on the receiver when the air pressure was reduced and of course an

---

[1]   The whole book is available on the web from Early English Books Online (http://eebo.chad-wyck.com/). This needs a subscription for access, but the online version is an accurate facsimile of the 1660 edition. A second edition, published in 1662 with some additional material, is available from Books on Demand (UMI, Ann Arbor, MI). The contents of the text are identical but the pagination is different. The 1660 book is also reprinted in volume 1 of *The Works of Robert Boyle* (16). This is the easiest version to read.

[2] There are at least two full-size modern reconstructions of the Boyle-Hooke pump. One is in the Museum of the History of Science in Oxford, UK, and the other is in the Science Museum, Kensington, London, UK. Neither are probably on display but can be seen by appointment. Both are well worth a visit partly because they emphasize the great size of the pump and particularly the receiver.

Fig. 4.3 Engraving of the air pump devised by Robert Boyle and Robert Hooke (1635–1703). The complete pump is shown at center, and some of the disassembled parts are at right. Various small pieces of equipment that were used in the experiments are also shown. See text for details

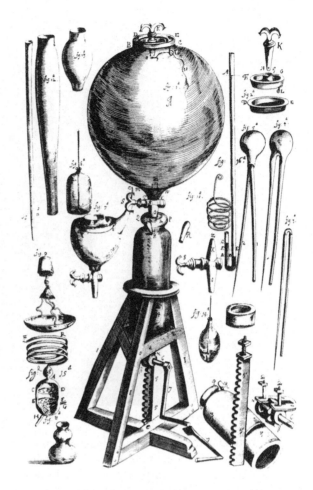

implosion could have been disastrous. In fact, in one of the experiments, a glass vial containing air at normal pressure exploded and cracked the receiver.

At the top of the receiver, there was a round hole ~4 in. in diameter with a lip of glass almost 1 in. high. This allowed relatively large objects to be introduced into the receiver. The orifice was closed with a brass ring that was cemented in place. The brass ring had a smaller hole in it ~0.5 in. in diameter through which smaller objects could be introduced into the receiver. This hole was closed with a ground brass stopper, which could be rotated so that a string attached to the bottom could control equipment in the receiver. In addition, this stopper could be easily removed and replaced. By contrast, the 4-in. brass ring had to be re-cemented every time it was removed. Incidentally, Boyle makes occasional comments that the design could be improved, and at this point he suggests that a better design for the 4-in. hole would be a ground glass taper "in case your Lordship should have such another Engine made for you."

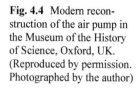

Fig. 4.4 Modern reconstruction of the air pump in the Museum of the History of Science, Oxford, UK. (Reproduced by permission. Photographed by the author)

At the bottom of the receiver was a brass stopcock similar to designs used today. The connection between the stopcock and the receiver was a challenge. The solution was to form a piece of "tin" with a conical shape and fill the space between the tin and the receiver with cement made of pitch, resin, and wood ashes "well incorporated." To prevent the cement from plugging the hole in the receiver during this process, a cork was placed in the hole and withdrawn through the top of the receiver with a string after the cement had set. Modern glassmakers are able to form the glass to fit the stopcock, and in fact this is what was done in the modern reconstruction at the Museum of the History of Science, Oxford, UK (Fig. 4.4).

The second part of the pump was the hollow cylinder together with the piston, which was driven by a rack and pinion. The cylinder was a piece of cast brass about 14 in. long with a hole 3 in. in diameter bored within it. Machining a hole of accurate constant diameter would have been a challenge, but boring large holes in metal was a well-known skill in making cannons.

One of the most critical parts of the pump was the piston or "sucker" as Boyle referred to it. Unfortunately, the description is less detailed than we would like. Boyle simply says that it consisted of two parts, one (marked 44 just above the rack in Fig. 4.3) "somewhat less in Diameter than the cavity of the cylindre, upon which is nail'd a good thick piece of tan'd shoe Leather, which will go so close to the Cylindre, that it will need to be very forcibly knock'd and ram'd in, if at any time it be taken out." It is not clear from this description exactly how the piston is constructed. Presumably, the first part [44] is made of wood to accept the nails, but where the thick piece of leather was placed is uncertain. Conant [12] in his discussion of the pump shows the leather seal as nailed to the upper side of the wooden piece, but this would not work well to form an airtight seal as the piston was pulled down because the leather would rise up at the periphery. Much better would be to place the leather on the bottom of the wooden piston. This is the arrangement used in reverse in modern bicycle pumps. The reconstruction of the pump in the Science Museum in London interpreted Boyle's description differently, and there the strip of leather is nailed around the periphery of the wooden piston. It is uncharacteristic of Boyle to be so vague about this critically important element of the pump because, of course, the ability of the pump to reduce the pressure in the receiver depended critically on the fit of the piston in the cylinder.

The piston was pulled down using a rack and pinion, which is clear from the engraving. Finally, the cylinder had a small valve at the top consisting of a tapered hole with a well-fitting brass plug that could be easily removed. To operate the pump, the stopcock at the base of the receiver was closed, the plug valve was opened, and the piston was cranked to the top of the cylinder. Then the plug valve was closed, the stopcock was opened, and the piston was cranked down. This process was then repeated as necessary. In some experiments, for example *experiment 37*, the piston was drawn down before the stopcock was opened.

At the end of this description of the pump, Boyle apologized to his nephew, stating that the description may seem to be unnecessarily prolix. However, he then went on to say that he included many details because of the great difficulty of getting the pump to perform satisfactorily. He had many problems with air leaks through the cement at the top and bottom of the receiver, and we can assume that the fit of the piston in the cylinder was always a serious problem. To reduce leaks, he poured into the top of the receiver a little "sallad oyl" to make the stopcock more airtight. Also a "pretty store of oyl" was poured into the cylinder to lubricate the piston. Interesting (and surprising to Boyle) was the fact that adding some water to the oil improved the operation of the pump.

Because of the problem of leaks and the resulting loss of the vacuum in a relatively short time, Boyle divided possible experiments into two types. The first was those that could be carried out in a short time, and the book concentrates on this group. The second type of experiment, such as studies of the preservation of animal or other bodies in a vacuum, or the germination and growth of vegetables, required a sustained partial vacuum over a long period of time, and Boyle conceded that his pump could not provide this.

Despite these imperfections, the pump was a major technological and engineering advance. As such, it owed much to the ingenuity of Hooke. The result was that, for the first time, it was possible to subject various materials and processes to a partial vacuum while setting up an identical control experiment in the air alongside the pump. Alternatively, the experiment could be performed in the receiver at normal atmospheric pressure and again when the air was removed. Boyle himself was quite aware of how innovative all this was, and the modern-day reader of the 1660 book senses the excitement and novelty of it.

How successful was the pump in developing a partial vacuum? As we shall see in *experiments 17–19*, Boyle reported that the level of mercury in a Torricellian barometer could be brought down to 1 in. above the surface of the mercury trough, indicating that the pressure had been reduced to ~3 % of its normal value (corresponding to an altitude of ~23 km). Again, in *experiment 33*, Boyle pumped down to a good vacuum and found that it then required ~150 lbs. of weight on the piston to pull it down. If we take the diameter of the piston as 3 in., this represents essentially full atmospheric pressure.

4.5   The Experiments

As indicated above, Boyle reported 43 separate experiments [13, 16]. However, some of these are clearly related, and it is convenient for us to divide them into seven groups. Because the work was done nearly 350 years ago, it is not surprising that some experiments are more interesting than others. In particular, physiology was in its infancy in 1660, and the experiments reported here precede the important studies of Boyle himself and those of Hooke, Lower, and Mayow later in the century. It should be remembered that Boyle's book appeared only 32 years after William Harvey's groundbreaking *De Motu Cordis* [15], which ushered in a revolution in physiology. Interestingly, Boyle was one of Harvey's patients because of his weak eyes.

Spring of the air (experiments 1–9, 32, 33, and 36). Boyle began his account of the experiments by discussing some of the most obvious properties of the air based on everyday use of his "pneumatical engine" or pump. He wrote, "I hold it not unfit to begin with what does constantly and regularly offer it self to our observation, as depending upon the Fabrick of the Engine it self, and not upon the nature of this or that particular Experiment which 'tis employed to try." He then described how the force necessary to pull down the piston became greater with successive strokes as the amount of air in the receiver was reduced. He attributed this to the fact that "the Particles of the remaining Air, having more room to extend themselves in, will less press out one another." Therefore, the pressure of the normal outside air resisted the downward action of the plunger more. He noted that if the air in the receiver was greatly rarified and the stopcock was opened, the piston was forcibly carried to the top of the cylinder.

Boyle then has a short digression on what he believed was the mechanical basis of the spring of the air. He likened air to a fleece of wool that can be greatly compressed, but if the compressing force is removed, it expands rapidly. He compared this explanation with that given by Rene Descartes (1596–1650), who had an alternative hypothesis, i.e., that the air consisted of particles in restless agitation so that each corpuscle endeavored to beat off all others coming close to it. This is similar to the modern concept of the kinetic theory of gases. Boyle did not choose one model over another, although he thought that the first was easier to understand.

Boyle then argued that the spring of the air is related to its pressure, which is high near the surface of the Earth because of the weight of the air. Here, he was echoing Torricelli's statement cited earlier. Boyle described how it was possible to demonstrate that air has weight by weighing a dry lamb's bladder containing air which was compressed by tying thread around it, and comparing this with the weight of the empty bladder when the air was removed by pricking holes in the bladder. Then, based on the density of air and the pressure at the surface of the Earth, he referred to calculations by others that the atmosphere must be at least 50 miles high in some places.

Another demonstration of the spring of the air in the daily operation of his pump was that if the piston was drawn down with the stopcock closed, it was extremely difficult to remove the small valve at the top of the cylinder. In fact, he stated that people who tried to remove the valve under these conditions were led to believe that there was some large weight attached to the bottom of it.

A further way of demonstrating the spring of the air was to take a lamb's bladder about half full of air, secured at the neck with a string, and place it in the receiver, which was then evacuated. The volume of the bladder increased dramatically but returned to its former volume when the pressure in the receiver was returned to normal. A similar experiment had been performed by Gilles de Roberval (1602–1675) using a calf bladder exposed to low pressure by means of a Torricellian barometer. In fact, Boyle reported that he was able to distend the bladder until it ruptured with a "great report, almost like a Craker." He also showed that a partially distended bladder increased its volume greatly if it was held near a fire, and he suggested that this might be because of increased "Agitation of the Aërial Particles." Again, this explanation is consistent with the modern kinetic theory of gases.

In *experiment 9*, Boyle described how he took a glass vial (spelled viol in the book) partly filled with water but with the neck closed by means of a glass pipe that was cemented in (this is the small container marked fig. 14 in Fig. 4.3 just to the right of the center of the wooden stand). At the very first descent of the piston, a piece of glass flew out of the vial striking the receiver and cracking it in many places. Boyle was surprised that this glass vessel exploded while others made of thin glass remained intact when the pressure in the receiver was reduced. He surmised that differences in the quality of the glass and the shape of the containers might be responsible. Boyle also observed wryly that receivers "are more easily crack'd then procur'd" and described ways of cementing cracks in the receiver with plaster so that it could continue to be used.

Torricellian experiments (experiments 17–19). Boyle used this term to describe experiments in which he studied the effect of reducing the pressure around the mercury in the dish of a barometer described by Torricelli. He made it clear that these experiments were some of the most important that he planned for his pneumatical engine, and he began his account of *experiment 17* as follows:"Proceed we now to the mention of that Experiment, whereof the satisfactory tryal was the principal Fruit I promis'd my self from our Engine.... I considered that, if the true and onlye reason while the Quick-silver falls no lower, be, that at the Altitude the Mercurial Cylinder in the Tube, is in Aequilibrium with the Cylinder of Air, supposed to reach from the adjacent Mercury to the top of the Atmosphere:then if this Experiment could be try'd out of the Atmosphere, the Quick-silver in the Tube would fall down to a levell with that in the Vessel, since then there would be no pressure on the Subjacent, to resist the weight of the Incumbent Mercury." Accordingly, Boyle arranged to have the mercury dish of a Torricellian barometer placed in his receiver while the tube of the barometer passed up through the hole at the top, which was carefully sealed with cement.

When the pressure in the receiver was reduced by cranking down the piston, Boyle was very satisfied to see that the level of the mercury in the barometer tube fell. By hard pumping, it was possible to reduce the height of the column of mercury to ~1 in., but despite laborious pumping the level would go no lower. Boyle attributed this to the fact that it was impossible to prevent small leaks of outside air into the receiver particularly through the cement that was used in several places. Incidentally, Boyle was obviously very pleased with this experiment and repeated it in the presence of three friends, "Famous Mathematick Professors, Dr. Wallace, Dr. Ward and Mr. Wren." This is the same Christopher Wren who built St. Paul's Cathedral in London and was involved in many scientific discussions.

Boyle noted that the extent to which the barometer level fell with the first descent of the piston was greater than with subsequent pumpings, and he correctly surmised that, as the pressure of the air in the receiver fell, less of it could be removed with each exsuction, as he called it. In fact, he argued that if one knew the volume of the receiver and the displaced volume of the piston in the pump, it should be possible to calculate the extent to which the mercury fell. Here, he was close to describing Boyle's Law, that is the inverse relationship between volume and pressure, but this was not done in the 1660 book. Instead, it was part of an addendum to the second edition of the book published in 1662, when he responded to various criticisms of his work, particularly one by Franciscus Linus [20]. Boyle also carried out the opposite experiment in which he raised the pressure inside the receiver with the pump and noted that the level of the mercury in the Torricellian barometer rose.

At this point, Boyle has a digression in which he describes variations in the height of the mercury in his Torricellian barometer when it was set up by the window of his bedroom so that he could observe it over a long period of time. He states that, over a period of ~5 wk, the height of the mercury varied by 2 in., although, interestingly, he thought it fell in warm weather and rose in cold weather. He was not able to explain the variations in height, although he wondered whether the phases of the moon and the tides had some effect. He also made a few experiments with a water-filled barometer and noted that it was impossible to reduce the level of this below ~1 ft above the surface of the water in the trough, and he recognized that this was because of the much lower density of water than mercury.

*Burning of candles and other substances (experiments 10–15).*When Boyle placed an ordinary tallow candle in the receiver and gradually reduced the pressure, he found that the appearance of the flame changed and that it was eventually extinguished. He stated that after the first two or three strokes of the pump, the flame got smaller, became blue in color, and moved further up the wick until it was only at the very top. The candle then went out. Boyle repeated this experiment with other flammable materials including "Coals, in which it seemed there had remained some little parcels of Fire," matches, and even a pistol containing gun powder! Boyle's interpretation of why these flammable materials went out when the air was rarified emphasizes the early state of knowledge in 1660. For example, he expected the fire to burn more brightly when the air was removed because this would allow greater space for the products of combustion. He stated it thus:"Whereby it seem'd to appear that the drawing away of the ambient Air made the Fire go out sooner than otherwise it would have done; though that part of the Air that we drew out left the more room for the stifling steams of the Coals to be received into." Of course Boyle knew nothing about oxygen at this time; Priestley was not to isolate it until over 100 years later. In the experiment with the pistol, Boyle found that the force of the flintlock was not apparently altered, and that sparks were produced just as in air at normal pressure. However, the gunpowder was not ignited in most of the experiments, although in one the flame appeared to expand more than expected in normal air. Boyle also attempted to ignite combustible material within the receiver by concentrating the sun's rays with a burning glass and succeeded in making smoke, but the results of these experiments were unsatisfactory he said because the thickness of the glass of the receiver impeded the sun's rays.

Physiological observations (experiments 40 and 41). As indicated earlier, measurements on animals form only a small part of the book. Probably one of the main reasons for this was that Boyle was so intrigued by the physical consequences of rarifying air that studies of living things took second place. He subsequently wrote other books and articles that included some physiology including "New experiments concerning the relationship between light and air (in shining wood and fish)" [7], "New pneumatical observations upon respiration" [9], *Spring of the Air, First Continuation* [8], and *Spring of the Air, Second Continuation* [10].

In *experiment 40*, Boyle studied the behavior of winged insects in his receiver as he removed the air. A large fly was introduced, and it dropped down from the side of the receiver where it was walking when the pressure was reduced. A bee fell down from a flower within the receiver when the pressure was lowered, although Boyle was unclear whether this was because the air was too rarified for it to fly, or whether it was weakened by the low pressure. In a footnote, he mentions that a white butter fly at first fluttered up and down but after the pressure was reduced it "fell down as in a swoon, retaining no other motion then some little trembling of the Wings."

These observations led to a much longer account in *experiment 41* on studies on the nature of respiration. A lark was placed in the receiver and sprang to a good height on several occasions when the pressure was normal. But when air was removed, it began to "droop and appear sick, and very soon after was taken with as violent and irregular Convulsions as are wont to be observ'd in Poultry, when their heads are wrung off." Another experiment was carried out on a hen-sparrow, and the bird seemed to be dead ~7 min after the pump was employed. However, when the air was restored, the bird revived and nearly escaped through the top cover, which had been removed. But when the air was removed a second time, the bird convulsed and died. A mouse inserted into the receiver behaved in a similar way, being very active initially but when the pressure was reduced appeared giddy and staggered before falling down unconscious. Again, the animal was revived when fresh air was let in.

In interpreting these experiments, Boyle posed the same questions as he did about the burning of candles referred to above, and which indicate the early state of knowledge about both combustion and respiration. The question was whether "the death of the fore-mention'd Animals preceded from the want of Air, then [than] that the Air was over-clogg'd by the steams of their Bodies, exquisitely pent up in the glass." Boyle argues that his experiments support the former possibility because, when he removed air with his pump, the animals clearly suffered, whereas when he readmitted air they revived.

Boyle then embarks on a long digression on the nature of respiration. He discusses the movement of the lungs and notes that, since they have no muscles themselves, they must be moved by the diaphragm and intercostal muscles. Interestingly, he states that "the Diaphragme seems the principal Instrument of ordinary and gentle Respiration."

There is a long section in which he discusses some of the theories of respiration in the literature, including the belief that its main function is to cool the blood. Much of this is of little relevance today. However, there is an interesting passage where Boyle considers "whether or no, if a Man were rais'd to the very top of the Atmosphere, he would be able to live many minutes, and would not quickly dye for the want of such Air as we are wont to breathe here below." He then referred to the experience of Joseph de Acosta in the high mountains of Peru who, when indisposed by the high altitude, stated "I therefore perswade my self, That the Element of the Air is there so subtle and delicate, as it is not proportional with the breathing of Man, which requires a more gross and temperate Air." The first edition of Acosta's famous book was published in 1590 [1], and it is interesting that Boyle was familiar with it. Boyle concluded from all this that there is a special portion of the air that is essential for life and that when this is removed what remains will not support life.

Boyle found that all living things, including eels that were placed in his receiver, could not survive a prolonged reduction of pressure. However, an exception was house snails, which did not seem to be affected, presumably because the amount of air that they needed was so small.

Effect of rarifying the air on some physical phenomena (experiments 16, 26, 27, and 37). The great curiosity of Boyle (and presumably Hooke) generated a series of interesting observations on physical phenomena, but they will only be referred to briefly. In *experiment 16*, a magnetized needle was placed in the receiver, and Boyle showed that it could still be deflected by a magnet brought to the outside of the receiver after evacuation. This raised the question of how the magnetic influence could be propagated in the absence of air. A similar experiment (*experiment 27*) studied the propagation of sound when a watch was suspended in the receiver by a piece of thread. Boyle found that the sound of the ticking disappeared when the air pressure was lowered. By contrast the sound made by a bell when it was struck was quieter when the air was removed but did not disappear altogether. However, a technical problem here was that the material that suspended the bell may have conducted the sound. A further issue was that since not all the air could be removed by the pump, the residual air might have propagated the sound. Of course, Boyle recognized that light was transmitted through the evacuated receiver because it was still possible to see objects in it.

These experiments raised the question of whether there was another medium called "ether" that continued to be present in the receiver even if all the air was evacuated. Many scientists of the time thought that to transmit light or magnetism through a space required the existence of some medium. They argued that the air pump might able to remove the air but not the ether. Experiments on the possible existence of ether were not followed up in this 1660 book but are described in Boyle's *Spring of the Air, First Continuation* [8].

A particularly ingenious experiment was *experiment 26* on the swing of a pendulum. Boyle, or more likely Hooke who had done extensive studies on pendulums, expected that a pendulum would swing slightly more slowly and that the motion would last longer before decaying in air as opposed to a partial vacuum because the air would impede the motion. Accordingly, two identical pendulums were constructed and set in motion, one inside the receiver, which was evacuated, and one outside. To the surprise of the observers, no consistent differences could be measured between the two pendulums. An interesting sequel to this is that it is now accepted that the viscosity of a gas is independent of its pressure, a finding that is counterintuitive to most people. One of the standard textbooks of fluid dynamics when discussing this refers the reader to Boyle's experiment done 350 years ago [11].

Another surprising finding described in *experiment 37* occurred on some occasions immediately after the piston was drawn down and the stopcock was then opened. The reduction of pressure in the receiver was accompanied by a "kinde of Light in the Receiver, almost like a faint flash of Lightning in the Day-time, and almost as suddenly did it appear and vanish." The phenomenon was not always seen but only when the engine was "in a good humour" as Boyle put it. The "Apparition of Light" could be seen by both candlelight and daylight, and following its appearance the sides of the receiver seemed to be darkened as if some "whitish Steam" adhered to the wall. Boyle was at a loss to understand this phenomenon, but it was probably a transient condensation of water vapor produced by the sudden reduction in air pressure.

Behavior of liquids (experiments 20–25, 28, 35, 42, and 43). When a liquid such as water was placed in the receiver within a vial that was tightly closed with a cork, no changes in the appearance of the liquid were seen. However, if the water was suddenly exposed to the low pressure, for example by breaking the glass neck of the vial, the water immediately started to bubble vigorously. In one experiment, pieces of red coral covered by vinegar were placed in the receiver. The result was the formation of a small number of bubbles at normal pressure, but when the air was rarified with the pump the bubbling increased greatly. In a further experiment, water that had been "boyl'd a pretty while" was placed inside the receiver, and this time no bubbles appeared with the first three descents of the piston. However, when the pressure was further reduced, the water suddenly appeared to boil in the vial "as if it had stood over a very quick Fire."

The results of these experiments are not surprising to us today, but to Boyle they raised the whole question of the nature of gases and liquids and particularly how it was possible for a liquid to contain a gas.

Miscellaneous experiments (experiments 29–31, 34, 38, and 39). Boyle's curiosity and ingenuity resulted in a number of other experiments, which are now of generally minor interest either because the methods were flawed or because the objectives are of little relevance today. He studied the behavior of smoke in a partial vacuum because some philosophers had argued that "Steams or Exhalations" ascend because of "positive levity," i.e., negative weight. The confusion here was the failure to recognize that warming a gas reduced its density. In another experiment, Boyle wondered whether two "exquisitely polish'd" flat pieces of marble would be pressed against each other by atmospheric pressure in the same way as Guericke's hemispheres. However, he was unable to obtain two pieces so exactly ground that they remained in contact for more than 1 or 2 min, which was the time it took him to prepare the experiment. Because he thought that the surfaces were not completely flat, he wondered whether moistening them with "Spirit of wine" would help, but again the experiment was unsuccessful and, in any event, Boyle was not aware of the phenomenon of surface tension.

Experiment 34 was ingenious in that he wondered whether he could demonstrate the buoyancy of air by comparing the weight of a half-filled bladder tied securely at the neck at normal pressure and after reducing the pressure. The weight was obtained by placing the bladder on a balance with metal weights at the other end. The results of this experiment were apparently confused by changes in the weight of the bladder caused by moisture in the air, and Boyle did not reach any clear conclusion.

An experiment in which Boyle introduced a mixture of snow and common salt into the receiver did not give any clear results, but it allowed Boyle to digress on how ice that formed when water freezes can exert enormous pressures resulting in the breaking of stones and other structures. Boyle was unaware of the fact that water expands as ice is formed.

4.6   Conclusion

The Boyle-Hooke pneumatical engine not only was a triumph of engineering, but it opened up a whole new field of study. In its day, it was famous and, for example, it was often displayed to visitors who came to the Royal Society. Interestingly, Boyle did not immediately continue his experiments on reduced air pressure after 1660 but, for example, in 1661 published a book on chemistry, *The Sceptical Chymist* [5] and, in a completely different area, *Style of the Scriptures* [4]. However, in 1662, he published a second edition of the 1660 book with an addendum *Whereunto is Added a Defence of the Authors Explication of the Experiments, Against the Obiections of Franciscus Linus and Thomas Hobbes* [6]. Linus had argued that, although the column of mercury in the Torricellian barometer was partly raised by atmospheric pressure, there was another structure called a "funiculus" within the vacuum that helped to support it. Boyle's vigorous response was that Linus' hypothesis of a funiculus was "partly precarious, partly unintelligible, and partly insufficient, and besides needless." But the most enduring feature of his response was the demonstration of the inverse relationship between the pressure and volume of a gas, which we now know as Boyle's Law. Hobbes's argument was more philosophical and had to do with the penetrative characteristics of air as an ethereal fluid, and Boyle contended that the experimental evidence was against this. Boyle's 1660 book makes satisfying reading even today, although some modern readers may be put off by the occasional bizarre spelling and typography. For that reason, a modern version of the text [16] is easier to read. At any event, modern students who are interested in high-altitude physiology should be aware of this classic book.

References

1. Acosta I. Historia Natural y Moral de las Indias. Seville: Iuan de Leon; 1590.
2. Bert P. La Pression Barométrique. Paris: Masson; 1878. English translation by Hitchcock MA and Hitchcock FA. Columbus: College Book; 1943.
3. Boyle R. New experiments physico-mechanicall, touching the spring of the air, and its effects (Made, for the most part, in a new pneumatical engine). Written by way of letter to the Right Honorable Charles Lord Vicount of Dungarvan, Eldest Son to the Earl of Corke. Oxford: H. Hall; 1660.
4. Boyle R. Some considerations touching the style of the H. Scriptures. London: H. Herringman; 1661.
5. Boyle R. The sceptical chymist, or, chymico-physical doubts paradoxes. London: J. Cadwell; 1661.
6. Boyle R. New experiments physico-mechanicall, touching the air: where unto is added a defence of the authors explication of the experiments, against the obiections of Franciscus Linus and Thomas Hobbes. Oxford: H. Hall; 1662.
7. Boyle R. New experiments concerning the relationship between light and air (in shining wood and fish). Phil Trans. 1667;2:581–600.
8. Boyle R. A continuation of new experiments, physico-mechanicall, touching the spring and weight of the eir, and their effects. Oxford: H. Hall; 1669.
9. Boyle R. New pneumatical observations about respiration. Phil Trans. 1670;5:2011–31.
10. Boyle R. A continuation of new experiments physico-mechanicall, touching the spring and weight of the air, and their effects. London: Miles Flesher; 1682.
11. Chapman S, Cowling TG. The mathematical theory of non-uniform gases. 3rd ed. Cambridge: Cambridge University Press; 1970.
12. Conant JB. Harvard case histories in experimental science, volume 1. Cambridge: Harvard University Press; 1957.
13. Frank R. Harvey and the Oxford physiologists. Berkeley: University of California Press; 1980.
14. Galileo G. Dialogues concerning two new sciences [1638], translated by Crew H and de-Salvio A. New York: MacMillan Publishing; 1917.
15. Harvey W. Exercitatio Anatomica de Motu Cordis et Sanguinis in Animalibus. Frankfurt a. M.: Guilielmi Fitzeri; 1628.
16. Hunter M, Davis EB. The works of Robert Boyle, 14 vols. London: Pickering Chatto; 1999.

17. Schott GP. Gasparis Schotti Mechanica Hydraulico-Pneumatica. Francofurti ad Moenum: Sumptu Heredum Joannis Godefridi Schönwetteri Bibliopol. Francofurtens: excudebat Henricus Pigrin, Typographus Herbipoli; 1657.
18. Stirling W. Some apostles of physiology. London: Waterlow and Sons; 1902.
19. Torricelli E. Letter of Torricelli to Michelangelo Ricci, [1644] (English translation of relevant pages in high altitude physiology, edited by West JB). Stroudsburg: Hutchinson Ross; 1981.
20. West JB. The original presentation of Boyle's Law. J Appl Physiol. 1999;87:1543–5.

Chapter 5
The Original Presentation of Boyle's Law

Abstract The original presentation of what we know as Boyle's law has several interesting features. First, the technical difficulties of the experiment were considerable, because Boyle used a glass tube full of mercury that was nearly 2.5 m long, and the large pressures sometimes shattered the glass. Next, Boyle's table of results contains extremely awkward fractions, such 10/13, 2/17, 13/19, and 18/23, which look very strange to us today. This was because he calculated the pressure for a certain volume of gas by using simple multiplication and division, keeping the vulgar fractions. Boyle was not able to express the numbers as decimals because this notation was not in common use at the time. Finally, his contention that pressure and volume were inversely related depended on the reader's comparing two sets of numbers in adjacent columns to see how well they agreed. Today we would plot the data, but again orthogonal graphs were not in general use in 1662. When Boyle's data are plotted by using modern conventional methods, they strongly support his hypothesis that the volume and pressure of a gas are inversely related.

Every student of physiology is familiar with Boyle's law, which states that the pressure and volume of a gas are inversely related (at constant temperature). In fact, Robert Boyle (1627–1691) did not refer to a law as such, but to a hypothesis which, he argued, was supported by experimental data. The way in which Boyle presented the data has some remarkable features.

Figure 5.1 shows the original table from his 1662 paper, *New Experiments Physico-Mechanical, Touching the Air: Whereunto is Added A Defence of the Authors Explication of the Experiments, Against the Obiections of Franciscus Linus, and, Thomas Hobbes* [3]. This was the second edition of a book of almost the same title which had been published 2 years earlier [2]. Linus had objected to Boyle's contention that the pressure of the atmosphere was sufficient to raise a mercury column by ~29 in., and he argued that something invisible above the mercury must be holding it up. Boyle countered the objection of Linus by showing that, if air was trapped in the small limb of a U tube and the long limb was filled with mercury, the compressed air was capable of generating a very high pressure, which could support a very long column of mercury

Boyle described the experiment in the text accompanying his table, shown in Fig. 5.1. With considerable difficulty, he procured a glass U tube, the longer leg of

© American Physiological Society 2015
J. B. West, *Essays on the History of Respiratory Physiology,*
Perspectives in Physiology, DOI 10.1007/978-1-4939-2362-5_5

A table of the condensation of the air.

A	*A*	*B*	*C*	*D*	*E*
48	12	00		29 2/16	29 2/16
46	11 1/2	01 7/16		30 9/16	33 6/16
44	11	02 13/16		31 15/16	31 12/16
42	10 1/2	04 6/16		33 8/16	33 1/7
40	10	06 3/16		35 5/16	35 - -
38	9 1/2	07 14/16		37	36 15/19
36	9	10 2/16		39 5/16	38 7/8
34	8 1/2	12 8/16		41 10/16	41 2/17
32	8	15 1/16		44 3/16	43 11/16
30	7 1/2	17 15/16		47 1/16	46 3/5
28	7	21 3/16		50 5/16	50 - -
26	6 1/2	25 3/16		54 5/16	53 10/13
24	6	29 11/16	Added to 22 1/8 makes	58 13/16	58 2/8
23	5 3/4	32 3/16		61 5/16	60 18/23
22	5 1/2	34 15/16		64 1/16	63 6/11
21	5 1/4	37 15/16		67 1/16	66 4/7
20	5	41 9/16		70 11/16	70 - -
19	4 3/4	45 - -		74 2/16	73 11/19
18	4 1/2	48 12/16		77 14/16	77 2/3
17	4 1/4	53 11/16		82 12/16	82 4/17
16	4	58 2/16		87 14/16	87 3/8
15	3 3/4	63 15/16		93 1/16	93 1/5
14	3 1/2	71 5/16		100 7/16	99 6/7
13	3 1/4	78 11/16		107 13/16	107 7/13
12	3	88 7/16		117 9/16	116 4/8

AA. The number of equal ſpaces in the ſhorter leg, that contained the ſame parcel of air diverſly extended.

B. The height of the mercurial cylinder in the longer leg, that compreſſed the air into thoſe dimenſions.

C. The height of the mercurial cylinder, that counterbalanced the preſſure of the atmoſphere.

D. The aggregate of the two laſt columns *B* and *C*, exhibiting the preſſure ſuſtained by the included air.

E. What that preſſure ſhould be according to the hypotheſis, that ſuppoſes the preſſures and expanſions to be in reciprocal proportion.

Fig. 5.1 Original table of data and calculations given by Boyle [3] to support his hypothesis that the pressure and volume of a gas are inversely related. The letter "s" appears to the modern reader to be an "f", and fractions (where given) in *column E* are difficult to read. They are as follows (*top* to *bottom*): 2/16, 6/16, 12/16, 1/7, 15/19, 7/8, 2/17, 11/16, 3/5, 10/13, 2/8, 18/23, 6/11, 4/7, 11/19, 2/3, 4/17, 3/8, 1/5, 6/7, 7/13, 4/8

which was nearly 8 ft. (2.44 m) long, while the shorter leg was some 12 in. (30.5 cm) long and was sealed at the end. He then prepared a narrow piece of paper, on which he marked 12 in. and their quarters, and he placed this in the shorter limb. A similar piece of paper, again divided into inches and quarters, was placed in the longer limb. Holding the U tube vertically, he then poured mercury into the long limb so that a column of air 12 in. long was trapped in the short limb, and the mercury levels in the two limbs were initially the same. This was the situation represented by the top row of numbers in Fig. 5.1. He then carefully added more mercury, little by little, to the long limb, and he observed the compression of the column of air in the short limb. For example, the second row of the second column of Fig. 5.1 shows that he stopped adding mercury for the second set of readings when the length of the air column was 11 1/2 in. The third column (*B*) shows that, at this time, the additional height of the mercury in the longer limb was 1 7/16 in. Additional mercury was then added until the air column was 11 in. high (*row 3*), at which time the additional height of the mercury in the long column was 2 13/16 in. Although the paper strip in the long limb was only divided into quarters of an inch, Boyle was able to interpolate and measure the height of the mercury to one-quarter of each small division (that is, 1/16 of an inch). In all, Boyle added mercury 24 times, until the length of the column of air was reduced to 3 in. (bottom row, *second column A*) and the additional height of the mercury was 88 7/16 in. (bottom row, *column B*).

Boyle added some interesting details on how he carried out this experiment. He had trouble with the breaking of the glass tubes because of the high pressures developed by the long column of mercury, so the lower part of the tube was placed in a square wooden box. This allowed him to catch the valuable mercury. As indicated above, the mercury was poured in very slowly because, as Boyle noted, it was "far easier to pour in more, than to take out any, in case too much at once had been poured in." The long tube was so tall that the experiment was carried out in a stairwell. Boyle also used a small mirror behind the tube to help him measure the height of the mercury accurately.

As indicated above, the second column of the table (A) and third column (B) show, respectively, the length of the trapped column of air and the additional height of the mercury in the long limb (both in inches). The first column of the table (also headed A) is simply the number of quarter-inches occupied by the trapped air. In other words, it is simply the second column multiplied by 4, and it is proportional to the volume of the gas (assuming a constant cross-sectional area of the tube). In the fourth column, headed C, Boyle states "added to 22 1/8." This is actually a misprint. The correct value is 29 1/8 in., which Boyle took to be the height of a mercury column supported by the normal atmospheric pressure. Therefore, the fifth column (headed D)is the sum of *column B* and 29 1/8 in., except that all the fractions in *column D* are given in 16ths. This column shows the pressure to which the bubble of gas was subjected. The last column (E) is the calculated pressure for the volume shown in the first column (A) according to Boyle's hypothesis that volume and pressure are inversely related. The fractions are difficult to read and are listed in the Fig. 5.1 legend.

The first thing that strikes today's reader is the extremely awkward fractions such as 2/17, 10/13, and 18/23. How did Boyle end up with such strange numbers? The answer is that he used simple multiplication and division and kept the vulgar fractions. As an example of how the pressures in *column E* were calculated, consider *row 6*, where the value in the first *column (A)* is 38. The hypothesis is that $P_1 V_1 = P_2 V_2$ or $P_2 = P_1 V_1 / V_2$, where P is pressure and V is volume. The first row shows that P_1 is 29 2/16 while V_1 is 48. Because V_2 is 38, the expression is (29 2/16 × 48)/38 which gives $P_2 = 36$ 15/19 in. of mercury (see Appendix).

Finally, Boyle invited the reader to compare the measured pressure in *column D* with the calculated pressure in *column E*. The agreement between the two was close over a large range of pressures, from 29 1/8 to 117 9/16 in. of mercury, which corresponds to a range of volumes from 48 to 12 (that is, a factor of 4).

There are additional interesting points about Boyle's table. First, why did Boyle not convert the values to decimals rather than keep the very awkward fractions such as 15/19? As an example, it is not immediately clear to most of us to what extent 107 13/16 differs from 107 7/13. However, if we put these two numbers into a decimal format as 107.813 and 107.538, we can immediately see that the difference between the two is 0.3 %.

The explanation of why Boyle did not use decimal notation is that this was not in general use in 1662. Some notations that bear some resemblance to modern decimals can be found in ancient China and medieval Arabia [1, 4]. Simon Stevin (1548–1620) is often credited with the first use of decimals as we know them today, and these were introduced into England in about 1608. Napier's book on logarithms [8] used decimals in the modern notation. However, it was not until the eighteenth century that decimal notation become standardized, and, therefore, it is not surprising that Boyle did not make use of decimals.

Another surprising feature for the modern reader about Boyle's treatment of his data is why he did not plot the results in a graph. One problem with the table shown in Fig. 5.1 is that the reader has to inspect each row of *columns D* and *E* to get a feeling for how much the observed pressures differed from the pressures calculated from the hypothesis. In addition, we need to compare the differences in the top rows with the bottom rows to see if the differences depend on the magnitude of the pressure or volume. Finally, we need to scan the table to make sure that the whole range of pressures/volumes is covered more or less uniformly.

All this information is quickly seen if the results are plotted as shown in Fig. 5.2. Because the hypothesis is that pressure and volume are inversely related, we could plot either the reciprocal of volume against pressure or the reciprocal of pressure against volume to test whether the points lie on a straight line. Both give essentially the same information, but the latter (shown in Fig. 5.2) is preferred, because Boyle added mercury for each measurement until the volume reached an exact one-quarter inch; therefore, volume can be regarded as the independent variable. Figure 5.2 immediately shows, to most people's satisfaction, that the reciprocal of pressure is linearly related to volume, that the deviations of the data points from the straight line of best fit do not depend on the magnitude of the volume, that the whole range

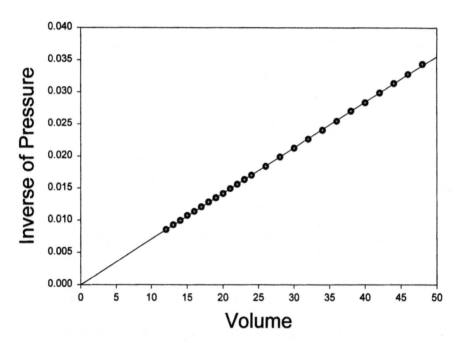

Fig. 5.2 Plot of Boyle's data, showing the reciprocal of pressure against volume. Units on vertical axis are (inches of mercury)-1, and units on the horizontal axis are (cross-sectional area of the tube/4) in cubic inches. The points lie close to the line of best fit, and the intercept on the vertical axis is very small. See text for details. The analysis supports Boyle's hypothesis

of volumes is adequately covered, and that the straight line passes very close to the origin. In formal terms, the coefficient of determination (R^2), the most common measure of how well a regression model describes the data, is >0.9999, the square root of the sum of squares of the residuals is only 3.77×10^{-4}, and the y-axis intercept is 2.474×10^{-5}.

Again, the explanation for why Boyle did not plot his data in this way is that graphs to depict data were not in general use in 1662. For example, in "La Géométrie" which René Descartes (1596–1660) wrote as an appendix to his major work of 1637 [6], few orthogonal coordinate systems are used for graphs. In fact, a number of the graphical methods for depicting data that we use today, such as pie graphs, were not introduced until the nineteenth century.

There are other ways of treating Boyle's data to determine how well it fits his hypothesis. One was an analysis by Geary [7], who used Boyle's data to derive the equation

Log P = 1.00404 Log V + C where C is a constant. Geary concluded that "with only 25 pairs of observations, it is evident that the numerical coefficient is not significantly different from unity." The implication is that pressure X volume (P X V) is constant.

Of course, pressure X volume is constant only if temperature is constant. Boyle was aware of this, although he did not pursue it. In the text near the table shown in Fig. 5.1, he described how he heated the trapped air in the small limb of the U tube with the flame of a candle "so that we scarce doubted, but that the expansion of the air would, notwithstanding the weight that opprest it, have been made conspicuous, if the fear of unseasonably breaking the glass had not kept us from increasing the heat" [3].

Finally, we need not go into the issue of to what extent Boyle's work was based on the work of others and, indeed, whether Boyle's name should even be attached to the law. This has been discussed extensively elsewhere [5, 9, 10]. Suffice it to say that at least five other people have some claim to be considered, including Viscount Brouncker, Robert Hooke, Edmé Mariotte, Henry Power, and Richard Towneley. In fact, in France the law is known as "la loi de Mariotte." However, many of us will be satisfied that Boyle's law remains as a reminder of this remarkable scientist.

Appendix

For people whose arithmetic is rusty, here is the calculation keeping the vulgar fractions.

The original expression is (29 2/16 × 48)/38.
Converting the numbers inside the brackets to 16ths gives
(29 × 16 + 2) x 48 or 22368
Dividing this by 16 gives 1398.
Dividing this by 38 gives 36 + 30/38 or 36 15/19

References

1. Boyer CB A history of mathematics. New York: Wiley; 1968.
2. Boyle R New experiments physico-mechanicall, touching the spring of the air, and its effects. Oxford: H. Hall for T. Robinson; 1660.
3. Boyle R New experiments physico-mechanical, touching the air: whereunto is added a defence of the authors explication of the experiments, against the obiections of Franciscus Linus, and, Thomas Hobbes. Oxford: H. Hall for T. Robinson; 1662.
4. Cajori F A history of mathematical notations. Chicago: Open Court; 1928.
5. Cohen IB Newton, Hooke, and 'Boyle's Law'. Nature. 1964;204:618–21.
6. Descartes R Discours de la méthode pour bien conduire sa raison chercher la verité dans les sciences. Leiden: Jan Maire; 1637.
7. Geary RC Accuracy of Boyle's original observations on the pressure and volume of a gas. Nature. 1943;151:476.
8. Napier J Mirifici Logarithmorum Canonis Descriptio. Edinburgh: Andrew Hart; 1614.
9. Webster C Richard Towneley and Boyle's law. Nature. 1963;197:226–8.
10. Webster C The discovery of Boyle's law, and the concept of the elasticity of air in the seventeenth century. Arch Hist Exact Sci. 1965;2:441–502.

Chapter 6
Robert Hooke: Early Respiratory Physiologist, Polymath, and Mechanical Genius

Abstract Robert Hooke (1635–1703) was a polymath who made important contributions to respiratory physiology and many other scientific areas. With Robert Boyle, he constructed the first air pump that allowed measurements on small animals at a reduced atmospheric pressure and this started the discipline of high altitude physiology. He also built the first human low pressure chamber and described his experiences when the pressure was reduced to the equivalent of an altitude of about 2400 m. Using artificial ventilation in an animal preparation he demonstrated that movement of the lung was not essential for life. His book *Micrographia* describing early studies with a microscope remains a classic. He produced an exquisite drawing of the head of a fly showing the elaborate compound eye. There is also a detailed drawing of a flea, and Hooke noted how the long, many-jointed legs enable the insect to jump so high. For 40 years he was the curator of experiments for the newly founded Royal Society in London and contributed greatly to its intellectual ferment. His mechanical inventions covered an enormous range including the watch spring, wheel barometer and the universal joint. Following the Great Fire of London in 1666 he designed many of the new buildings in conjunction with Christopher Wren. Unfortunately Hooke had an abrasive personality which was partly responsible for a lack of recognition of his work for many years. However during the last 25 years there has been renewed interest and he is now recognized as a brilliant scientist and innovator.

6.1 Introduction

Among the early respiratory physiologists, Robert Hooke (1635–1703) stands out as a polymath who excelled in an extraordinary range of areas. With Robert Boyle (1627–1691) he built the first air pump that allowed physiological measurements to be made on small animals at low barometric pressures. This marked the beginning of the study of high altitude physiology and clarified the physiological effects of hypoxia. Hooke went on to make one of the first studies of artificial ventilation showing how the lung could function as a gas exchanger in the absence of its movement. Another first was the building of a human decompression chamber in which

© American Physiological Society 2015
J. B. West, *Essays on the History of Respiratory Physiology,*
Perspectives in Physiology, DOI 10.1007/978-1-4939-2362-5_6

he described his own experiences during acute hypobaria. In a different field, he exploited the newly invented microscope to clarify structure and function in small animals. His book *Micrographia* contains many exquisite drawings and remains a classic. Most of this work was carried out while he was the full-time curator of experiments for the Royal Society and the records of that early institution dramatically describe the intellectual ferment in England during the middle of the seventeenth century [2].

Hooke was a prolific and ingenious inventor with a host of novel ideas covering a wide range of objects including telescopes, watch springs, pendulums and gears. He formulated the law of elasticity that states that strain is proportional to stress. Following the Great Fire of London in 1666 that destroyed much of the city, Hooke was responsible for planning many of the new buildings and he had a reputation as a talented architect. For example, he is credited with assisting Christopher Wren (1632–1723) with the design of the dome of St. Paul's Cathedral. The Monument to the Great Fire can be considered as a monument to Hooke.

Unfortunately Hooke's personality was abrasive and this may be one reason why history has not treated him as well as his brilliance deserves. However during the last 25 years, and particularly following the tercentenary of his death in 2003, there has been a resurgence of interest in Hooke and the publications of several books [1; 6; 7; 8; 14; 15; 16]. For an introduction, Inwood [14] is recommended. Hooke's diary [18] also received increased attention, as did his many contributions set out in the History of the Royal Society of London [2]. Few medical and graduate students today are aware of Hooke's genius and this essay may help to remedy this.

6.2   Brief Biography

Hooke was born in the village of Freshwater in the Isle of Wight just off the south coast of England. His father was the minister of the local parish and had modest means. Hooke was a sickly boy but at an early age he exhibited an interest in mechanical things. When he was 14 he was sent to London where he briefly studied art with the portraitist Peter Lely (1618–1680) and this presumably helps to explain his later beautiful drawings. He then entered Westminster School where again he showed an aptitude in mechanics. At the age of 18 he moved to Christ Church College, Oxford as a chorister and this appointment was probably helped by the fact that he had previously shown an ability to play the organ.

Once at Oxford his interest in science flowered. He first worked as an assistant to Thomas Willis (1621–1675) who was a brilliant scientist and one of the founding members of the Royal Society. Willis described a number of diseases for the first time but is best known for the Circle of Willis, that is the anastamosis of blood vessels at the base of the brain. More important for Hooke was that he became an assistant to Robert Boyle, a wealthy aristocrat and productive scientist. During the period from about 1657 to 1662 Hooke worked with Boyle on the development of the first air pump, and the ensuing experiments on the physiological effects of low atmospheric pressure (see below). Hooke made many important friends in Oxford including the celebrated architect Christopher Wren, the clergyman and polymath John Wilkins (1614–1672) and the astronomer Seth Ward (1617–1689). This illustrious group were among those responsible for founding the Royal Society of London in 1660.

In 1662 Hooke was chosen to be the first curator of experiments for the Royal Society where his responsibilities were to provide three or four experiments at each meeting. This onerous task was not always fulfilled but Hooke was extremely active with many demonstrations and commentaries. The meetings of the Society were recorded every week and eventually printed [2] where they make fascinating reading even today. Hooke showed great enthusiasm and industry in this position which he held for 40 years, and the range of topics covered by the Society was extraordinary. His book *Micrographia* [10] was published when he was only 30 years old and was the first major publication of the Royal Society. With its beautiful copper plate engravings it was a bestseller at the time and remains enjoyable reading today.

Hooke's position as curator in the Royal Society gave him a high profile but was unpaid. However in 1664 Hooke was appointed as lecturer in Gresham College in London. As a result he became what was probably the first professional research scientist in the area of natural history. He published a series of lectures that he had delivered at Gresham College and these are known as the Cutler Lectures named after the endowment. They covered a great variety of topics including studies of the motion of the earth by observation of a bright star, improvements in the construction of telescopes, and the properties of springs. This last included what we now know as Hooke's law, that is that in a spring, strain is proportional to stress. Hooke also claimed some precedence in proposing that gravitation obeys an inverse square law, and this resulted in acrimonious correspondence with Newton who never gave Hooke credit for the idea and indeed generally belittled Hooke's contributions. In 1677 Hooke was elected as one of the two secretaries of the Royal Society and he was active in publishing its discussions.

The Great Fire of London occurred in 1666 and Hooke played an important role in planning the reconstruction of the city. In fact less than a week after the end of the fire, he had produced a model for rebuilding the burned area. He was one of the three official surveyors for the project and he designed a number of buildings in conjunction with Christopher Wren. One of these is the Monument to the fire which remains today. He acquired a reputation as a skilled architect although in many instances it is impossible to separate his contributions from those of Wren's.

Hooke's character had serious flaws which may have diminished his place in history. He was involved in a series of bitter disputes with a number of people including Newton and Henry Oldenburg who preceded Hooke as secretary of the Royal Society. Hooke was affected by ill health in his last years and he died in London in 1703. His grave is not marked. It is extraordinary that no authenticated likeness of Hooke has survived and this is perhaps unique among distinguished scientists of his time.

6.3 Air Pump and the First Experiments on Rarified Air

In the field of respiratory physiology, one of Hooke's major achievements was constructing the first air pump designed for physiological research, and that resulted in Robert Boyle's landmark book "New experiments physico-mechanicall, touching the spring of the air and its effects" [4]. Hooke's pump and the experiments that Boyle and he carried out are described in detail in Chap. 4. In fact Hooke is barely mentioned in the book although later Boyle acknowledged his critical contributions. There is no doubt that Hooke was responsible for the design and construction of the pump which was far ahead of its time. Indeed Fig. 4.3 in Chap. 4 with its emphasis on the mechanical details of the construction shows that this was made by someone who had special technical skills.

6.4 Artificial Ventilation

In the account of the history of the Royal Society [2] an entry for October 24, 1667 reads "Mr. HOOKE's account of the experiment of keeping a dog alive by blowing into his lungs, and even without the motion of his lungs, only by keeping them extended with a constant supply of fresh air, was read, and ordered to be registered". "Registered" meant that it was to be written up and included in the journal *Philosophical Transactions of the Royal Society of London* where it can be read today [11]. Hooke begins his account with "And because some Eminent Physicians have affirmed that the Motion of the Lungs was necessary to life upon the account of promoting the Circulation of the Blood, and that it was conceiv'd, the Animal would immediately be suffocated as soon as the Lungs should cease to be moved, I did…make the following additional Experiment…" [11]. Here Hooke is referring to authorities such as William Harvey (1578–1657) who wrote "It is manifest that the blood passes through the lungs, not through the septum [in its course from the right to the left side of the heart], and only through them when they are moved in act of respiration, not when they are collapsed and quiescent" [9].

Hooke then went on to describe how a dog's chest was opened and slits were made on the surface of the lungs with a "slender point of a very sharp pen-knife". Two sets of bellows were employed to create a continuous flow of air via the trachea, and this air escaped through the slits. Hooke wrote "This being continued for a pretty while, the dog, as I expected, lay still, as before, his eyes being all the time very quick, and his Heart beating very regularly: But, upon ceasing this blast, and suffering the Lungs to fall and lye still, the Dog would immediately fall into Dying convulsive fits; but he as soon reviv'd again by the renewing the fulness of his Lungs with the constant blast of fresh Air…

"Towards the end of this Experiment a piece of the Lungs was cut quite off; where 'twas observable, that the Blood did freely circulate, and pass thorough the Lungs, not only when the Lungs were kept thus constantly extended, but also when they were suffered to subside and ly still. Which seemed to be Arguments that as the *bare* Motion of the Lungs *without fresh Air* contributes nothing to the life of the Animal, he being found to survive as well when they were not mov'd, as when they were; so it was not the subsiding or movelessness of the Lungs, that was the immediate cause of Death, or the stopping of the Circulation of the Blood through the Lungs, but the *want* of a sufficient supply of *fresh Air*" [11].

Following this experiment, an objection was raised that it was not the fresh air that kept the dog alive but the motion of the bellows. Hooke with his colleague Dr. King therefore carried out another experiment which was reported to the Society on May 9, 1668 [2]. A brass tube was tied into the trachea of a dog and the animal was allowed to rebreathe into a large bladder attached to the tube. The report states "After about three or four minutes, the dog began to struggle violently, and to repeat his endeavors for breath very frequently... yet, after about six minutes, his strength failed a-pace... and then he began to be convulsed; and at the end of about eight minutes, we could see no signs of life". The investigators then removed the bladder and ventilated the dog with fresh air using a bellows and reported "Within less than a minute, the dog, by our moving the thorax first and continually blowing, recovered motion in his breast". The cannula was then removed and it is reported that the dog then walked away. The report continued "And he [the dog] presently fell to licking of himself, as not much concerned".

Hooke also suggested an experiment to determine whether blood changed its color as a result of its passage through the lung, but to do this he needed the assistance of a colleague, Richard Lower (1631–1691) to carry out the surgery. However Lower was not available at the time, but subsequently did the experiment and famously showed that the blood's color was changed from dark to bright red in its passage through the lung. He reported this in his book Tractatus de Corde [17] where he gave credit to "the very famous Master *Robert Hooke*".

Incidentally artificial respiration had previously been used to keep animals alive, for example by Vesalius and Leonardo da Vinci but Hooke's ingenious experiment was designed to answer a very specific question, that is whether motion of the lungs was necessary for life.

6.5 Human Decompression Chamber

We have seen that Hooke's air pump provided the first opportunity to subject small animals to a reduced barometric pressure in the laboratory, and it is perhaps not surprising that Hooke with his extraordinary energy and ingenuity would want to do the same thing to humans. The best account of this comes from a series of entries in "The History of the Royal Society of London" [2].

The first mention is in an entry for January 12, 1670/1. (This curious date stems from the fact that the calendar had recently changed.) Incidentally the entries for that date highlight the great variety of interests of the Royal Society and the lively intellectual activity. For example Hooke showed a curiosity sent to the Society by the Archbishop of Canterbury which was supposed to be several pieces of a hippo-potamus dug up at Chatham southeast of London! Another odd entry stated that the King had laid a wager of fifty pounds to five for the compression of air by water; and that it was acknowledged that his Majesty had won the wager. A further entry stated that Hooke had produced an engine for grinding glasses of a true both ellipti-cal and hyperbolical figure. Finally there is a note, again by Hooke, stating that "he proposed a new way of making a vessel for extracting the air, so large, that man might fit in it, and so contrived, as to rarify the air to a certain degree, and to supply the person sitting in it with fresh air. He was desired to get such a vessel made".

The next relevant entry on February 9 reads "Mr. Hooke being asked, whether the air-vessel for a man to fit in was yet ready, answered, that it was, and that he now intended to make some experiments in it, and to report them at the next meeting. He added, that the chief design of this vessel was to find what change the rarefac-tion of the air would produce in man, as to respiration, heat, etc. Being asked, how it was contrived, he said, that it consisted of two tuns [large barrels], one included in the other; the one to hold a man, the other filled with water to cover the former, thereby to keep it stanch [airtight]; with tops to put on with cement; or to take off; one of them having a gage, to see to what degree the air is rarified; as also a cock to be turned by the person, who sits in the vessel, according as occasion shall require, etc.". It was then agreed that anyone who wished to see the experiment should meet at Gresham College later in the week.

The next report is dated March 2 and states "Mr. Hooke made a report of the suc-cess of the experiment made in the vessel for rarifying the air, viz. that himself had been in it, and by the contrivance of bellows and valves blown out of it one tenth part of the air (which he found by a gage suspended within the vessel) and had felt no other inconvenience but that of some pain in his ears at the breaking out of the air included in them, and the like pain upon the readmission of the air pressing the ear inwards".

Finally on March 23 we read "Mr. Hooke brought in a report of the experiment, which he had again made in the air vessel; which was, that he had blown out one fourth of the air that was in the vessel, estimated by a gage; and that he had contin-ued in it somewhat above a quarter of an hour without any other inconvenience than feeling some pain in his ears, and finding himself deaf, while the straining of the air upon him in blowing out the air; which pain and deafness he likewise found upon the forcible rushing in again of the air into his ears; but that when he came out, and had walked a little while up and down, his hearing returned. He added, that having taken a candle burning with him into this vessel, the candle went out long before he had felt any of that inconvenience in his ears".

Assuming the gauge was correct, Hooke subjected himself to a pressure of about 570 mm Hg which is equivalent to an altitude of about 2400 m. This experiment did not produce any remarkable results but again it exemplifies Hooke's ingenuity.

6.5.1   *Micrographia*

This is the most famous book by Hooke and indeed is one of the best known books in the history of microscopy. Readers who are not aware of it are in for a pleasant surprise.

The history of microscopy is a little unclear but many authorities attribute the invention of the compound microscope, that is one with both an objective and eyepiece lens, to Hans and Zaccharias Janssen in the Netherlands at the end of the sixteenth century. Galileo (1564–1642) apparently used a compound microscope in 1609 but it took some time before it was exploited for scientific research. Marcello Malpighi (1628–1694) was one of the first to use a compound microscope in 1660. However Hooke's *Micrographia* published in1665 had an enormous influence because of its superlative illustrations. Other early microscopists interested in biology were Nehemiah Grew (1641–1712) who was a botanist, Jan Swammerdam (1637–1680) who studied insects, and Antonie van Leeuwenhoek (1632–1723) who was one of the first scientists to see algae and bacteria.

A feature of *Micrographia* is its magnificent engravings based on the hand drawings by Hooke. Here we should remember that Hooke showed an early interest in illustration and that his first apprenticeship when he went to London was with Peter Lely in his art studio. The great detail and elegance of Hooke's drawings are additional evidence of the extraordinary abilities of this man.

The full title of the book is *Micrographia: or some Physiological Descriptions of Minute Bodies Made By Magnifying Glasses with Observations and Inquiries thereupon*. Notice Hooke's use of the term physiological. In addition to his minute studies of the anatomy of various animals he discusses some of the physiological implications of his findings. In his long preface Hooke emphasizes the importance of careful observation, memory and reason, and warns against excessive speculation and philosophy. He then points out that the power of human sight is limited and that much can be gained by increasing its ability using magnifying lenses.

The title of Hooke's book "*Micrographia: or some Physiological Descriptions of Minute Bodies...* might suggest that it deals only with small animals. However this is not the case. The first part describes microscopic appearances of familiar objects such as the point of a needle, the edge of a razor, and various types of cloth including silk. Perhaps Hooke's objective here was to point out how much additional information is afforded by the microscope even when we look at familiar objects. For example although a razor is usually thought to have a sharp clean edge, the microscope reveals that it has many indentations and imperfections.

Another early section is on what Hooke calls glass drops. These are formed when drops of molten glass are rapidly cooled by allowing them to fall into cold water. Hooke was intrigued by the fact that when one of these solid drops was broken the "whole bulk of the drop flew violently, with a very brisk noise, into multitudes of small pieces". He realized that the rapid cooling resulted in a solid body under very large stresses. Interestingly in the discussion of the drops, Hooke stated that "*Heat is a property of a body arising from the motion or agitation of its parts*". He is there-

Fig. 6.1 Hooke's drawing of
the microscopic appearance
of cork in *Micrographia*. He
was the first person to use the
word cell. From [10]

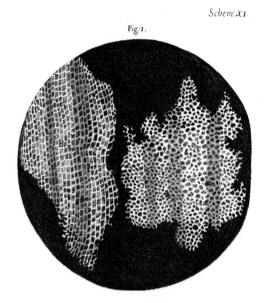

fore getting close to the modern notion of the relation between temperature and the kinetic energy of the molecules of a substance.

Later there is a section on the microscopic appearance of charcoal. The description of the appearance is not particularly interesting and there is no illustration. However Hooke then speculates on combustion stating that some component of the air combines with part of the combustible material. Here he is on a path that could have led to the discovery of oxygen.

Turning to his microscopical studies of plants and animals, a famous section is devoted to the texture of cork. He cut a small piece with a very sharp pen knife and was surprised to find "that it had very little solid substance, in comparison of the empty cavity that was contained between …". He went on to say "in that these pores, or cells, were not very deep, but consisted of a great many little Boxes…". His drawing is shown in Fig. 6.1 and it is often pointed out that this is the first time that the word cell was used in the description of biological tissue. Hooke then goes on to describe the appearance of many leaves of plants including beautiful illustrations of the barbs of stinging nettles.

However the most interesting parts of *Micrographia* deal with his drawings of small animals. One of the most striking is the flea shown in Fig. 6.2. The draftsmanship and attention to detail is outstanding. Of course we should appreciate that what we see here is an engraving based on Hooke's drawing. But it is a superb illustration. Hooke goes on to relate structure and function when he states "For its strength, the *Microscope* is able to make no greater discoveries of it than the naked eye, but only the curious contrivance of its leggs and joints, for the exerting that strength, is very plainly manifest, such as no other creature, I have yet observ'd, has anything like it". In other words he infers the jumping ability of the flea from its anatomy.

Fig. 6.2 Hooke's drawing of a flea in *Micrographia*. He pointed out that the many-jointed structure of the legs explained the extraordinary ability of the flea to jump. From [10]

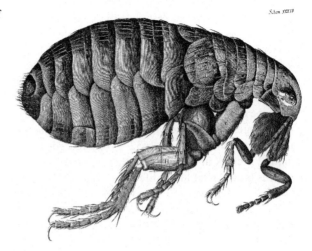

But the climax of Hooke's drawings is perhaps the head of a grey drone-fly with the compound eyes as shown in Fig. 6.3. What a magnificent example of draftsmanship this is. Hooke concedes that he has little notion of the function of the various structures but he is clearly obsessed with the sheer beauty of the insect.

Towards the end of *Micrographia* Hooke has a section on "A new Property in the Air…". The link between this discussion and the earlier microscopic images seems

Fig. 6.3 Head of a grey drone-fly (*Eristalis tenax*) drawn by Hooke in *Micrographia*. The compound eye shows exquisite draftsmanship. From [10]

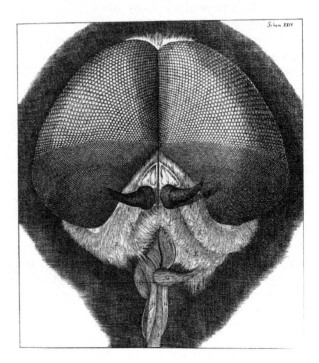

tenuous. However Hooke describes the inverse relationship between pressure and volume in a gas that we normally know as Boyle's law, and of course he was working with Boyle on the air pump when the relationship was first described. He uses this to speculate on the weight of the atmosphere. He then goes on to describe the large number of stars discoverable by the telescope, and also some features of the moon. The only obvious link between this discussion and the rest of the book is that a telescope uses glass lenses just as does a microscope.

The images in *Micrographia* can be revisited many times with pleasure. The famous diarist, Samuel Pepys (1633–1703), recorded that in January 1665 he stayed up until 2 o'clock in the morning reading *Micrographia* which he called "the most ingenious book I read in my life". A convenient version of the book was produced by Project Gutenberg and is now on the Web at http://www.gutenberg.org/ebooks/15491. This includes fine reproductions of the illustrations.

6.6   Mechanical Inventions

One of the most extraordinary features of Hooke was the prolific number of inventions that are attributed to him. These are referred to in his diary [18], the History of the Royal Society [2], and Hooke's various writings. Some authorities state that his inventions number over 1,000. It was noted above that Hooke showed an interest in engineering from an early age, and Fig. 4.3 in Chap. 4 with its drawings of the components of the air pump clearly shows Hooke's interest in the technical features of the machine.

Timekeeping. This was a major interest of Hooke. First he worked extensively on the pendulum which was also an interest of other scientists at the time such as the Dutchman Christiaan Huygens (1629–1695). Hooke studied the isochronicity of the simple pendulum, that is the conditions under which its period remained constant. For example, he studied the effect of altitude and gravity on the period of swing. He also worked extensively on the conical pendulum in which the bob describes a circle. A particularly interesting invention was Hooke's so-called long pendulum. The entry for May 6, 1669 in the History of the Royal Society [2] states that "Mr. Hooke produced a new kind of pendulum of his own invention, having a great weight appendant to it, and moved with a very small force; viz. by such a contrivance, that a pendulum of about fourteen feet long, so as a single vibration of it is made in two seconds, with an excussion [excursion] of half an inch or less, having a weight of three pounds hanging on it, and moved by the sole force of a pocket watch... shall go 14 months...". It seems very counterintuitive that such an enormous pendulum could be maintained swinging for so long by means of the minute power of a pocket watch but this is the case. There is a working model of this pendulum in the Science Museum in London, and the author made a smaller model 12 feet long which is swinging in his study.

Of more practical importance was Hooke's invention of a pocket watch in which the mechanism was a coiled spring attached to the arbor of the balance. However this became a topic of great controversy. The history is complicated but Hooke apparently considered taking out a patent early on but held back because of a technical issue. Some 15 years later Huygens had the same idea and did acquire a patent which annoyed Hooke immensely. There was a great deal of controversy which even involved the king himself. In fact many scholars now believe that Hooke did have priority with this critically important development in timekeeping. One of the reasons for its importance was that it could be used to determine the longitude of a sailing ship. This depended on having an accurate timepiece so that the time when the sun reached its zenith could be accurately determined. From this the longitude was easily determined. Clocks in which the rate was determined by a pendulum were generally not accurate on a ship because of the rolling motion.

Hooke's interest in the properties of springs has already been alluded to. He is credited with inventing the spring balance which is still extensively used. Here the weight in the pan determines the extension of the spring, that is, as the weight so the extension (in Latin, ut pondus, sic tensio). Later this was modified to Hooke's law, that is the extension of the spring is determined by the force (ut tensio, sic vis).

Telescopes. Hooke had a great interest in telescopes and clearly this topic is related to microscopy. He proposed a 60-foot long telescope to be erected in the quadrangle of Gresham College and made a sketch of this. He also invented a lens grinding machine which he claimed could be employed for very large lenses although this was apparently never built and others argued that it would not have the power to complete the task. He proposed a design of a folded telescope with mirrors that would allow a very long focal length but not be as cumbersome. An important invention was an equatorial clock drive for a telescope. This device is universally used today to maintain the alignment of a telescope with the stars by turning the mounting to counteract the rotation of the earth. He also invented the iris diaphragm that is used today in cameras.

A fascinating project that was years ahead of its time was an attempt to determine the diameter of the orbit of the earth around the sun by measuring the apparent change in position, or parallax, of a star using a vertical or zenith telescope. Hooke chose gamma Draco, the brightest star in that constellation, which is directly above London. He cut one-foot square holes in the roof and floor of one of his rooms in Gresham College, mounted a 10-foot long tube with a lens with a focal length of 36 feet, and made observations from the room below. The orientation of the telescope was adjusted using plumb lines. He was unsuccessful because the star which is about 11 light years distant would only change its position by about 0.3 arcseconds (one arcsecond is 1/3600 of a degree). This minute angle was far less than he could measure. Hooke also made another audacious attempt using the 62 meters high Monument to the Great Fire as a zenith telescope (see below) but vibrations caused by traffic thwarted the parallax measurement.

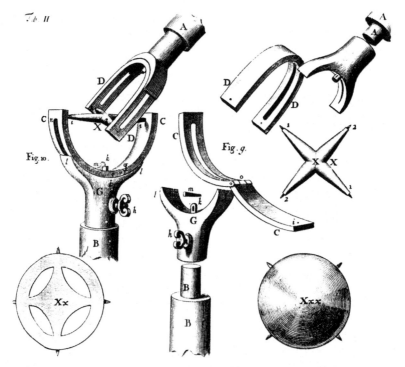

Fig. 6.4 Hooke's invention of the universal joint. This is used in automobiles today. From [12]

Meteorology. Hooked invented the wheel barometer in which the height of the mercury column is indicated on a large dial by a pointer. The mercury is contained in a U tube, and the small limb has a float on the surface of the liquid that is connected by a thread to a wheel on the shaft holding the pointer. This instrument is still frequently found today in the hallway of large houses. Hooke also invented a hygrometer for measuring the humidity of the air. This was based on the coiling or uncoiling of the beard or awn of a wild oat when its moisture alters. Hooke was successful in linking the beard to a pointer as he did in the wheel barometer and he was delighted to find that simply breathing on it resulted in a complete revolution of the pointer [13]. He also developed a hydrometer for measuring the specific gravity of liquids, using a float that submerged to a variable extent.

Gears. Hooke developed the worm gear that is now frequently used for turning a large toothed wheel. This has a large mechanical advantage so that a small torque on the worm drive results in a great deal of force on the toothed wheel. He also invented the universal joint shown in Fig. 6.4. This is a component of the drive train of most automobiles today. He also described a wheel cutting system for making gear wheels. In this the cutter is automatically advanced by a known number of degrees using a slotted wheel and this basic design is frequently used today.

6.7 Architecture

Christopher Wren was one of the first friends that Hooke made when he moved to Oxford, and the two collaborated extensively following the Great Fire of London which destroyed much of the city in September 1666. It gutted over 13,000 houses and nearly 90 churches including St. Paul's Cathedral. Hooke was intimately involved with the reconstruction that followed. In fact he put forward a plan based on a grid system only a week after the fire. Although this system was never used, he helped with the design of many of the replacement buildings.

Hooke assisted Wren in the design of St. Paul's Cathedral that stands today. A feature of this is the very large dome and Hooke was responsible for the basic engineering design of this based on a catenary, that is the shape taken by a chain when it is suspended between two points. He stated "as hangs the flexible line, so but inverted will stand the rigid arch" [12]. In other words in a suspended chain, all the elements (links of the chain) are in tension with no other forces operating. In an arch or dome, it is important that all the elements are exposed only to compressive forces, with no shearing stresses for example. Both the hanging chain and the arch must be in equilibrium, and the forces are simply reversed. For a fuller analysis see [3]. Hooke designed many other buildings including a large Bethlem Royal Hospital for the insane but this building no longer exists. One of Hooke's buildings that is still extant is the church of St. Mary Magdalene in Willen, Milton Keynes, UK.

One of the most interesting structures designed by Hooke with Wren is the Monument to the Great Fire which is still a tourist attraction in London (Fig. 6.5). In fact this is one of the best memorials of Hooke that we have. As noted earlier there is no authenticated portrait of him. But the Monument which was also designed as an enormous zenith telescope 62 m high is vintage Hooke. The top of the column has an urn with a hinged lid that can be opened to allow the telescope to be used. Each of the 311 steps is exactly 6 inches high to facilitate studies of the relation between barometric pressure and altitude. This was a feasible study because the fall in barometric pressure on ascending 62 m above sea level is about 5 mm Hg which would be easily measured using his wheel barometer. Hooke also planned experiments on the effects of altitude on the period of a pendulum. However this experiment was doomed to failure. The altitude of 62 m increases the distance from the center of the earth by only about one part in a million, and gravity which obeys the inverse square law is therefore essentially unchanged. Since the period of a pendulum is inversely proportional to the square root of gravity this will also not be measurably changed.

Incidentally Hooke made a good deal of money by working on the reconstruction of London following the Great Fire and when he died his estate was estimated to exceed £10,000 [7]. This was a remarkable change in fortune for a man who had been born in a family of modest means and whose salary during most of his working life was small.

Fig. 6.5 Monument to the Great Fire of London of 1666 which still stands. It was designed by Hooke and Christopher Wren and also served as a zenith telescope. By permission of the London Metropolitan Archives, City of London

6.8   Diary

Hooke kept a detailed diary from 1672 to 1680 [18]. It was not meant for publication, and the style is very different from that of the diary of his contemporary, Samuel Pepys which is a readable narrative. Hooke's diary consists of a large number of very brief entries about his daily activities and experiences. It was written in microscopic longhand which is almost impossible to read but has been deciphered. For the serious student it makes fascinating reading. One of the interesting features is the large number of friends and colleagues who are mentioned. He also briefly described some of his experiments and there are occasional sketches of proposed inventions. For the prurient, it notes his sexual encounters with the astrological sign of Pisces.

In conclusion, Robert Hooke was one of the most colorful figures in the early history of respiratory physiology. Together with Robert Boyle he pioneered studies on the effects of rarified air and hypoxia. His book *Micrographia* remains a classic. As curator of the Royal Society for 40 years he was involved in an enormous number of new findings. He built the first human decompression chamber and carried out an important experiment on artificial ventilation. He was a prolific inventor and his engineering insights continue to give pleasure. He was heavily involved with rebuilding London after the Great Fire of 1666 although few examples of his architecture remain. Unfortunately he had a difficult personality and was involved in acrimonious arguments with other prominent scientists such as Newton and Huygens. This characteristic was partly responsible for his eclipse for many years but there has been a renaissance of interest during the last 25 years and his brilliance is now generally accepted.

References

1. Bennett JA, Cooper M, Hunter M, Jardine L. London's Leonardo: the life and work of Robert Hooke. Oxford: Oxford University Press; 2003.
2. Birch T. The history of the Royal Society of London; 1756–1757.
3. Block P, DeJong M, Ochsendorf J. As hangs the flexible line: equilibrium of masonry arches. Nexus Network J. 2006;8:13–24.
4. Boyle R. New experiments physico-mechanicall, touching the spring of the air, and its effects (made, for the most part, in a new pneumatical engine) written by way of letter to the Right Honorable Charles Lord Vicount of Dungarvan, eldest son to the Earl of Corke. Oxford: H. Hall; 1660.
5. Boyle R. New experiments physico-mechanical, touching the air: whereunto is added a defence of the authors explication of the experiments, against the objections of Franciscus Linus and Thomas Hobbes. Oxford: H. Hall; 1662.
6. Chapman A. England's Leonardo: Robert Hooke and the art of experiment in Restoration England. Proc Royal Inst Gt Br.1996;67:239–75.
7. Chapman A. England's Leonardo: Robert Hooke and the seventeenth-century scientific revolution. Philadelphia: Institute of Physics; 2005.
8. Cooper M, Hunter M. Robert Hooke: tercentennial studies. Burlington: Ashgate; 2006.
9. Harvey W. An anatomical disquisition of the motion of the heart and blood in animals. Cambridge, 1649. In: The works of William Harvey (translated by R. Willis). Philadelphia, University of Pennsylvania Press; 1939:136.
10. Hooke R. Micrographia: or some physiological descriptions of minute bodies made by magnifying glasses with observations and inquiries thereupon. Royal Society of London; 1665.
11. Hooke R. An account of an experiment made by Mr. Hook, of preserving animals alive by blowing through their lungs with bellows. Phil Trans R Soc Lond. 1667;2:539–40.
12. Hooke R. A description of helioscopes and some other instruments. London. TR for John Martyn; 1676.

13. Hooke R. Microscopic observations, or Dr. Hooke's wonderful discoveries by the microscope. London: print for Robert Wilkinson; 1780.
14. Inwood S. The forgotten genius: the biography of Robert Hooke 1635–1703. San Francisco: MacAdam/Cage; 2003.
15. Kent P, Chapman A. Robert Hooke and the English renaissance. Leominster: Gracewing; 2005.
16. Keynes G. A bibliography of Dr. Robert Hooke. Oxford: Clarendon Press; 1960.
17. Lower R. Tractatus de Corde. Item de motu & colore sanguinis et chyli in eum transitu, translated by K.J. Franklin. London; 1669. In: Early Science in Oxford, Vol. XI, edited by R.T. Gunther. Oxford; 1935.
18. Robinson, HW, Adams, W. editors. The diary of Robert Hooke, M.A., M.D., F.R.S., 1672–1680. London: Taylor & Francis; 1968.
19. Schott GP. Gasparis Schotti Mechanica *Hydraulico-Pneumatica*. Francofurti ad Moenum: Sumptu Heredum Joannis Godefridi Schönwetteri Bibliopol. Francofurtens, excudebat Henricus Pigrin, Typographus Herbipoli; 1657.
20. West JB. The original presentation of Boyle's Law. J Appl Physiol. 1999;87:1543–5.
21. West JB. Robert Boyle's landmark book of 1660 with the first experiments on rarified air. J Appl Physiol. 2005;98:31–9.
22. West JB. Torricelli and the ocean of air: the first measurement of barometric pressure. Physiology (Bethesda). 2013;28:66–73.

Chapter 7
Marcello Malpighi and the Discovery of the Pulmonary Capillaries and Alveoli

Abstract Marcello Malpighi (1628–1694) was an Italian scientist who made outstanding contributions in many areas including the anatomical basis of respiration in amphibia, mammals and insects, and also in the very different fields of embryology and botany. He was one of the first biologists to make use of the newly-invented microscope, and is best known as the discoverer of the pulmonary capillaries and alveoli. However he also discovered the spiracles and tracheae that enable respiration in insects. His studies of the embryology of the chicken were far ahead of his time, and he then turned to the anatomy of plants where he made important contributions. Indeed in some articles he is referred to as the father of embryology, and in other publications as one of the fathers of plant anatomy. His work on the lung was chiefly carried out on the frog and he referred to this animal as the "microscope of nature" because it allowed him to see structures that were not visible in larger animals such as mammals. He also argued that nature undertakes its great works in larger animals after a series of attempts in lower animals. For breadth of interest, innovation and productivity, it is not easy to think of his equal in the field of life sciences.

7.1 The Man

Marcello Malpighi (1628–1694) was born in Crevalcore near Bologna into a family that was comfortably off (Fig. 7.1). An interesting tidbit about his date of birth is that this was the year of publication of William Harvey's De motu cordis describing the circulation of the blood, and in a sense Malpighi completed Harvey's missing link on the pulmonary circulation. Little is known about Malpighi's childhood and youth except that he was schooled in "grammatical studies" in Bologna. Readers who become interested in Malpighi are fortunate that Adelmann [1] has written an exhaustive study in five large volumes which is not only extremely detailed but also highly readable.

Malpighi studied philosophy for a few years but in 1653 he turned his attention to anatomy at the University of Bologna, and this was the beginning of an extraordinarily productive career in this science. In 1656 he was invited to be professor of theoretical medicine at the University of Pisa where he began a long

© American Physiological Society 2015
J. B. West, *Essays on the History of Respiratory Physiology*,
Perspectives in Physiology, DOI 10.1007/978-1-4939-2362-5_7

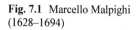

Fig. 7.1 Marcello Malpighi (1628–1694)

friendship with Giovanni Borelli (1608–1679). This man was an eminent mathematician and naturalist who was active in the Accademia del Cimento, one of the earliest scientific societies, and Malpighi became a member. However in 1659 Malpighi decided to return to Bologna and the subsequent 2 years were very productive with extensive discoveries about the lung. He moved again in 1662 to a professorship in medicine at the University of Messina, Sicily, and in fact throughout his life he spent periods at both Pisa and Messina although he always regarded Bologna as his home. His time in Messina also was very productive but he returned to Bologna in 1667.

In 1668 Malpighi received a letter from the Royal Society in London inviting him to send manuscripts to this prestigious group. Malpighi was honored by the invitation and as a result he carried out a classical study on the anatomy of the silkworm. He sent the manuscript to the Society which published it in 1669 [7] and made him a member. He maintained strong links with that group throughout the rest of his life and most of his anatomical studies were published by the Society. The admiration was mutual. On one occasion the secretary wrote to him "… to no one … observing the structure of the human body does Nature seem to have revealed her secrets as fully as to her beloved Malpighi… So too our Royal Society embraces no one with greater affection" [1].

Unhappily Malpighi was involved in a number of bitter disputes about his scientific discoveries throughout his life and he also suffered from ill health in his later years. In 1684 there was a disastrous fire at his home which destroyed many of his manuscripts and much of his equipment. In 1691 he was invited to Rome by Pope Innocent XII to be the Pope's personal physician which was a high honor. Malpighi died of a stroke in 1694 and is buried in the church of Santi Gregorio e Siro in Bologna where there is a memorial.

Malpighi was an extraordinarily productive scientist. As we shall see he was the first person to describe the pulmonary capillaries and the alveoli. In addition he was the first to describe the anatomical basis of insect respiration as a result of his studies on the silkworm. Many of his discoveries initially bore his name and some still do. He discovered the renal glomeruli (initially called the Malpighian corpuscles), the Malpighian corpuscles in the spleen, the Malpighi layer in the skin and the Malpighian tubules in the excretory system of insects. He then went on to make extensive botanical studies, and finally did such extensive work on morphogenesis that he is sometimes called the father of embryology based on his work on the chicken embryo.

7.2 Discovery of the Pulmonary Capillaries

Much of Malpighi's research was made possible by the recent invention of the compound microscope. Magnifying spectacles using one lens go back a long way and were in use in the thirteenth century. However the compound microscope, that is one with both an objective and an eyepiece lens, appeared much later. Some authorities attribute the invention to Hans and Zacharias Janssen in the Netherlands at the end of the sixteenth century. Galileo developed a compound microscope in 1609 but it was some time before this was exploited for scientific research. Certainly Malpighi was one of the first to use a compound microscope in 1660, and Robert Hooke's Micrographia published in 1665 contained beautiful illustrations. Nehemiah Grew (1641–1712) and Antonie van Leeuwenhoek (1632–1723) made further important advances. However many of Malpighi's discoveries were apparently made using a single magnifying lens

Malpighi's historic description of the pulmonary capillaries was made in his Second Epistle to Borelli published in 1661 with the title De Pulmonibus [5]. Early in this letter Malpighi beautifully described how he came to use the frog for his dissections. He first studied sheep and other mammals but in spite of enormous efforts the results were disappointing. In fact Malpighi frequently emphasized the tediousness and inadequate results of most of his dissections.

However when he eventually used the frog he was jubilant and in a striking passage referred to the frog as the "microscope of nature". By this he meant that he was able to visualize with a relatively small magnification, features as minute as the capillary network that had eluded him in mammals because the structures were so small that they could not be seen under his microscope. He went on to say that nature is accustomed "to undertake its great works only after a series of attempts at lower levels, and to outline in imperfect animals the plan of perfect animals". In other words he is alluding to the fact that, as he sees it, evolution has tried out its advances in "imperfect animals" by which he means frogs before using the same structures in a more advanced form in the so-called "perfect animals" by which he means mammals. He goes on "For Nature is accustomed to rehearse with certain large, perhaps baser, and all classes of wild [animals], and to place in the imperfect the rudiments of the perfect animals".

A little later Malpighi makes a droll statement about the amount of labor the work has taken him. "For the unloosing of these knots [that is elucidating these problems] I have destroyed almost the whole race of frogs, which does not happen in the savage Batrachomyomachia of Homer". Here he is referring to the imaginary fierce battle between frogs and mice which is a parody of the Iliad and which, incidentally, was probably not written by Homer.

Now turning to Malpighi's actual observations, after his abortive efforts on mammals such as sheep he first looked at the living lung in the frog. However although he could clearly see the blood moving rapidly through small arteries, he could not determine what eventually happened to it. He then dried the lung of a frog and wrote as follows, "I could not extend the power of the eye any further in the living animal, hence I believe that the mass of blood poured into an empty space and was re-collected by the outgoing vessels and the structure of their walls... however, the dried lung of a frog resolved my doubts. In a very small portion of it ... there may be seen, with a perfect glass no broader than the eye, the points which are called "Sagrino" [dark spots on the surface of the lung of a frog] forming the membrane, but mixed with looped vessels. So great is the branching of these vessels, after they extend out hither and thither from the vein and artery, that no larger system of vessels will be served, but a network appears, formed by the offshoots of the two vessels. This network not only occupies the whole floor [of the airspace] but is extended to the walls and adheres to the outgoing vessel, just as I could observe more abundantly, but with greater difficulty in the oblong lung of a tortoise, which is likewise membranous and translucent. Here it lies revealed to the senses that, as the blood passes out through these twisting divided vessels, it is not poured into spaces, but is always passed through tubules and is distributed by the many windings of the vessels" (Translation from [12]).

The two letters to Borelli contained two illustrations shown in Figs. 7.2 and 7.3. Malpighi's drawing of the capillary mesh is shown in the bottom part of Fig. 7.2. This depicts one of the vesicles (alveoli) with its base and various sides that has been opened up so that the dense network of capillaries can be seen in all the walls. The upper part of Fig. 7.2 shows the two lungs of the frog. On the left we can see the alveoli and on the right is another depiction of the capillaries.

Malpighi's discovery of the pulmonary capillaries was momentous. In fact it was the first description of capillaries in any circulation. Harvey in De motu cordis in 1628 had supposed, as had Galen and Columbus before him, that the blood found its way from the right ventricle through the parenchyma of the lungs into the pulmonary vein and left ventricle. However Harvey could not see the capillary vessels but called them "pulmonum caecas porositates et vasorum eorum oscilla", that is "the invisible porosity of the lungs and the minute cavities of their vessels". It was Malpighi's great triumph that he was the first person to see and describe them.

Fig. 7.2 Malpighi's drawing
of the pulmonary capillaries
and alveoli. **I** Shows the two
lungs with the alveoli on the
left and the capillaries on the
right. **II** Shows the pulmonary capillaries in a diagram
of an alveolus which has been
opened up. (From [5])

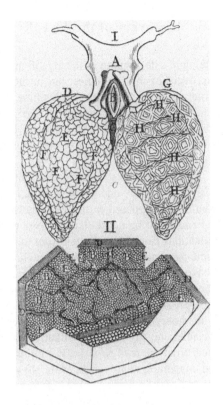

Fig. 7.3 Malpighi's drawing of the alveoli. **I** and **II**
Show two different views of
alveoli. **III** Diagram showing
the branching of the trachea
ending up in the alveoli.
(From [5])

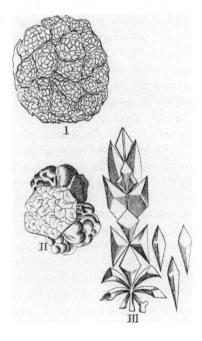

7.3   Discovery of the Alveoli

Malpighi's description of the alveoli may not have been quite so momentous as his discovery of the capillaries, but in fact it completely changed perceptions of the structure of the lung. Prior to his time the structure and function of the lung were a mystery. Vesalius following the writings of Galen held that the lungs had been formed from solidified bloody foam. Harvey compared the substance of the lung with that of the kidney or liver and argued that one of its principal purposes was to cool the animal.

Malpighi refers to these earlier notions near the beginning of his Epistle I of "De Pulmonibus" addressed to Borelli. He stated "The substance of the lungs is commonly supposed to be fleshy because it owes its origin to the blood, and it is believed to be not unlike the liver or the spleen …". (This and the subsequent quotations from Malpighi are from the translation by Young [14].) But then in the same paragraph Malpighi goes on to drop his bombshell. "By diligent investigation I have found the whole mass of the lungs, with the vessels going out of it attached, to be an aggregate of very light and very thin membranes, which, tense and sinuous, form an almost infinite number of orbicular vesicles and cavities, such as we see in the honey-comb alveoli of bees, formed of wax spread out into partitions. These [vesicles and cavities] have situations and connection as if there is an entrance into them from the trachea, directly from the one into the other; and at last they end in the containing membrane". Malpighi goes on to refer to one of the illustrations at the end of the Epistles which is shown in Fig. 7.3. In referring to the left-hand part of the figure he states that "with greatest diligence I have been able to make out, those membranous vesicles seem to be formed out of the endings of the trachea, which goes away at the extremities and sides into ampulus cavities". Parts I and II of the figure show the vesicles (alveoli) and Part III is a diagrammatic representation of the final branching of the prolongations of the trachea. He adds "Seeing that the air which rushes from the trachea into the lungs requires a continuous path for easy and rapid ingress and egress, whence possible this internal tunic of the trachea, ends in sinuses and vesicles, makes a mass of vesicles like an imperfect sponge so to speak". Although the illustrations are from the frog lung, Malpighi also refers to somewhat similar observations in a dog's lung.

So Malpighi had discovered that air entering the lung is conducted down a series of what we now call airways into the tiny alveoli, and also that the surface of the alveoli is covered with a rich network of blood vessels as shown in image II at the bottom of Fig. 7.2. He seems to be very close to an understanding of the primary function of the lungs which is the exchange of gases between the alveolar space and the blood in the capillaries. However this eludes him. Further on in the first Epistle he states "Concerning the use of the lungs I know that many views are held from the ancients onwards, and about them there is very much dispute—especially about the cooling, which is taken to be the principal purpose, when it strives with the imagined excessive heat of the heart which may require eventation; wherefore these things have made me diligently inquisitive in the investigations of another purpose and from these things which I subjoin I can believe that the lungs are made by Nature for mixing the mass of blood". So Malpighi is poised to understand the gas exchange function of the lung but at the last moment steps away.

7.4   Insect Respiration

It is fascinating that in addition to discovering the pulmonary capillaries and alveoli in amphibia and by implication mammals since he made similar observations in the dog, Malpighi also was the first person to describe the mode of respiration in insects. Admittedly this was not his primary purpose in his study of the silkworm. Nevertheless he was the first person to clearly describe the spiracles and tracheae that allow oxygen to reach the body tissues of insects.

Malpighi's work on the silkworm was stimulated by a letter from Oldenburg who was the secretary of the Royal Society in London. We have already seen that Malpighi's reputation as a scientist was so great in 1667 that he was invited to become a correspondent of the Society and send his scientific findings to them for publication. Oldenburg told Malpighi that the Society was particularly interested in the silkworm, and indeed Malpighi had previously done some work on this animal. However stimulated by the invitation, he embarked on a major project to describe the anatomy of the silkworm. This was presented to the Society in a manuscript titled "Dissertatio epistolica de bombyce" which appeared in the Philosophical Transactions of the Royal Society [7].

Adelmann quotes Cole [2] who described the contents of De Bombyce and excerpts are summarized here. Malpighi anatomized all phases of the species, but, apart from his very remarkable and accurate observations on the genitalia of the moth, the larva claimed the greater part of his attention. The head of the caterpillar is described in detail and the arrangement of the legs and their function in crawling is analyzed. However the most important part from our point of view is that Malpighi described for the first time the spiracles of insects, and the system of vessels associated with them (Fig. 7.4). There are nine pairs of spiracles in the larva of bombyces and each spiracle has its own bundle of vessels [airways], two of which anastamose with corresponding vessels in front and behind. The result is a longitudinal spiracular trunk on each side which stretches from head to tail. There also transverse anastamoses across the body. Like the arteries, the tracheae from the spiracles go on dividing and diminishing in size until they can hardly be seen even with a microscope. Malpighi described these air tubes as tracheae and compared them with the lungs of vertebrates. He carried out additional studies in the cicada, stag-beetle, locust, wasp and bee and identified expansions of the tracheae which he identified with the air chambers of lungs. He combined physiological studies with his anatomical observations. For example he occluded the spiracles with oil and immediately the animal developed convulsions and died "intra Dominicae orationis spatium" [during the time it takes to recite the Lord's Prayer]. He went on to show that if the anterior spiracles are blocked only the corresponding part of the body is

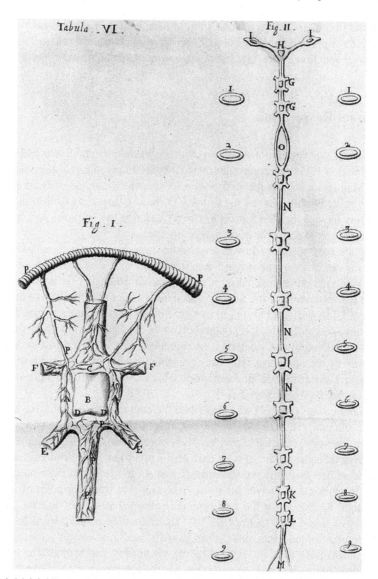

Fig. 7.4 Malpighi's drawing of the spiracles, tracheae and nerve trunk of the silkworm. **I** Shows the small tracheoles ending near a nerve ganglion. **II** Shows the nerve trunk in the center and the nine pairs of spiracles that form the openings of the tracheae on the body surface. (From [7])

affected, and the animal subsequently recovers. Malpighi also described the nerve trunk and ganglia, the silk glands, the pulsating blood vessel extending the whole length of the worm which is the primitive heart, and the urinary tubules that are still referred to by his name. Figure 7.4 shows Malpighi's illustration of a silkworm larva with the spiracles clearly identified. The nerve trunk is also shown with a ganglion on the left and its supply of tracheae.

This brilliant project enhanced Malpighi's reputation even more. Adelmann lists some of the accolades which begin with the statement that Malpighi's book was generally admitted to be the first really great contribution to insect morphology. One scientist referred to it as a "tissue of discoveries" and went on to describe it as "a treatise from which one can learn more about the wonderful internal structure of insects than from all earlier works combined". Another stated that "Despite the few months at his disposal for carrying out his anatomical researches, Malpighi obtained prodigious results and saw more with eyes armed with a simple lens than many others who came later, thought they possessed all the means subsequently offered by the perfected physical sciences".

7.5 Embryological Studies

Malpighi's skills with the microscope prompted him to embark on another major scientific enterprise and the result was a description of the development of the chicken embryo in far more detail than had been previously possible. In fact some authorities argue that his morphogenesis studies were his most important scientific contributions. For example four of the five large volumes in Adelmann's exhaustive study of Malpighi are devoted to his embryological work.

Malpighi was influenced by a philosophical movement at the time of emphasizing the similarity of biological functions to those of machines. For example the French philosopher René Descartes (1596–1650) had referred to man as an earthy machine (machine de terre). This was also part of the attitude of Galileo (1564–1642) whom Malpighi never met but for whom he had an enormous regard. Malpighi's friend Borelli shared the same notions and his famous book "De motu animalium" contains many examples. Malpighi's most important book on morphogenesis was "Dissertatio epistolica de formatione pulli in ovo" (Discourse letter on the formation of the chicken in the egg) [8] in which Malpighi draws an analogy with an artisan who "in building machines must first manufacture the individual parts, so that the pieces are first seen separately, which must then be fitted together". Elsewhere he writes "Nature in order to carry out the marvelous operations [that occur] in animals and plants has been pleased to construct their organized bodies with a very large number of machines, which are necessarily made up of extremely minute parts so shaped and situated as to form a marvelous organ, the structure and composition of which are usually invisible to the naked eye without the aid of the microscope…".

Malpighi's philosophy here has some similarities with that described earlier where he stated that Nature is accustomed "to undertake its great works only after a series of attempts at lower levels, and to outline in imperfect animals the plan of perfect animals". Now in De formatione pulli in ovo he states that the study "of the first unelaborated outlines of animals in the course of development" is particularly useful because the artisan Nature forms them separately before combining them with one another. He gives as an example the miliary glands (embryonic hepatocytes) of the embryo which will merge to form the liver but are still distinguishable as individual groups of cells which in crustaceans remain distinct.

Malpighi's studies of the chicken embryo were very extensive. He was particularly interested in the early development of the heart including the primitive cardiac tube and its segmentation, the aortic arches, and the somites. He also worked on the development of the nervous system including the neural folds, the neural tube, the cerebral vesicles, and the optic vesicles. He studied the heart within 30 h of incubation and noticed that it began to beat before the blood became red. Malpighi's drawings of the heart in an embryo 2 days old are shown in Fig. 7.5. Incidentally he was also apparently the first person to identify red blood cells although he did not understand their significance. In an observation that became controversial, he stated that he had seen the complete form of the embryo in an unincubated egg. This notion of "preformation" was taken up by subsequent scientists and of course was an error but Malpighi only alluded to it tangentially.

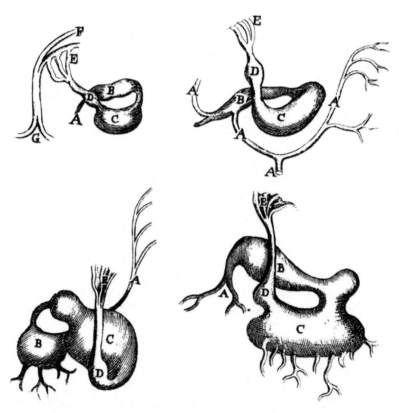

Fig. 7.5 Drawings of the developing heart in the chick embryo. The blood comes from the veins *A* and passes into the auricle [atrium] *B*. Sometimes there is a brief intermediate channel as the blood is pushed into the right ventricle *C* and from there into the left ventricle *D*. Finally the blood enters the arteries *E* and from them moves either to the head *F* or to the umbilical vessels *G*. (From [8])

7.6 **Botanical Studies**

Malpighi's enormous energy and scientific curiosity are evidenced by the fact that, as we have seen, he not only made fundamental advances on the anatomy of the lungs, but in addition he carried out extensive studies on the silkworm, and embryology of the chicken. His boundless enthusiasm is further indicated by the following quotation when he referred to his interest in the investigation of higher animals and added "… but these, enveloped in their own shadows, remain in obscurity; hence it is necessary to study them through the analogues provided by simple animals. I was therefore attracted to the investigation of insects; but this too has its difficulties. So, in the end, I turned to the investigation of plants so that by an extensive study of this kingdom I might find a way to return to earlier studies, beginning with vegetant Nature. But perhaps not even this will be enough, since the simpler kingdom of minerals and elements should take precedence. At this point the undertaking becomes immense, and absolutely out of all proportion to my strength" [1].

Fortunately for the progress of science he resisted the temptation to take on minerals and elements. But Malpighi turned to the study of plants with extraordinary success. The results were published in Anatomes Plantarum Pars Prima [9] followed by Anatomes Plantarum Pars Altera [10]. In fact Malpighi is commonly regarded as the founder of the microscopic study of plant anatomy along with his English contemporary Nehemiah Grew (1641–1712).

Malpighi's interest in the anatomy of plants began in Messina when he was walking in the garden of his patron Visconte Ruffo. He noticed that when a branch of a chestnut tree was snapped, what appeared to be vascular bundles projected from the open end. These were the tubes of the xylem that transport water and soluble mineral nutrients from the roots to the leaves. They are known as tracheids and the nomenclature is interesting: trachea in mammals, tracheae in insects, and tracheids in plants.

Malpighi was immediately reminded of the tracheae of insects and this aroused his interest in the nutrition of plants. He then took up the topic with his customary enthusiasm and systematically studied all aspects of plant anatomy including the stem, leaf, root, bud, flower, seed and seedling. When he examined the fine tubes in the xylem of the stem that reminded him of the insect tracheae, he found that these formed annular rings in some plants but scattered bundles in others. Here he was inadvertently referring to a fundamental difference between the two great families of plants, the dicotyledons and the monocotyledons although these were not described until much later.

Malpighi recognized that the leaves of a plant constituted a laboratory which produced sap that was essential for plant nutrition. He showed for example that when he removed one of the two leaves of a germinating seedling its growth was stunted and if he removed both leaves the plant died. He also described that when the bark of a tree was removed in a ring around the trunk, a swelling appeared above the ring because of obstruction to the sap descending from the leaves in the phloem which is located in the periphery of the trunk just under the bark, and the periphery of a stem just under the epithelium.

Malpighi was a talented artist and sketched beautiful sections of plants such as Nigella that he noticed produced honey in the depths of the flowers. Examples of his drawings are shown in Fig. 7.6. In keeping with his interest in morphogenesis that he had demonstrated in his studies of the chick embryo, he studied the early stages of development of plants such as the bean (Leguminosae) with elegant drawings. Another interest was galls which are abnormal growths on the trunks or branches of some trees. He demonstrated that these were caused by the deposition of insect eggs.

With Nehemiah Grew, Malpighi is known as a founder of the microscopic study of plant anatomy. Both of them communicated their publications to the Royal Society in the same period and they were on cordial terms. Grew became the secretary of the Society after the retirement of Oldenburg.

7.7   Other Studies

It is impossible to summarize all the scientific contributions of Malpighi. However some others should be briefly mentioned. He described the papillae of the tongue and postulated their role in taste. In studies of the skin he described a layer of cells that now bears his name. Additional studies were made on the brain with particular reference to the white matter that he showed consisted of bundles of fibers.

His book De viscerum structura exercitatio anatomica [6] describes the histology of the kidney, spleen and liver. He saw the glomeruli in the kidney and for a time these were named after him. Small groups of lymphatic bodies in the spleen were known as Malpighian corpuscles. In the liver he saw small lobules which he concluded represented the fundamental unit of that organ.

7.8   Malpighi's Difficulties

It would be natural to think that a man as original, industrious and productive as Malpighi would live a satisfying life. Unhappily this was not the case. First he was unfortunate in that he suffered from ill health for most of his life. More important was the savage criticism of much of his scientific work. In part this was the result of professional jealousy. However in addition there was criticism of Malpighi's scientific approach. Although we marvel at how much he achieved with an early microscope, many of his detractors argued that this type of research was useless and it would not lead to improvements in medical treatment.

A tragedy in 1659 had a lasting influence on Malpighi. His brother Bartholomew got into an argument with the eldest son of Durolamo Sbiraglia and fatally stabbed him. The result was a smoldering feud between the two families and a lifelong bitter

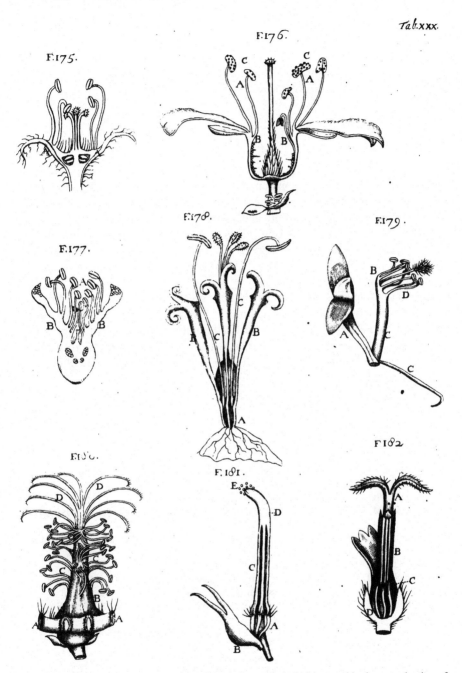

Fig. 7.6 Botanical illustrations by Malpighi. He was particularly interested in the reproduction of plants and many drawings show the sex organs including the stamens, anthers, pistils, stigmas and styles. (From [9])

criticism of Malpighi's work by Sbiraglia who was also a scientist. Perhaps there are echoes of the feud of the two families in Shakespeare's *Romeo and Juliet*. Malpighi was finally driven to publish a long bitter response to his critics in his Opera Posthuma published in 1697 [11]. However history will always regard Malpighi as one of the most outstanding biologists.

7.9   Note on Sources

By far the best source of information is the five-volume work by Adelmann, Marcello Malpighi and the Evolution of Embryology [1]. The first volume is a general account of Malpighi's life and work and in spite of its length makes excellent reading. A review of the book by Wilson [13] is a good introduction. Also useful is E.J. Cole's History of Comparative Anatomy [2], and the chapter on Malpighi in Michael Foster's Lectures on the History of Physiology [4]. The latter contains another translation of part of De Pulmonibus.

Access to Malpighi's publications has been greatly improved because of the Internet. Publications with reference numbers 5, 7, 8, 9, and 10 are on the Web. Some of the original publications are in specialized libraries especially that of the Royal Society in London. There are English translations of De Pulmonibus by Young [14], De polypo cordis by Forrester [3] which forms part of reference 6, and Pulli in Ovo by E. Corti on the Web.

References

1. Adelmann HB. Marcello Malpighi and the evolution of embryology (5 vols). Ithaca: Cornell University Press; 1966.
2. Cole FJ. A history of comparative anatomy. London: Macmillan and Co.; 1949.
3. Forrester JM. Malpighi's De Polypo Cordis: an annotated translation. Med Hist. 1995;39:477–92.
4. Foster M. Lectures on the history of physiology. Cambridge, UK: Cambridge University Press; 1924.
5. Malpighi M. De Pulmonibus. London: Phil Trans Roy Soc; 1661.
6. Malpighi M. De viscerum structura exercitatio anatomica. Bologna: Jacopo Monti; 1666.
7. Malpighi M. Dissertatio epistola de bombyce. London: Phil Trans Roy Soc; 1669.
8. Malpighi M. Dissertatio epistolica de formatione pulli in ovo. London: Phil Trans Roy Soc; 1673.
9. Malpighi M. Anatomes Plantarum Pars Prima. London: Phil Trans Roy Soc; 1675.
10. Malpighi M. Anatomes Plantarum Pars Altera. London: Phil Trans Roy Soc; 1679.
11. Malpighi M. Opera Posthuma. London: Churchill; 1697.
12. Wilson LG. The transformation of ancient concepts of respiration in the seventeenth century. Isis. 1960;51:161–72.
13. Wilson LG. Malpighi and seventeenth-century embryology: an essay review. J Hist Med Allied Sci. 1967;22:190–8.
14. Young J. Malpighi's "De Pulmonibus". Proc R Soc Med. 1929;23:1–11.

Chapter 8
Stephen Hales: Neglected Respiratory Physiologist

Abstract Stephen Hales was an eminent early eighteenth century scientist and minister of the parish of Teddington near London. He is well known for his early work on blood pressure. However, he made many contributions to respiratory physiology. He clarified the nature of the respiratory gases, distinguishing between their free (gaseous) and fixed (chemically combined) forms, demonstrated that rebreathing from a closed circuit could be extended if suitable gas absorbers were included (to remove carbon dioxide), suggested a similar device as a respirator for noxious atmospheres, invented the pneumatic trough for collecting gases, measured the size of the alveoli, calculated the surface area of the interior of the lung, calculated the time spent by the blood in a pulmonary capillary, invented the U-tube manometer, and measured intrathoracic pressures during normal and forced breathing. Hales's work is remarkable for its emphasis on the "statical" method, i.e., meticulous attention to detail in measurement and careful calculations. In his later life he made important contributions in the area of public health. He was a trustee of the new colony of Georgia and willed his own library of books to the colony though their whereabouts is unknown. He deserves more recognition in the history of respiratory physiology.

The Reverend Stephen Hales was a remarkable man even for the eighteenth century. He was the full-time minister of the parish of Teddington near London but in addition was an eminent scientist and physiologist. He was born in 1677 when England was riding on the crest of a wave of scientific discovery. William Harvey had published his De Motu Cordis in 1628 just 50 years before, and in that 50 years scientists at London and Oxford, including Boyle, Hooke, Lower, and Mayow, had transformed many aspects of the physical and biological sciences and had, in the process, laid the foundations of modern respiratory physiology [5].

The scientific revolution had begun to gather momentum at the turn of the century. Harvey attended the University of Padua from 1600 to 1602 at the same time that Galileo was there. Indeed it seems likely that when Harvey was doing his first experiments on the circulation of the blood, Galileo was training his telescope on Jupiter and its moons. What Galileo saw proved to anyone who was willing to listen that Copernicus and Kepler were correct about the movements of the earth and planets around the sun. Thus these observations finally abolished the 1,500-year-old Ptolemaic system just as Harvey's experiments proved the circular movement of the

© American Physiological Society 2015
J. B. West, *Essays on the History of Respiratory Physiology,*
Perspectives in Physiology, DOI 10.1007/978-1-4939-2362-5_8

Fig. 8.1 Stephen Hales, D.D., F.R.S. (1677–1761)

blood and finally ousted the old Galenical order that had dominated medicine for so long.

But Galileo's new ideas ran counter to the established teaching of the Church, which had been the repository of knowledge for so many years. His last book had to be published in the Netherlands far from the Inquisition, and the center of gravity of science moved from Italy where the Renaissance had begun to northern Europe. Galileo himself died under house arrest in 1642, and in that same year on Christmas Day Isaac Newton was born.

When Stephen Hales entered Cambridge in 1696, Newton had already left to become Warden of the Mint in London, but his influence was still much in evidence. Indeed the young Hales became interested in astronomy, and in 1705 he built an orrery, i.e., a machine to show the movements of the planets. As a consequence Hales is cited in histories of astronomy. This was the year in which Edmund Halley, using Newton's laws of motion, first predicted the return of a comet 53 years later; its next apparition is in 1985. Newton's influence must have been a factor in Hales's analytical quantitative approach in which, as he stated, "We must in all reason… number, weigh and measure." Hales gave the name "staticks" to this numerical approach - thus the names of his two principal books *Haemastaticks* [7] and *Vegetable Staticks* [6].

Hales was remarkably versatile. In addition to fulfilling his responsibilities as vicar of Teddington (Fig. 8.1), he was one of the most influential scientists of the first half of the eighteenth century, and he has an assured place in the history of botany and chemistry. However, his considerable contributions to respiratory physiology have generally been overlooked. As examples, the chapter by Perkins [10] on the history of respiratory physiology at the beginning of the American Physiological Society's *Handbook of Physiology* devotes only one short paragraph to Hales, and he gets short shrift in Michael Foster's [4] *Lectures on the History of Physiology*, where he is dismissed with the opinion that he did not make "any definite special contribution to our knowledge of respiration."

Stephen Hales is best known in physiology for his work on blood pressure, which is described in his book *Haemastaticks* [7]. A modern facsimile reprint is available [9]. This includes an extensive analysis of flows and pressures in keeping with the mechanical Newtonian flavor of his work. Most of his material on respiratory physiology is in his other great book *Vegetable Staticks* [6]. A modern reprint is available [8]. This, as its name implies, is chiefly about botanical research including an explanation for the rise of sap in plants. The title of this book presumably partly explains why Hales's contributions to respiratory physiology are so little known.

Hales was a great experimentalist, with an inexhaustible curiosity and drive. This is shown in the first page of *Haemastaticks* in which he plunges into an experimental account of how he first measured blood pressure in the horse, "In *December*, I caused a Mare to be tied down alive on her Back, she was fourteen Hands high, and about fourteen Years of Age, had a *Fistula* on her Withers, was neither very lean, nor yet lusty: Having laid open the left crural Artery about three Inches from her Belly, I inserted into it a brass Pipe whose Bore was one sixth of an Inch in Diameter." His questioning thrusting inquisitive attitude was typical of the time as evidenced by the stimulating accounts of the meetings of the Royal Society [2]. Hales was elected a Fellow in 1718, and his *Vegetable Staticks*, published in 1727, had the imprimatur of Isaac Newton who was the President of the Royal Society at that time.

Hales worked in a number of areas of respiratory physiology. First, he was interested in the nature of the respiratory gases, although there was bound to be some confusion in this area until Black, Scheele, Priestley, Cavendish and Lavoisier had identified oxygen, carbon dioxide and nitrogen later in the century. Hales was fascinated by the fact that gases could exist in what he called the elastic and fixed states, i.e., in the gaseous form and as nongaseous compounds. In one experiment he rebreathed air from a bladder for a minute at the end of which he had to stop because of a feeling of suffocation (Fig. 8.2). He showed that during this time about one-thirteenth of the total volume of air was transformed from the elastic to the fixed state, i.e., had been absorbed by the body [6] (experiment CVIII).

He was able to show that the period of rebreathing could be extended if pieces of moist cloth soaked in vinegar or sal tartar (probably made from the sediment of wine casks) were included in the breathing circuit. These presumably absorbed some of the carbon dioxide and this observation prompted Joseph Black to carry out his experiments on caustic lime which led to the discovery (or actually the rediscovery) of carbon dioxide. Hales argued that the increase in weight of these absorbers reflected the amount of gas changed from the elastic to the fixed state. The breathing circuit that he used for these experiments included two one-way valves and would look familiar today (Fig. 8.2). Hales went on to suggest that a similar closed-circuit breathing device could be employed as a respirator by someone who had to enter a noxious atmosphere for a short time.

In connection with these experiments Hales invented the pneumatic trough for collecting gases, and this was to be used extensively by later chemists including Priestley and Lavoisier. This trough can be considered a derivative of the upturned jar which was employed so effectively by the Oxford physiologists, e.g., Mayow, in their experiments on respiration and combustion.

Fig. 8.2 Upper part shows closed circuit for rebreathing with 2 one-way valves (*r, b*). *n* is an elastic bladder containing hoops on which were placed pieces of moist cloth to absorb gases. Lower part shows pneumatic trough invented by Hales for collecting gases

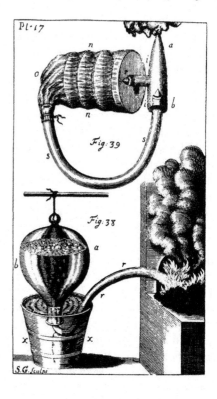

Hales's observations on the absorption of gases by the lung led him to studies of the morphometry of the lung and its implications for function [6] (experiment CIX). He measured the size of the alveoli in the calf, obtaining a value of one hundredth of an inch or 250 μm. The actual value is about 100 μm [11], so Hales's measurement using the crude equipment of that time was impressively accurate. He calculated the surface area of the interior of the lungs to be 289 ft² (about 30 m²) and pointed out that this was 10 times the surface area of a man's body. He recognized the importance of this large surface area for gas exchange and remarked that the "partitions" in the lung are very thin to allow a "continued succession of fresh air" to be absorbed by the blood. In other experiments he measured the amount of gas in blood by heating the latter in a closed container connected to a gas reservoir and reported that "a cubick inch of *Hog's blood*... produced thirty three cubick inches of Air."

While studying the pulmonary circulation in *Haemastaticks* [7], (experiment X) he noted that the flow of blood in the pulmonary veins was pulsatile. He also calculated that the time spent by the blood in the pulmonary capillary was a little over 2 s for a capillary, which he calculated to be about 0.000672 in. (or 17 μm) in diameter. Hales was apparently the first person to use the term capillary.

Hales also invented the U-tube manometer, which he used on many occasions for measuring the pressure developed by sap in vines and trees (Fig. 8.3). He was intrigued by pressures of all kinds, including of course blood pressure, and he

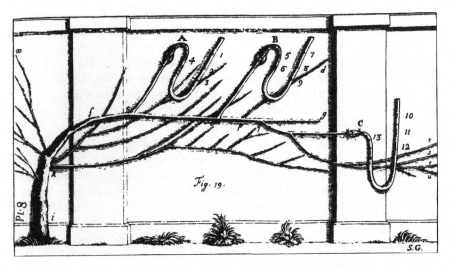

Fig. 8.3 U-tube manometers invented by Hales and used in this instance for measuring pressure developed by sap in a vine

is rightly given credit for its first measurement. One prevalent theory of muscle movement at this time was that it was generated by changes of pressure in small blood vessels in the muscle. Hales was able to disprove this by calculating the force that could be derived from a given pressure. Indeed he went on to suggest that electrical impulses along nerves might be involved in muscle contraction.

He measured mouth pressures in man during respiratory efforts and also intrathoracic pressure in dogs by inserting a small tube through an intercostal space [6] (experiment CXIII). He recorded, for example, that a pressure of 6 in. of water could be developed by the animal during normal breathing, but the manometer rose to 24–30 in. during "great and laborious inspirations." He recognized that the development of even a small intrathoracic pressure would result in a large total force on the diaphragm because of its large surface area. Thus he argued that if contraction of the diaphragm could develop a pressure of 2 in. of mercury the total force across the diaphragm must equal the weight of a cylinder of mercury with a base equal to the area of the diaphragm and a height of 2 in.

Throughout Hales's work there is a strong emphasis on the "statical" method, i.e., a meticulous attention to detail in measurement and sometimes laboriously long calculations. An amusing example of this is Hales's calculation of the water flow in a new drain installed in the churchyard at Teddington. Hales wrote that the "quantity of water which ran from the springs was estimated by fixing in clay at the mouth of the covered drain a small trough, and placing inside it a vessel containing two quarts which was filled in three swings of a pendulum beating seconds, which pendulum was 39 + 2/10 in. long from the suspending nail to the middle of the plumbet or hub. By which means it was found that more than sixty tuns of water ran there in 24 h, allowing 53 gal. Winchester measure to the hogshead."

This emphasis on numbers, weights, flows, and pressures might suggest that Hales was so obsessed with details of the function of animals and plants that he ignored the whole organism. But this is not so; he was keenly aware of how all aspects of physiology contributed to the "animal oeconomy." He was full of admiration for the overall design, as he saw it, and on one page of *Haemastaticks* after a long section in which he calculates the number of capillaries in man to be 3,541,713 he exclaims, "So curiously are we wrought, so fearfully, and wonderfully are we made!"

His two most important books *Vegetable Staticks* and *Haemastaticks*, published in 1727 and 1733, respectively, established his reputation among the scientists of his day, and in 1739 he was awarded the Copley medal of the Royal Society. His interests then moved to public health, particularly to the advantages of using mechanical ventilators to improve the quality of the air in con-fined spaces such as the living quarters of ships and prisons. One of his ventilators was installed on the top of Newgate Prison in London, and it apparently reduced the dreadful mortality from jail fever. His ventilators were especially effective on slave and transport ships and they were also used in hospitals. Hales was interested in other problems affecting the merchant navy, including methods for making fresh water from seawater by distillation and a device for the automatic measurement of sea depth. Other interests in the area of public health included substances to dissolve renal stones (a serious problem at that time) and efforts to reduce the consumption of gin, which was a scourge of the working classes.

Hales's scientific work directly influenced Joseph Black, Priestley, Lavoisier, and Humphry Davy, and there are many references in scientific books published later in the eighteenth century to the writings of the "learned Dr. Hales." It is interesting that Hales was never attracted to the erroneous phlogiston theory, which generally had a large negative influence in the middle of the eighteenth century and dominated the thinking of even such great chemists as Priestley.

Hales was Perpetual Curate of the parish of Teddington, then a small village on the Thames 15 mi. west of London. Later in his life he also became Rector of Farringdon in the west of England. A happy feature of an interest in Hales is that his church at Teddington (St. Mary's) is still in use and contains many mementos of his period there. During a visit to the church about five years ago I was shown the parish records that were kept in the church safe and were written in Hales's own hand. They are full of interesting details that are the daily concern of a conscientious vicar and, incidentally, show that Hales was a stern disciplinarian as far as parish morals were concerned. For example, it is recorded that several parishioners were required to do public penance for fornication, a practice that would be difficult to institute today. The stone tower that Hales built to replace an old wooden one remains as a monument to him, and his grave with its commemorative stone is beneath this. The church now has a fine stained-glass window commemorating its famous minister.

Hales has an important link with early America because he was a trustee of the new colony of Georgia. Though he never came to the New World, he was a conscientious trustee and there are numerous references to his work in the State Library in

Atlanta. Hales willed his own library to the new colony as follows, "All my bound Books at Teddington, which my Executor hereafter named, shall think proper to be sent to Georgia, I give and bequeathe for a public parochial Library to such Town or Parish in Georgia in America, the Governor shall think fit to appoint. The carriage of them thither to be at the expense of my Executor and Executrix hereafter named." However, the whereabouts of these books is unknown. Indeed there seems to be no direct evidence that the library ever reached the colony. If it did, it may have been damaged when the original state capital, Savannah, was captured by the British in 1778 and the seat of government moved west. A hunt for the missing library would make a nice project for a respiratory physiologist in his declining years.

There are two biographies of Hales. The first [3] is by the physician A. E. Clark-Kennedy and was published in 1929 and reprinted in 1965. It was an extension of a memorial address at Corpus Christi College, Cambridge, celebrating the 250th anniversary of Hales's birth. The most recent biography [1] is by D. G. C. Allan and R. E. Schofield and was published in 1980. However, neither biography does justice to Hales's considerable interests in respiratory physiology.

Hales was much admired by his contemporaries as a scientist and humanitarian. He lived to be 84 and is commemorated by a monument in Westminster Abbey, though his bones remain at Teddington. With his contributions to respiratory physiology in the areas of gas exchange and mechanics and his intriguing links with early America, he deserves to be better known.

References

1. Allan DGC, Schofield RE. Stephen Hales: scientist and philanthropist. London: Scholar; 1980.
2. Birch T. The history of the royal society of London. Vols. 1–4. New York: Johnson Reprint; 1968.
3. Clark-Kennedy AE. Stephen Hales D.D., FRS. London: Cambridge University Press; 1929.
4. Foster M. Lectures on the history of physiology. London: Cambridge University Press; 1924.
5. Frank RG. Harvey and the Oxford physiologists. Berkeley: University of California Press; 1980.
6. Hales S. Vegetable staticks. London: Innys & Innys; 1727.
7. Hales S. Statical essays containing haemastaticks. London: Innys & Manby; 1733.
8. Hales S. Vegetable staticks. London: Scientific Book Guild; 1961.
9. Hales S. Statical essays containing haemastaticks. New York: Hafner; 1964.
10. Perkins JF, Jr. Historical development of respiratory physiology. In: Handbook of physiology. Respiration. Washington, DC: Am Physiol Soc. 1964, Sect. 3, vol. 1, chap. 1, p. 1–62.
11. Tenney SM, Remmers JE. Comparative quantitative morphology of the mammalian lung: diffusing area. Nature. 1963;197:54–6.

Chapter 9
Joseph Black, Carbon Dioxide, Latent Heat, and the Beginnings of the Discovery of the Respiratory Gases

Abstract The discovery of carbon dioxide by Joseph Black (1728–1799) marked a new era of research on the respiratory gases. His initial interest was in alkalis such as limewater that were thought to be useful in the treatment of renal stone. When he studied magnesium carbonate, he found that when this was heated or exposed to acid, a gas was evolved which he called "fixed air" because it had been combined with a solid material. He showed that the new gas extinguished a flame, that it could not support life, and that it was present in gas exhaled from the lung. Within a few years of his discovery, hydrogen, nitrogen and oxygen were also isolated. Thus arguably Black's work started the avalanche of research on the respiratory gases carried out by Priestley, Scheele, Lavoisier and Cavendish. Black then turned his attention to heat and he was the first person to describe latent heat, that is the heat added or lost when a liquid changes its state, for example when water changes to ice or steam. Latent heat is a key concept in thermal physiology because of the heat lost when sweat evaporates. Black was a friend of the young James Watt (1736–1819) who was responsible for the development of early steam engines. Watt was puzzled why so much cooling was necessary to condense steam into water, and Black realized that the answer was the latent heat. The resulting improvements in steam engines ushered in the Industrial Revolution.

9.1 Introduction

Joseph Black (1728–1799) has a special place in the history of respiratory physiology because his research on the nature of carbon dioxide marked the beginning of the rapid advance in knowledge of the respiratory gases (Fig. 9.1). Carbon dioxide had actually been briefly described about 100 years before by van Helmont (1580–1644) but Black was responsible for first elucidating its properties. His work influenced subsequent investigators such as Cavendish (1731–1810), Priestley (1733–1804), Scheele (1742–1786) and Lavoisier (1743–1794), and over the ensuing 20 years after Black's publication, all the respiratory gases including oxygen, hydrogen, nitrogen and water were characterized.

Black's pioneering work on carbon dioxide assures him a major place in the history of physiology. However he did much more. In fact historians of science some-

© American Physiological Society 2015
J. B. West, *Essays on the History of Respiratory Physiology,*
Perspectives in Physiology, DOI 10.1007/978-1-4939-2362-5_9

Fig. 9.1 Joseph Black (1728–1799). (From http://en.wikipedia.org/wiki/File:Joseph_Black_b1728.jpg)

times ignore his work on carbon dioxide and instead emphasize his major advances in the area of heat. He was the first person to recognize latent heat, that is the heat added or lost in the change in state of a substance. An example is water when it is converted into steam or ice. He also described the specific heats of various substances. His work influenced his friend James Watt (1736–1819) who made critical advances in the design of the steam engine. This had been invented by Thomas Newcomen (1664–1729) but was inefficient. Watt's modification greatly improved its performance and was a major factor in the development of the Industrial Revolution which had an enormous influence in history.

9.2   Brief Biography

Joseph Black was born in Bordeaux, France where his father was a wine merchant who himself had been born in Belfast but was of Scottish origin. Joseph's mother was also from Scotland and it was she who taught her children to read English. At the age of 12 Joseph was sent to a private school in Belfast. There he was reported to have been an excellent scholar [7]. Four years later in 1744 he entered the University of Glasgow where he took the arts curriculum. However there is a note that in his fourth year he studied physics and was the favorite pupil of the professor of natural philosophy [7].

At the end of his arts course he studied medicine under Dr. William Cullen (1710–1790). This man was one of the most illustrious professors of medicine in the English-speaking world at the time and an important figure in the Scottish Enlightenment. When he moved to Edinburgh he was the physician of David Hume

(1711–1776), the eminent philosopher, and had a wide circle of friends including Adam Smith (1723–1790), the economist. Cullen was a firm believer in the experimental method and he employed Joseph Black as his assistant in the laboratory who later reported that Cullen treated him "with the same confidence and friendship... as if I had been one of his own children" [13].

In 1752 Black moved to Edinburgh University to continue his medical studies. At the time, this institution had the finest medical education in the United Kingdom. It is not always appreciated that the Scottish universities at that time were far ahead of the better known English universities such as Oxford and Cambridge in the field of medicine.

Black was required to write a thesis for his M.D. degree in Edinburgh and he became interested in the properties of limewater which was thought to be valuable in the cure of kidney stone, a common ailment at the time. However it transpired that the action of limewater was a contentious subject between two of the major professors and Black therefore decided to work on a related topic, the properties of *magnesia alba* (magnesium carbonate). A feature of this research was Black's use of accurate balances and in fact Black is credited with inventing the first accurate analytical balance. His experimental work on *magnesia alba* and his subsequent elucidation of the properties carbon dioxide are described below.

In 1756 Black returned to Glasgow where he was appointed lecturer, and later became professor. Remarkably, his research interests changed considerably. He became interested in the heat transfer that occurs particularly in a change of state, for example the transition from water to ice or water to steam. He had noticed that the change in state can occur over a prolonged period of time when a substance is heated or cooled without a change in temperature. For example snow at near freezing point gradually melts to form water over a period of time without a change in temperature. He introduced the term "latent heat" to refer to this phenomenon. He also discovered that different liquids have different capacities to take up heat, and he introduced the concept of "specific heat".

James Watt (1736–1819), who was one of Scotland's most famous engineers, came to Glasgow at the age of 18 and became an instrument maker to Black. The latter remarked "I soon had occasion to employ him to make some things... and found him to be a young man possessing most uncommon talents for mechanical knowledge and practice... which often surprised and delighted me in our frequent conversations together" [4]. Watt was influenced by Black's work on latent heat and applied this knowledge to improve the steam engine. This had been invented by Thomas Newcomen (1663–1729) and was extensively used to pump water from mines but was very inefficient. The improvements developed by Watt played a critical role in the Industrial Revolution that made Britain a leader in industry.

In 1766 Black returned to Edinburgh where he concentrated on teaching. His lectures became famous and were read for many years. Later in life he had episodes of hemoptysis, presumably caused by pulmonary tuberculosis, and he died in Edinburgh in 1799. To many peoples' surprise, his will indicated that he was quite wealthy.

For readers who want additional information, Ramsay wrote an early, readable biography of Black [13] and more extended accounts with corrections are by Guerlac [8, 9] and Donovan [7]. A collection of Back's lectures was compiled by Robison [15] and an extensive series of letters between Black and Watt is available [14].

9.3   The Chemistry of Alkalis and Carbon Dioxide

The circumstances leading to Black's work on alkaline chemicals were bizarre. Renal stones were apparently more common in the eighteenth century than they are now and they were a therapeutic challenge. "Cutting for stone", that is operating to remove a renal stone or gravel from the bladder, was frequently described and in the period before anesthesia was a very painful and dangerous operation. As a result there was much interest in possible medical treatments. In 1739 a Mrs. Joanna Stephens invented a concoction that seemed to be helpful, and the English parliament voted her the sum of £5000 (an enormous amount in those days) for the recipe. This turned out to be a strange mishmash of eggshells, snails, soap and various other unlikely constituents, and as a result various medical people in the University of Edinburgh became interested in the properties of limewater which was assumed to play a role. A dispute developed between two professors, and although limewater interested Black, he thought it best to stay out of the controversy. He explained in a letter to his father "I found it proper to lay aside limewater which I had chosen for the subject of my Thesis. It was difficult and would have appeared presumptuous in me to have attempted settling some points about which two of the Professors themselves are disputing" [8]. He therefore chose to study a similar substance, *magnesia alba* (magnesium carbonate $MgCO_3$) and for his MD thesis he wrote a dissertation titled *De humore acido a cibis orto, et magnesia alba* (On the acid humour originating from food, and on magnesia alba). A year later he read a modified version of his dissertation as a paper titled "Experiments upon magnesia alba, quicklime, and some other alcaline substances" to the Philosophical Society of Edinburgh, and in 1756 this appeared in the second volume of the journal "Essays and Observations; Physical and Literary" of the Society [2] which later became the Royal Society of Edinburgh. The original paper is now very difficult to obtain but was reprinted by the Alembic Club in 1944 [3] (Fig. 9.2).

In early experiments Black added acid to *magnesia alba* and showed that it effervesced and lost weight. He used both distilled vinegar and oil of vitriol (sulfuric acid). In modern nomenclature the reaction was

$$MgCO_3 + H_2SO_4 = MgSO_4 + H_2O + CO_2$$

He also found that when *magnesia alba* was heated in a furnace, it also lost weight but the resulting material which he called *magnesia usta* did not lose weight when acids were added. Here the reaction was

$$MgCO_3 \rightarrow MgO + CO_2$$

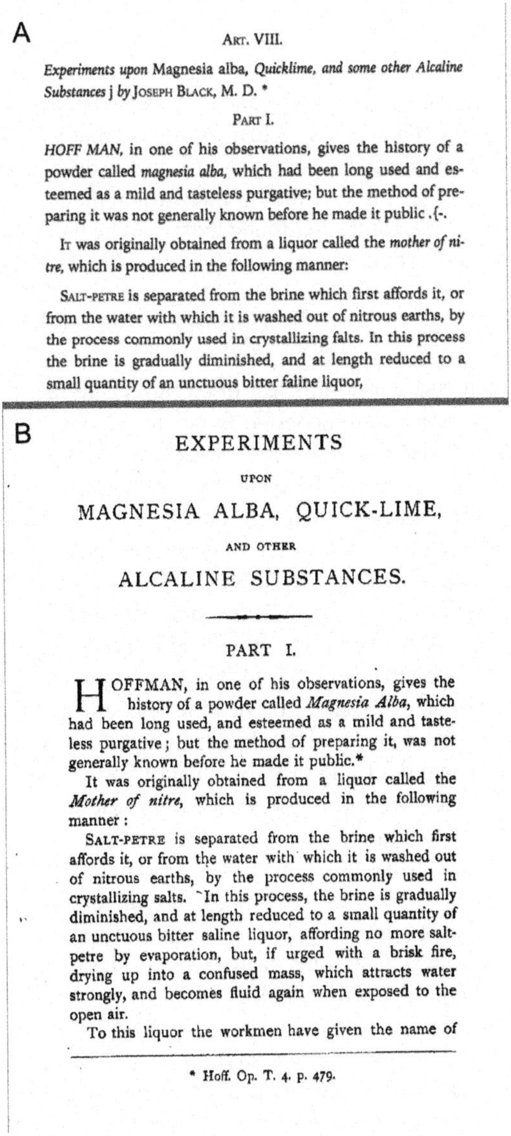

Fig. 9.2 Black's publication on alkalis and the discovery of carbon dioxide. This was his only major publication in English. A shows the original article which is now almost unobtainable. B shows a later reprint. A is from [2]. B is from [3]

A

Art. VIII.

Experiments upon Magnesia alba, *Quicklime, and some other Alcaline Substances* j *by* Joseph Black, M. D. *

Part I.

HOFF MAN, in one of his observations, gives the history of a powder called *magnesia alba*, which had been long used and esteemed as a mild and tasteless purgative; but the method of preparing it was not generally known before he made it public .{-.

It was originally obtained from a liquor called the *mother of nitre*, which is produced in the following manner:

Salt-petre is separated from the brine which first affords it, or from the water with which it is washed out of nitrous earths, by the process commonly used in crystallizing falts. In this process the brine is gradually diminished, and at length reduced to a small quantity of an unctuous bitter faline liquor,

B

EXPERIMENTS

upon

MAGNESIA ALBA, QUICK-LIME,

and other

ALCALINE SUBSTANCES.

PART I.

HOFFMAN, in one of his observations, gives the history of a powder called *Magnesia Alba*, which had been long used, and esteemed as a mild and tasteless purgative; but the method of preparing it, was not generally known before he made it public.*

It was originally obtained from a liquor called the *Mother of nitre*, which is produced in the following manner:

Salt-petre is separated from the brine which first affords it, or from the water with which it is washed out of nitrous earths, by the process commonly used in crystallizing salts. In this process, the brine is gradually diminished, and at length reduced to a small quantity of an unctuous bitter saline liquor, affording no more saltpetre by evaporation, but, if urged with a brisk fire, drying up into a confused mass, which attracts water strongly, and becomes fluid again when exposed to the open air.

To this liquor the workmen have given the name of

* Hoff. Op. T. 4. p. 479.

He realized that the action of heat on *magnesia alba* was the same as heating limestone which is calcium carbonate $CaCO_3$. Again the reaction was

$$CaCO_3 \rightarrow CaO + CO_2$$

Black then looked at the properties of the "air" given off when *magnesia alba* was either treated with acid or heated. He found that it was not the same as atmospheric air. For example he reported in a letter to Cullen "I mixed together some chalk [CaCO$_3$] and vitriolic acid... The strong effervescence produced an air or vapour, which, flowing out at the top of the glass, extinguished a candle that stood close to it; and a piece of burning paper immersed in it, was put out as effectually as if it had been dipped in water". He also showed that it was toxic to animals that breathed it. For example sparrows "died in it in ten or eleven seconds" although "they would live in it for three or four minutes when the nostrils were shut by melted suet" [15] (page 231). He further found that when he bubbled his new gas through limewater it formed a white precipitate. This was calcium carbonate and the reaction was

$$Ca(OH)_2 + CO_2 = CaCO_3 + H_2O$$

When Black used this test in a brewery he found that the same gas was given off in the process of alcoholic fermentation. He called the gas "fixed air" because in his experiments on alkalis the gas had been combined with a solid material. This was the first demonstration that gas was a weighable constituent of a solid body. As noted earlier, Black had developed very accurate chemical balances and these enabled him to show that when magnesium carbonate was heated and "fixed air" was liberated, there was a loss of weight. He also realized that it was the same gas as that described about 100 years earlier by Jan Baptist van Helmont who called it *gas sylvestre* and who produced it by adding acid to limestone. He also recalled that Stephen Hales (1677–1761) had suggested that the loss of weight of *sal tartar* (potassium carbonate) on heating was due to the loss of "elastic fluid" which he called fixed air [10]. Black also thought that it might be the same gas produced in the Grotto del Cano in Italy where it was known that people could survive if they were standing but dogs perished because the noxious gas being heavy remained close to the ground.

Black also showed that the gas exhaled from the lung contained fixed air. He did this by bubbling expired gas through limewater and noting the white precipitate of calcium carbonate. In his "Lectures on the Elements of Chemistry" [15] which were collected after his death (Fig. 9.3) he stated "And I convinced myself, that the change produced on wholesome air by breathing it, consisted chiefly, if not solely, in the conversion of part of it into fixed air. For I found that by blowing through a pipe into limewater, or a solution of caustic alkali, the lime was precipitated, and the alkali was rendered mild" [15] (p. 231). He carried out a particularly colorful experiment in Glasgow in the winter of 1764–5 when he placed a solution of lime-water that dripped over rags in an air duct in the ceiling of a church where it is said that 1500 people remained at their devotions for 10 h. Apparently the result was the formation of a substantial amount of calcium carbonate although further details of this study are not available [15] (p. 231).

Black's discovery of carbon dioxide was the first major advance in the discovery of the respiratory gases. At the time the topic was known as "pneumatic chemistry". Black's critical publication was in 1757 and it was only nine years after this that Henry Cavendish (1731–1810) reported the discovery of hydrogen. Six years after this in 1772, a pupil of Black, Daniel Rutherford (1749–1819), isolated nitrogen.

LECTURES

ON THE

ELEMENTS OF CHEMISTRY,

DELIVERED

IN THE UNIVERSITY OF EDINBURGH;

BY THE LATE

JOSEPH BLACK, M.D.

PROFESSOR OF CHEMISTRY IN THAT UNIVERSITY,

PHYSICIAN TO HIS MAJESTY FOR SCOTLAND; MEMBER OF THE ROYAL SOCIETY OF EDIN-
BURGH, OF THE ROYAL ACADEMY OF SCIENCES AT PARIS, AND THE IMPERIAL
ACADEMY OF SCIENCES AT ST. PETERSBURGH.

NOW PUBLISHED FROM HIS MANUSCRIPTS,

BY

JOHN ROBISON, LLD.

PROFESSOR OF NATURAL PHILOSOPHY IN THE UNIVERSITY OF EDINBURGH.

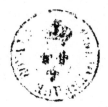

VOL. I.

EDINBURGH:

PRINTED BY MUNDELL AND SON,

FOR LONGMAN AND REES LONDON, AND WILLIAM CREECH EDINBURGH.

1803.

Fig. 9.3 Title page of Black's lectures on the elements of chemistry. These were very popular for many years. From [15]

Priestley produced oxygen in 1774 although Carl Wilhelm Scheele (1742–1786) had done this previously but reported it several years later. Finally Lavoisier in 1777 in a memoir to the Académie des Sciences was able to make the dramatic statement "Eminently respirable air [oxygen] that enters the lung, leaves it in the form of chalky aeroform acids [carbon dioxide]… in almost equal volume.…Respiration acts only on the portion of pure air that is eminently respirable… The excess, that is the mephitic portion [nitrogen] is a purely passive medium which enters and leaves the lung… without change or alteration" [12].

Lavoisier's research was the coup de grâce of the erroneous phlogiston theory originally championed by George Ernst Stahl (1659–1734). However it has been pointed out that an early blow to the phlogiston theory was actually delivered by Black because his experiments with magnesium carbonate and limestone in which carbon dioxide was released were incompatible with the phlogiston theory. Later in his life in 1796 he admitted his change in belief when he was able to state "After having, for between 30 and 40 years, believed and taught the chemical doctrines of Stahl, I have become a converter to the new views of chemical action; and subscribe to almost all Lavoisier's doctrines…" [6].

Black's paper that appeared in the second volume of the "Essays and Observations: Physical and Literary" of the Philosophical Society of Edinburgh was one of only three in English that Black wrote [2]. However he had a reputation as an inspiring speaker, and his lectures attracted a large number of students. A compilation of his lectures was published in 1803 [15] and is available on the Internet.

After his move to Edinburgh in 1766 Black apparently did not pursue his work on alkalis and carbon dioxide. However he continued to spend periods in his laboratory and was a consultant on various chemical problems mainly on industrial processes. For example he worked on methods of improving bleaching in the woolen industry, and he studied ways of converting seaweed into caustic potash. With others he also worked on the new design of furnaces for the production of iron and on material for coating ships' hulls to preserve them. He was also interested in sugar refining, brewing, and water analysis.

As indicated earlier, Black's interest in alkalis was stimulated by the medical problem of renal stone, and his two major publications, one in Latin and the other in English, stemmed from his work on this topic and earned him an MD degree. He practiced medicine although on a small scale in both Glasgow and Edinburgh. Although he submitted his work on alkalis in his MD thesis in 1754 he apparently saw his first patient in 1753. Among his patients was the famous philosopher David Hume (1711–1776), and he also advised the father of Walter Scott (1771–1832), the well-known author, that his son was in danger of developing tuberculosis because his nurse had the disease.

Black also had a number of administrative responsibilities in connection with medical institutions. He was a manager of the Royal Infirmary of Edinburgh which is still a major hospital, and he was president of the Royal College of Physicians of Edinburgh. Among his honors was his appointment as physician to "his majesty for Scotland" (George III who actually never visited the country), and his election as a member of the Imperial Academy of St. Petersburg. In fact Catherine the Great invited him to teach there at one time [1].

Black had some interesting friends in Edinburgh in 1766 to 1797. This 30 years was part of the period sometimes known as the Scottish Enlightenment which was characterized by a ferment of intellectual and scientific accomplishments. Higher education was particularly strong in Scotland at the beginning of the seventeenth century when the country could boast five universities compared with England's two. Black's friends included the philosopher David Hume and the economist Adam Smith. Various clubs existed to facilitate the exchange of ideas.

One of the most colorful was the Poker Club which derived its name not from the familiar card game but as a poker in a fireplace stirs up a flame. Many of the liveliest minds in Edinburgh attended. Much of the discussion occurred over dinner which began "soon after two o'clock, at one shilling a-head, the wine being confined to sherry and claret, and the reckoning to be called at six o'clock" [5]. Adam Smith (1723–1790) was a member and his treatise on economics "The Wealth of Nations" [16] had an enormous influence. David Hume, the philosopher who wrote "A Treatise of Human Nature" found the Poker Club a way of dispelling the depression that apparently developed when he pondered human nature. He wrote "Most fortunately it happens that since reason is incapable of dispelling these clouds... I dine... I converse, and am merry with my friends" [11]. Black was fortunate to be able to spend time with these inquiring minds.

9.4   Latent Heat

As indicated earlier, when Black returned to Glasgow in 1756 to be lecturer and later professor, his research interests changed to the topic of heat. Although this now seems to us as a major change in direction, he may not have seen it that way. After all his previous work had been on the effects of heat on substances, and particularly on the subsequent elimination of fixed air with a corresponding loss of weight. His new interest in heat may have been prompted by observations of Cullen who had been working in Glasgow. Cullen had noticed that when ordinary ether, a volatile liquid with a low boiling point, was exposed to a partial vacuum by means of an air pump, it began to boil, and the remaining liquid cooled markedly. Cullen published an account of this experiment in 1748, that is eight years before Black moved to Glasgow. However as indicated before, Cullen had a high opinion of Black and they probably discussed these topics.

Black was aware of the well-known fact that snow is very slow to melt after the air temperature has risen above the freezing point of water. This suggested that heat was necessary for the change of state from ice to water because the water that was formed after thawing had a temperature only slightly above freezing. Black therefore carried out some experiments to investigate this. In one of these, two containers were set up a large room where the temperature remained at 47 °F. (Black used the scale described in 1724 by Daniel Fahrenheit (1686–1736) in all his work.) The containers were located 18 inches apart and one of them contained five ounces of ice at 32 °F while the other contained the same weight of water at 33°. Black found that the water warmed to 40° in half an hour. However the ice took ten and a half hours to obtain the same temperature, or in other words 21 times as long as the water. Black therefore argued that the heat absorbed by the ice was (40–33) × 21, or 147 units of heat. By contrast the water had only absorbed 8 units of heat. Therefore 147–8 or 139 units of heat had been absorbed by the melting ice and were concealed, as it were, in the water [13].

In another experiment Black took a piece of ice at a temperature of 32 °F, weighed it, and added it to a known weight of water of known temperature. This allowed him to calculate the amount of heat required to melt the ice, and this came out to be that which would have heated an equal quantity of water by 143 °F. In a third experiment, a lump of ice at 32 °F was placed in an equal volume of water that had been heated to 176 °F and it was found that the water cooled to 32 °F. Therefore the heat required to melt the ice was 176–32 = 44 °F of heat. This means that the latent heat of fusion came out to be about 144 °F which corresponds to 80 °C. This is very near the currently accepted value although the units for latent heat in the SI system are now kilojoules per kilogram.

Black's calculations cited above refer to the latent heat of freezing water. He also measured the latent heat required to turn water into steam. To do this he compared the time required for a known weight of water to rise from a given temperature to boiling point on the one hand, and the time required to convert the same weight of water into steam. The number he got was 830 units on the Fahrenheit scale and the accepted figure today is 967.

Curiously Black never published his work on latent heat. Instead it was communicated to a group of professors at the University of Glasgow in 1762, and it formed part of the lectures that were given to his students and that are now available [15].

Black also studied the specific heats of various liquids. This term refers to the amount of heat required to increase the temperature of a particular substance by a known amount. For example Black described an experiment where a known weight of mercury at 150 °F was mixed with an equal weight of water at a temperature of 100 °F. He found that the temperature of the mixture at equilibrium was not 125° as might be expected, but 120°. In other words the mercury was cooled by 30° while the water was warmed to only 20° in spite of the fact that the quantity of heat gained by the water was the same as that lost by the mercury. Therefore, he said, the same quantity of heat has more effect in heating mercury than in heating an equal weight of water. Thus Black made a clear distinction between what we might call the intensity of heat on the one hand and the quantity of heat on the other. The intensity, or temperature, can be measured with a thermometer. However the quantity of heat transferred to a substance requires both a measurement of the change in temperature and the duration of the heat transfer.

Incidentally it is interesting to note the great contrast in the number of publications between Black and his near contemporary, Joseph Priestley, who came from Leeds a couple of hundred miles to the south. The taciturn Scot published only three papers in English [13] whereas the ebullient Priestley is credited with some 150 publications including about 50 books.

9.5   Joseph Black and James Watt

As mentioned earlier, Black was friendly with the young James Watt (Fig. 9.4) whom he met in Glasgow, and their friendship was apparently partly responsible for one of the most important innovations in the Industrial Revolution.

Watt was born in Greenock on the Firth of Clyde not far from Glasgow. His father was a shipwright and Greenock had a long tradition of shipbuilding. James showed an interest in engineering from an early age and when he was 18 went to London for a year to study instrument making. When he returned to Glasgow, he found a place in the University where he called himself a mathematical instrument maker. There he met John Robison who had an interest in astronomy and the two enjoyed working on engineering problems. Robison was later to publish a collection of Black's lectures [15]. When Black returned to Glasgow from Edinburgh in 1756 he met both Watt and Robison and the three became firm friends with an interest in engineering.

Apparently it was Robison who first interested Watt in early steam engines. These had been used for many years to pump water from mines but were ineffi-

Fig. 9.4 James Watt (1736–1819). (From the National Portrait Gallery, by permission)

Fig. 9.5 Diagram of the Newcomen steam engine which was the workhorse in the early eighteenth century. It was later modified by James Watt with help from Joseph Black with a resulting great increase in efficiency. (From http://railroad.linda-hall.org/essays/locomotives. html)

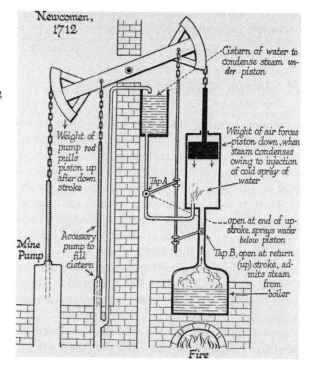

cient. The most important was the Newcomen engine shown in Fig. 9.5. At bottom right there is a furnace that heats water in a boiler to make steam. Above this is a cylinder with a piston. The beam at the top is balanced in such a way that the weight of the mine pump (bottom left) pulls the piston up filling the cylinder with steam. Then a valve is opened allowing cold water from the cistern to enter the cylinder as a spray thus condensing the steam. The result is a partial vacuum in the cylinder and the piston moves down because of the atmospheric pressure acting on it from above. When the piston is near the bottom of the cylinder the injection of water is stopped and the piston rises again filling the cylinder with steam. Note that the power stroke is the descent of the piston because of the partial vacuum, not the rise of the piston as a result of the injection of steam as might be expected. Since the downward movement of the piston that activates the pump is caused by atmospheric pressure, this was often called an atmospheric engine. This design had been used for a period of over 50 years with almost no change since it was first proposed by Thomas Newcomen to pump water from mines. Watt's interest was sharpened when he was asked to repair a Newcomen engine owned by the University and he realized that it was very inefficient.

On investigating the pump, Watt recognized that one of the problems was that the cylinder was cooled during each down stroke of the piston as the steam was condensed. Therefore its temperature changed greatly with each cycle with a consequent waste of heat. In addition he was surprised at the amount of water that was necessary to condense the steam. He then asked Black whether the amount of heat required to make steam was much greater than the heat required to raise water to its boiling point. Black realized that the key to this question was the latent heat of vaporization which he had shown to be very large. The upshot was that Watt decided to choose another design that did not involve heating and cooling the power cylinder. Watt later wrote that Black "told me that [the doctrine of latent heat] had long been a tenet of his and [he] explained to me his thoughts on the subject" [14].

The solution devised by Watt was to have a separate chamber apart from the piston in which the steam could be condensed. The result was that the temperature of the main cylinder was maintained and this was assisted by surrounding it with a steam jacket. Readers of today may wonder why steam from the boiler was not used for the power stroke as was the case in later designs. The reason was that at that time it was impossible to fabricate a cylinder with the close tolerances required by the piston, and that could tolerate the high pressures without being damaged. Later Watt joined Matthew Boulton (1728–1809) in Birmingham where precision engineering was available and "expansive" steam engines (that is, powered by steam above atmospheric pressure) were developed.

Watt's modified engine became extremely popular and was used for many years. It was one of the inventions responsible for the Industrial Revolution with a corresponding prodigious increase in industry in England and ultimately throughout the rest of the world. So it could be argued that Black's early work on latent heat had an important influence on the development of manufacturing and the resulting increase in the wealth of many countries.

In conclusion, Black clearly merits an important place in the history of respiratory physiology. His discovery, or more strictly re-discovery, of carbon dioxide started an avalanche of investigations of other gases of respiratory importance. Curiously some historians ignore his work on carbon dioxide but give much emphasis to his discovery of latent heat and specific heat. Arguably this influenced his friend James Watt, and ultimately was a factor in the development of the steam engine that ushered in the Industrial Revolution and had an enormous effect on the wealth of Britain and other nations. Yet in some ways Black is an enigma. He only published one major paper in English, and although his published lectures were read for many years, they soon became outdated. He later became a consultant on a series of projects in industrial chemistry and was surprisingly wealthy when he died [13]. He forms a striking contrast to his near contemporary, Joseph Priestley, who published prodigiously and discovered oxygen, the other major respiratory gas.

References

1. Anderson RGW. Black, Joseph (1728–1799). Oxford dictionary of national biography. Oxford, UK: Oxford University Press; 2004.
2. Black J. Experiments upon magnesia alba, quicklime, and some other alcaline substances. Essays Obs: Phys Lit. 1756;2:157–225.

3. Black J. Experiments upon magnesia alba, quicklime, and some other alcaline substances. Edinburgh: The Alembic Club reprints; 1944.
4. Buchanan WW, Brown DH. Joseph Black (1728–1799): Scottish physician and chemist. Practitioner. 1980;224:663–6.
5. Carlyle A. Autobiography of the Rev. Dr. Alexander Carlyle. Edinburgh: William Blackwood and Sons; 1860.
6. Cragg RH. Thomas Charles Hope (1766–1844) Med Hist. 1967;11:186–9.
7. Donovan A. James Hutton, Joseph Black and the chemical theory of heat. Ambix. 1975;25:176–90.
8. Guerlac H. Joseph Black and fixed air a bicentenary retrospective, with some new or little known material. Isis. 1957a;48:124–51.
9. Guerlac H. Joseph Black and fixed air: Part II. Isis. 1957b;48(154):433–56.
10. Hales S. Vegetable staticks. London: printed for W. and J. Innys and T. Woodward; 1727.
11. Hume D. A treatise of human nature. London: Henry Frowde; 1748.
12. Lavoisier A-L. Expériences sur la respiration des animaux, et sur les changemens qui arrivent à l'air en passant par leur poumon, 1777. Reprinted in Lavoisier, A-L Oeuvres de Lavoisier. Paris: Imprimerie Impériale; 1862–1893.
13. Ramsay W. The life and letters of Joseph Black. London: Constable; 1918.
14. Robinson E, McKie D, editors. Partners in science. Letters of James Watt and Joseph Black. Cambridge: Harvard University Press; 1970.
15. Robison J. Lectures on the elements of chemistry by Joseph Black. London: Longman and Rees; Edinburgh: William Creech; 1803. http://gallica.bnf.fr/ark:/12148/bpt6k62375d; https://archive.org/download/2543060RX2.nlm.nih.gov/2543060RX2.pdf.
16. Smith A. An inquiry into the nature and causes of the wealth of nations. London: W. Strahan and T. Cadell; 1776.

Chapter 10
Carl Wilhelm Scheele, the Discoverer of Oxygen, and a Very Productive Chemist

Abstract Carl Wilhelm Scheele (1742–1786) has an important place in the history of the discovery of respiratory gases because he was undoubtedly the first person to prepare oxygen and describe some of its properties. In spite of this, his contributions have often been overshadowed by those of Joseph Priestley and Antoine Lavoisier who also played critical roles in preparing the gas and understanding its nature. Sadly, Scheele was slow to publish his discovery and therefore Priestley is rightly recognized as the first person to report the preparation of oxygen. Having said this, the thinking of both Scheele and Priestley was dominated by the phlogiston theory, and it was left to Lavoisier to elucidate the true nature of oxygen. In addition to his work on oxygen, Scheele was enormously productive in other areas of chemistry. Arguably he discovered seven new elements, and many other compounds. However he kept a low profile during his life as a pharmacist, and he did not have strong links with contemporary prestigious institutions such as the Royal Society in England or the French Académie des Sciences. He was elected to the Royal Swedish Academy of Science but only attended one meeting. Partly as a result, he remains a somewhat nebulous figure in spite of the critical contribution he made to the history of respiratory gases, and his extensive researches in other areas of chemistry. His death at the age of 43 may have been hastened by his habit of tasting the chemicals that he worked on.

10.1 Introduction

Carl Wilhelm Scheele (1742–1786) (Fig. 10.1) was a Swedish-German chemist who has the distinction of being the first person to prepare oxygen and describe some of its most important properties. As such he would be expected to have a very prominent place in the history of respiratory physiology. However in fact he remains a somewhat enigmatic figure. One reason for this is that although he undoubtedly prepared oxygen before anybody else, he did not publish his discovery until several years later by which time Priestley had published his own independent

© American Physiological Society 2015
J. B. West, *Essays on the History of Respiratory Physiology,*
Perspectives in Physiology, DOI 10.1007/978-1-4939-2362-5_10

Fig. 10.1 Carl Wilhelm
Scheele (1742–1786). This
is an engraving by Evald
Waldemar Hansen. From [9]

discovery of the gas in very arresting prose [19]. It is also true that both Scheele and Priestley interpreted their findings in the light of the erroneous phlogiston theory. It was left to Lavoisier to elucidate the true nature of oxygen although, as we shall see, he benefited from a communication from Scheele and characteristically he did not acknowledge this. The upshot is that we have three important chemists all connected with the discovery of oxygen, and in many accounts Scheele is not given the credit to which he is due.

Another reason for Scheele's somewhat nebulous reputation is that he was rather self-effacing and worked in comparative obscurity as a pharmacist in Sweden. Added to this, his only book was not as accessible as the accounts of Priestley and Lavoisier. Having said this, Scheele had an extraordinarily productive research career with not only oxygen but many other chemical discoveries to his credit.

Consistent with the comments above, Scheele has not attracted anything like the attention that writers have given to Priestley and Lavoisier. However in 1892, A.E. Nordenskiöld published (in German) a comprehensive selection of the letters and notes of Scheele. This was a pioneering work because little attention had been given to Scheele's work since his death in 1786. Dobbin [16] published an English translation of collected papers of Scheele including his major book "Chemische Abhandlung von der Luft und dem Feuer" (Chemical treatise on air and fire) and a modern reprint is available as indicated in the list of references. Boklund [2] wrote a book on "Carl Wilhelm Scheele: His work and life" which contains a lengthy commentary on some of Scheele's papers including those in the so-called Brown Book which are almost undecipherable and were omitted by Nordenskiöld. An attractive profusely illustrated volume titled "The Apothecary Chemist: Carl Wilhelm Scheele" was written by Urdang [18] who was director of the American Institute of the History of Pharmacy. This contains a very readable introduction to Scheele's life and work. Other good accounts are Frängsmyr [7] and Boklund [3]. An English translation of Scheele's book "Chemical treatise on air and fire" including front matter not available in Dobbin [16], was made by J.R. Forster in 1780 and is available on the Internet at http://tinyurl.com/mnqp2v9 [14]. Partington [10] is authoritative on Scheele's chemistry discoveries.

10.2 Brief Biography

Scheele was born in Stralsund, a city in western Pomerania, now part of Germany. However at the time this area was under Swedish jurisdiction. The city is on the Baltic coast about 140 km due south of Malmö. Scheele's father was a rather unsuccessful brewer and corn-chandler. When Carl was 14 he moved to Gothenburg, Sweden to be an apprentice pharmacist. There he developed an interest in chemistry and apparently carried out experiments late in the night using the chemicals available in the pharmacy. He also read widely including the work of Georg Ernst Stahl (1659–1734) who was one of the main proponents of the phlogiston theory. Later Scheele moved to Malmö where he worked with C.M. Kjellström who had scientific interests. There Scheele also made contact with Anders Retzius who was a prominent chemist at Lund University.

A little later Scheele went to Stockholm to work in another pharmacy, but after two years he moved again, this time to Uppsala which was a celebrated academic center. Here he was the director of the laboratory of a large pharmaceutical company, and he became acquainted with Torbern Bergman(1735–1784) (Fig. 10.2) who was a professor of chemistry at the eminent university there. He was well known for his work on chemical affinities, that is the properties that allow dissimilar chemical species to form compounds. He also contributed to the theory of crystal structure

Fig. 10.2 Torbern Bergman (1735–1784). He was an eminent professor of chemistry at Uppsala University and a strong supporter of Scheele all his life. Lithograph by Otto Henrik Wallgren. From (https://commons.wikimedia.org/wiki/File:Torbern_Bergman-1849.jpg)

Fig. 10.3 Building of the pharmacy in Köping where Scheele did much of his later research. His laboratory was a wooden barn in the courtyard behind the house. From [18]

and worked on the crystal structure of some minerals. The uranium crystal, torbernite, is named after him. Bergman asked Scheele to help with a problem involving potassium nitrate and acetic acid and later this study led to the discovery of oxygen (see below). Bergman remained a strong advocate of Scheele and wrote a long introduction to Scheele's book "Chemical treatise on air and fire". Bergman later called Scheele his greatest discovery.

After five years in Uppsala, Scheele moved to Köping, a small town to the west of Stockholm, where he set up his own business as a pharmacist (Fig. 10.3). Much of his extensive research was done in this somewhat isolated setting. In fact, although the building shown in Fig. 10.3 looks modest, most of Scheele's chemical research was actually done in a wooden barn in the courtyard behind the house. However his abilities were recognized and he was elected as a member of the Royal Swedish Academy of Sciences. Scheele died at the comparatively early age of 43 and it is thought that his habit of tasting various dangerous chemicals may have been a factor in his demise (see below).

10.3 The Discovery of Oxygen

As indicated above, Scheele's studies that eventually resulted in the discovery of oxygen began when he was at Uppsala and he was asked by Torbern Bergman to clarify a problem that occurred when saltpeter (potassium nitrate) was heated with acetic acid and then produced a red vapor which was nitrogen dioxide. Apparently Bergman asked Scheele's advice because he was concerned about the purity of the saltpeter. When we start to track the events that led up to the discovery of oxygen at this time, a serious problem is that the information about Scheele's actual experiments comes only from notes and letters. These were subsequently published in German by Nordenskiöld [15] and an English translation of some of this material is available [16]. The fact that Scheele's work was only described in notes at this stage, some of which are difficult to decipher, makes the identification of the timeline for the discovery of oxygen challenging.

According to Partington [10], Nordenskiöld concluded from his study of the manuscripts that most of the work was completed in 1773, but some of the experiments, including the preparation of oxygen by heating potassium nitrate go back to 1770. Other authors claim that in the earliest experiments, oxygen was produced by heating manganese dioxide with sulfuric acid [4]. "Vitriol air" is an early term that was used by Scheele for oxygen, and this occurs in a manuscript referred to as number 52 from the Uppsala period with dates of 1770–1771.

Scheele produced oxygen by heating a variety of substances including mercuric oxide as did Priestley. Scheele also obtained the gas by heating potassium nitrate, silver carbonate, manganese nitrate and manganese oxide. He reported that the resulting gas was odorless, tasteless, and supported the combustion of a candle more than air. Figure 10.4 reproduces his account from his book (see below) which is here given in an English translation [14]. In a summary of Scheele's experiments written by Bergman in 1775, he stated that the gas formed by heating oxides of mercury, silver and gold supported both combustion and respiration better than common air [10]. Scheele's later name for oxygen, "Feuerluft" (fire-air) was first used in 1775.

It is therefore clear that Scheele's first experiments occurred two or three years before Priestley first produced oxygen because his date is well established [5]. His famous account reads "On 1st of August, 1774, I endeavored to extract air from *mercurius calcinatus per se* [mercuric oxide], and I presently found that... air was expelled from it very readily... but what surprized me more than I can well express, was, that a candle burned in this air with a remarkably vigorous flame... I was utterly at a loss how to account for it" [12]. Priestley first published his discovery in 1775 [11].

Scheele described his historic discovery of oxygen for the outside world in his only book "Chemische Abhandlung von der Luft und dem Feuer" (Chemical treatise on air and fire) [13]. The title page of this is shown in Fig. 10.5. The book was written in the autumn of 1775 and sent to Bergman early in 1776. However it was not published until 1777, that is two years after Priestley's first publication of the discovery of oxygen. Scheele blamed his publishers for the delay but of course Scheele was tardy in writing up his work for publication. Probably he did not concern himself greatly with the issue of priority in the early 1770s. There is some evidence that Scheele first learned of Priestley's work in November 1775 although it is also suggested that Scheele's first knowledge of Priestley's discovery was in a letter from Bergman of August 1776. An English translation of Scheele's book by J.R. Forster was published in 1780 (Fig. 10.6). The front matter includes a letter to

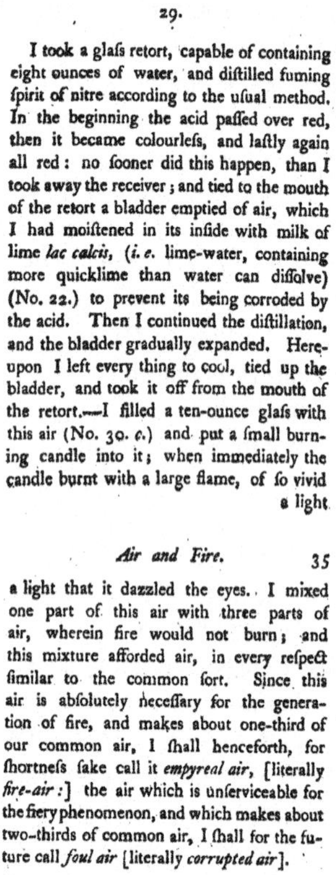

Fig. 10.4 Extract from the English translation of "Chemische Abhandlung von der Luft und dem Feuer". Note that when Scheele put a burning candle into his fire air it burnt with a flame so vivid that it dazzled the eyes. From [14]

29.

I took a glafs retort, capable of containing eight ounces of water, and diftilled fuming fpirit of nitre according to the ufual method. In the beginning the acid paffed over red, then it became colourlefs, and laftly again all red : no fooner did this happen, than I took away the receiver ; and tied to the mouth of the retort a bladder emptied of air, which I had moiftened in its infide with milk of lime *lac calcis*, (*i. e.* lime-water, containing more quicklime than water can diffolve) (No. 22.) to prevent its being corroded by the acid. Then I continued the diftillation, and the bladder gradually expanded. Hereupon I left every thing to cool, tied up the bladder, and took it off from the mouth of the retort.——I filled a ten-ounce glafs with this air (No. 30. *c.*) and put a fmall burning candle into it; when immediately the candle burnt with a large flame, of fo vivid
a light

Air and Fire.   35

a light that it dazzled the eyes. . I mixed one part of this air with three parts of air, wherein fire would not burn ; and this mixture afforded air, in every refpeƐ fimilar to the common fort. Since this air is abfolutely neceffary for the generation of fire, and makes about one-third of our common air, I fhall henceforth, for fhortnefs fake call it *empyreal air*, [literally *fire-air* :] the air which is unferviceable for the fiery phenomenon, and which makes about two-thirds of common air, I fhall for the future call *foul air* [literally *corrupted air*].

Priestley from the translator, and in the notes at the end of the book is a letter to Scheele from Priestley.

There is also a fascinating link with Lavoisier which is important in the context of the history of the discovery of oxygen. In 1774 Scheele wrote a letter to Lavoisier, a draft of which is now in the Centre for History of Science at the Royal Swedish Academy of Sciences in Stockholm [1]. It was recopied and sent to Lavoisier on September 30. Remarkably it was thought to have disappeared but was discovered in the Archives of the French Académie des Sciences in 1890 by Edouard Grimaux [8]. The letter is in French but a translation is as follows:

Sir,

I have received through Secretary Wargentin a book, which he says that you have had the goodness to give me (as a present). Although I do not have the honor of being known by you, I am taking the liberty of thanking you very humbly. I desire nothing with as much (passion) ardor as to be able to show you my gratitude.

For a long time I have wanted to be able to read an account of all the experiments that have been done in England, in France and in Germany on the many kinds of air. You have not only satisfied this wish, but by new experiments you have given scientists in the future the most beautiful opportunities to better examine fire and the calcination of metals.

During the past several years I have carried out experiments on several kinds of air, and I have also spent a good deal of time in discovering the singular properties of fire, but I have

Fig. 10.5 Title page of Scheele's only book "Chemische Abhandlung von der Luft und dem Feuer" published in 1777. This was his first public announcement of his discovery of fire air. From [13]

never been able to prepare ordinary air from fixed air: I have tried many times, following the opinion of M. Priestley, to produce an ordinary air from fixed air by a mixture of iron filings, sulfur, and water, but I have never succeeded because fixed air always united with the iron and made it soluble in the water. Perhaps you do not know a way to do this either. Because I do not have any large burning glass, I beg you to carry out an experiment (a trial) with yours in this way: Dissolve some silver in nitrous acid and precipitate it with alkaline tartrate, wash the precipitate, dry it, and reduce it with the burning glass in your machine, Fig. 8, but because the air in this bell jar (this receiver) is such that animals die in it and a part of the fixed air separates from the silver in this operation, it is necessary to place a bit of quick lime in the water where one has put the bell, so that this fixed air joins more quickly with the lime. This is the way that I hope that you will see how much air is formed during this reduction, and whether a lighted candle can keep burning and animals live in this air it. By this experiment you will do me a great favor. I would be infinitely obliged if you would inform me of the result of this experiment. I have the honor of remaining with great esteem, Monsieur, your very humble servant.

Uppsala, the _ September, 1774.

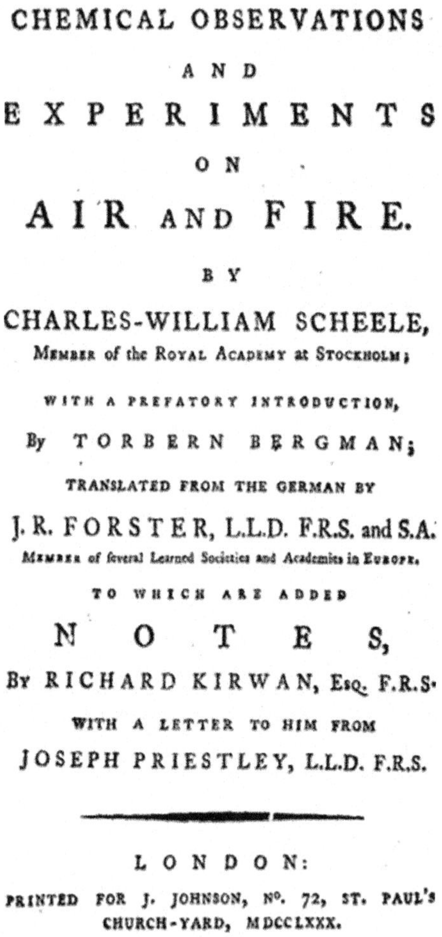

Fig. 10.6 English translation of "Chemische Abhandlung von der Luft und dem Feuer" published in 1780. In the notes at the end of the book there is a letter from Joseph Priestley. From [14]

CHEMICAL OBSERVATIONS

A N D

E X P E R I M E N T S

O N

A I R AND F I R E.

B Y

CHARLES-WILLIAM SCHEELE,

Member of the Royal Academy at Stockholm;

WITH A PREFATORY INTRODUCTION,

By T O R B E R N B E R G M A N;

TRANSLATED FROM THE GERMAN BY

J. R. F O R S T E R, L.L.D. F.R.S. and S.A.

Member of several Learned Societies and Academies in Europe.

TO WHICH ARE ADDED

N O T E S,

By R I C H A R D K I R W A N, Esq. F.R.S.

WITH A LETTER TO HIM FROM

J O S E P H P R I E S T L E Y, L.L.D. F.R.S.

L O N D O N:

PRINTED FOR J. JOHNSON, N°. 72, ST. PAUL'S
CHURCH-YARD, MDCCLXXX.

Note particularly the section "… I hope that you will see how much air is formed during this reduction, and whether a lighted candle can keep burning and animals live in this air". The original French reads: "j'espère que vous verrés combien d'air se produit pendant cette réduction, et si une chandelle allumée pouvait soutenir la flame, et les animaux vivre là-dedans".

This dramatic phrase essentially announces the discovery of oxygen, and the significance could hardly have escaped Lavoisier. However he never acknowledged receiving Scheele's letter and it has even been suggested that Lavoisier's wife hid the letter from her husband to allow him to claim credit for the discovery of oxygen [17]. This rather scandalous assertion was included in an entertaining play on the discovery of oxygen [6]. It should be added than an unfortunate characteristic of Lavoisier was his repeated failure to acknowledge contributions made by other scientists. The chemist Joseph Black among others accused him of this. It is also relevant here that Priestley visited Lavoisier in Paris in October 1774 and had dinner with him. At that time he told Lavoisier about his own experiments with oxygen.

Scheele's book "Chemische Abhandlung von der Luft und dem Feuer" is a substantial book consisting of 97 numbered paragraphs [13]. There is a long introduction by Bergman who reaffirms that Scheele's discovery of oxygen preceded Priestley's publications. In the preface, Scheele himself states that he had completed the greater part of his experiments before he obtained sight of Priestley's elegant observations.

The book begins with studies showing that atmospheric air is made up of two kinds of "elastic fluids". The first is "fire air" which we now know as oxygen. The second is variously translated as "spoiled" or "vitiated" or "foul" air. This is nitrogen. The proportion of the two components was given by Scheele as about 1 to 3. However in a supplement to the book published in 1779 he revised the proportion of fire air to total air as 9/33rds which is 27%. The actual value is 21%. Much of the latter part of the book is devoted to the properties of fire and light and is of less interest to us today. As mentioned earlier, both Scheele and Priestley interpreted all their results in the light of the phlogiston theory. It was left to Lavoisier to demolish this and thus describe the true nature of oxygen.

Although it is generally acknowledged that the trio of Scheele, Priestley and Lavoisier were responsible for the discovery of oxygen and understanding its nature, two previous investigators should be mentioned. One is John Mayow (1661–1679) who was a talented English experimentalist and stated that air contained a material that he named "nitro-aerial spirit". He argued that both animals and flames expired in a closed vessel "for want of nitro-aerial particles" but Mayow never generated oxygen. Another somewhat mysterious person whose claims have only recently been recognized was the Polish alchemist, Michael Sendivogius (1566–1636). He heated saltpeter (potassium nitrate) and referred to the gas produced, which included oxygen, as the "food of life". However this work was not apparently picked up by other chemists. More information is at www.sendivogius.pl.

10.4   Scheele's Other Discoveries

This essay concentrates on the discovery of oxygen because this is clearly Scheele's most important contribution in the context of the history of respiratory physiology. However Scheele made many other important discoveries in the field of chemistry and indeed in his short lifespan of 43 years, some authorities state that he discovered more elements (seven) than any other scientist. For example, it can be argued that Scheele discovered chlorine, oxygen, manganese, barium, molybdenum, tungsten and fluorine. However this assessment can be confusing. For example Scheele's chlorine was in fact a mixture with air, and the true nature of chlorine was first described by Humphry Davy. Indeed Davy certainly deserves an important place for "bean counters" since he discovered eight elements: lithium, boron, sodium, magnesium, potassium, calcium, strontium and barium. Little is to be gained by odious comparisons of these giants in the history of chemistry but the bottom line is that Scheele was enormously productive in spite of his low profile.

For a full account of Scheele's discoveries apart from oxygen, Partington [10] should be consulted. Scheele had a strong interest in mineralogy, and an early interest was a mineral known as black magnesia or pyrolusite, that is manganese dioxide. Scheele showed that this substance was a strong oxidizing agent and he also discovered manganese in plant ashes. Scheele's research on manganese led him to the discovery of chlorine when he dissolved the pyrolusite in hydrochloric acid and warmed it in a retort. He showed that the gas was a strong bleaching agent, it attacked many metals, and when combined with soda formed common salt. Scheele also worked on fluorspar, a mineral form of calcium fluoride. By distilling this material with sulfuric acid he produced hydrofluoric acid and recognized its powerful corrosive properties. He also prepared a number of its salts, that is fluorides. A further area of research was the chemical properties of bone and horn. By treating bone ash with sulfuric acid, he obtained phosphoric acid.

Another mineral studied by Scheele was molybdenite, that is molybdenum sulfide. By treating this in various ways with nitric acid and solvents he obtained molybdenum. Scheele also discovered tungsten trioxide which he obtained by boiling calcium tungstate, now known as scheelite. Scheele also worked on some of the properties of arsenic. He prepared copper arsenite which has a brilliant green color and this is now known as "Scheele's green".

An informative listing of Scheele's discoveries is as follows [18]: gases included oxygen, chlorine (though not in pure form), ammonia, hydrochloric acid gas. Inorganic acids included hydrofluoric, nitrosulfonic, nitrous, molybdic, tungstic and arsenic. Organic acids included lactic, gallic, pyrogallic, oxalic, tartaric, malic, mucic, uric. Scheele also isoloated glycerin and lactose, and determined the composition of borax and Prussian blue (ferric ferrocyanide). He invented new processes for preparing ether, calomel, magnesia and phosphorus. Incidentally his work on phosphorus was important in the development of the large Swedish match industry. His observation that different parts of the solar spectrum influence the decomposition of silver chloride to different degrees was important in the development of photography.

Partington finishes his section on Scheele's chemical discoveries with the following assessment "Every chemist who has attempted research will look over the record of Scheele's discoveries… with astonishment and admiration. Astonishment at the great volume of discoveries which he made in his short life in such disadvantaged circumstances; admiration of the way in which he carried out his work and the fundamental importance of it all".

10.5 Scheele's Death

In 1785, Scheele became ill with symptoms of renal disease. A short time later he developed a disease of the skin although its nature is not clear. However, remarkably, as his disease progressed, he decided to marry the widow of the man who had owned the pharmacy in Köping before he did. Her name was Margareta Pohl and they were married only three days before he died.

Because this series of events was so unusual, further information was sought. Apparently when Scheele moved to Köping in 1775, he became "provisor" of the pharmacy there. The pharmacist Pohl had died and the license thereby went to his widow, Margareta, born Sonneman. "Provisor" (a title used up to 1819) meant to manage a pharmacy without owning it. In 1776, another apothecary who was wealthy offered widow Pohl such a profitable price that she decided to accept it and sell the store. However Scheele bought the license from the widow promising to marry her. Such an arrangement was called "to preserve the widow" and was mostly used within the church, when a vicar in a parish had died. But Scheele apparently did not have the time or inclination to take this important step. Shortly before he died, however, he married widow Pohl, and by this arrangement the license went over to her again. This information was given to me in a letter from Tore Frängsmyr. His biography of Scheele has already been referred to [7].

Possible reasons for Scheele's early death have been discussed. Scheele had developed a habit of tasting various chemicals that he worked on. Since these included arsenic, lead, and other toxic materials, it has been suggested that these were a factor in his demise. In the event he died at the early age of 43 in May 1786 at his home in Köping.

In summary Scheele remains something of an enigma. There is no doubt that he was the first person to produce oxygen and describe some of its important properties. He was also enormously productive in other areas of chemistry. However all this was done by a man of low profile who moved a number of times during his professional life and who never developed a recognizable laboratory. His life forms a dramatic contrast to those of Priestley and Lavoisier who were the other members of the trio who discovered oxygen and clearly understood what it was.

Perhaps there is a lesson here about the importance of not delaying publication of important discoveries. The relative obscurity of Scheele is due to the fact that although he clearly was the first person to prepare oxygen, he was slow to report it, and as a result everybody knows about the contributions of Priestley and Lavoisier. As indicated earlier, Scheele was rather self-effacing and worked in comparative obscurity, and perhaps did not give a lot of thought to priority of publication. In this "publish or perish" age, his experience might serve as a wake up call for young scientists with the message that timely publication is essential.

References

1. Boklund U. A lost letter from Scheele to Lavoisier. Lychnos. 1957–1958;17:39–62.
2. Boklund U. Carl Wilhelm Scheele. His work and life. Stockholm: Roos Boktryckeri AB; 1968.
3. Boklund U. Scheele, Carl Wilhelm. Complete dictionary of scientific biography. Detroit: Charles Scribner's Sons; 2008.
4. Cassebaum H, Schufle JA. Scheele's priority for the discovery of oxygen. J Chem Educ. 1975;52:442–4.

5. Conant JB. The overthrow of the phlogiston theory. Cambridge: Harvard University Press; 1948.
6. Djerassi C, Hoffmann R. Oxygen: a play in two acts. New York: Wiley-VCH; 2001.
7. Frängsmyr T. Carl Wilhelm Scheele (1742–1786). Chemica Scripta. 1986;26:507–11.
8. Grimaux E. Une lettre inédite de Scheele à Lavoisier [An unpublished letter from Scheele to Lavoisier]. Revue générale des sciences pures et appliqués. 1890;1:1–2.
9. Krook A. Carl Wilhelm Scheele. Svenska Familj-Journalen. 1874;13:325–6.
10. Partington JR. A history of chemistry. London: Macmillan; 1962.
11. Priestley J. An account of further discoveries in air, in letters to Sir John Pringle. J Philos Trans. 1775;65:384–94.
12. Priestley J. Experiments and observations on different kinds of air. Vol. II. London: J. Johnson; 1776.
13. Scheele CW. Chemische Abhandlung von der Luft und dem Feuer: nebst einem Vorbericht. Upsala: Verlegt von Magn. Swederus; 1777.
14. Scheele CW, Forster JR (trans.). Chemical observations and experiments on air and fire. London: J. Johnson; 1780.
15. Scheele CW, Nordenskiöld AE. editor. Nachgelassene Briefe und Aufzeichnungen. Stockholm: P.A. Norstedt & söner; 1892.
16. Scheele CW, Dobbin L. editor. The collected papers of Carl Wilhelm Scheele. London: G. Bell & Sons Ltd.; 1931. (This was reprinted by the Kraus Reprint Co., New York; 1971).
17. Severinghaus JW. Priestley, the furious free thinker of the enlightenment, and Scheele, the taciturn apothecary of Uppsala. Acta Anaesthesiol Scand. 2002;46:2–9.
18. Urdang G. The apothecary chemist Carl Wilhelm Scheele. Madison: The American Institute of the History of Pharmacy; 1942.
19. West JB. Joseph Priestley, oxygen, and the enlightenment. Am J Physiol Lung Cell Mol Physiol. 2014;306:L111–9.

Chapter 11
Joseph Priestley, Oxygen, and the Enlightenment

Abstract Joseph Priestley (1733–1804) was the first person to report the discovery of oxygen and describe some of its extraordinary properties. As such he merits a special place in the history of respiratory physiology. In addition his descriptions in elegant eighteenth-century English were particularly arresting, and rereading them never fails to give a special pleasure. The gas was actually first prepared by Scheele (1742–1786) but his report was delayed. Lavoisier (1743–1794) repeated Priestley's initial experiment and went on to describe the true nature of oxygen which had eluded Priestley who never abandoned the erroneous phlogiston theory. In addition to oxygen, Priestley isolated and characterized seven other gases. However most of his writings were in theology because he was a conscientious clergyman all his life. Priestley was a product of the Enlightenment and argued that all beliefs should be able to stand the scientific scrutiny of experimental investigations. As a result his extreme liberal views were severely criticized by the established Church of England. In addition he was a supporter of both the French and American Revolutions. Ultimately his political and religious attitudes provoked a riot during which his home and his scientific equipment were destroyed. He therefore emigrated to America in 1794 where his friends included Thomas Jefferson and Benjamin Franklin. He settled in Northumberland, Pennsylvania although his scientific work never recovered from his forced departure. But the descriptions of his experiments with oxygen will always remain a high point in the history of respiratory physiology.

11.1 Introduction

For many of us interested in the history of respiratory physiology, Joseph Priestley (1733–1804) has a special place (Fig. 11.1). Not only was he was the first person to describe the gas we now call oxygen, but his description of his discovery was so arresting that rereading it always gives special pleasure. As we shall see, Carl Wilhelm Scheele (1742–1786) actually produced the gas earlier, but his description appeared after Priestley's and his style was more laconic. Antoine Laurent

© American Physiological Society 2015
J. B. West, *Essays on the History of Respiratory Physiology,*
Perspectives in Physiology, DOI 10.1007/978-1-4939-2362-5_11

Fig. 11.1 Joseph Priestley
(1733–1804). (From the
National Portrait Gallery, by
permission)

Lavoisier (1743–1794) repeated Priestley's critical experiment and then went on to clarify the nature of oxygen which eluded Priestley who adhered to the erroneous phlogiston theory. But Priestley has the distinction of first describing some of the properties of this gas from which much of respiratory physiology devolves. He was a talented writer with an enormous output of papers and books. His descriptions in late eighteenth-century English remain a delight to read, reflecting as they do his enthusiasm and wonder. Revisiting his reports of his great discovery never fails to evoke a certain frisson.

While Priestley's discoveries in the area of oxygen and a number of other important gases are naturally the focus of this article, it should be made clear at the outset that most of his writings were in an entirely different area. First and foremost he was a minister of religion with a lifelong emphasis on a humanist philosophy that resulted in him being a fierce nonconformist. This led to severe criticism from the established church in England. He was an ardent liberal who strongly supported both the French and American revolutions, and it was these political attitudes that ultimately led to his undoing in England, and his subsequent emigration to America. By far the largest part of Priestley's writings are on theology, and for example, in the definitive biographies such as those of Schofield [15, 16], Priestley's work on the respiratory gases occupies only a small section. He was also a man of prodigious energy, and apart from his studies of religion and oxygen and other gases, he made important contributions to the topic of electricity, particularly its history. He produced some 50 books and when once asked how many he had written he replied "Many more, Sir, than I should like to read" [8].

Not surprisingly, there is a large literature on Priestley, especially in relation to his work on oxygen. The chief purpose of this article is to give graduate and medical students (and perhaps their mentors) some insight into the color and excitement of his discoveries. In addition it is revealing to compare Priestley with his contemporary, Lavoisier, because of their very different personalities and writings. Finally there is a discussion of Priestley as a product of the Enlightenment that had such an enormous influence in Europe in the eighteenth century and beyond, that resulted in Priestley's interactions with Benjamin Franklin (1706–1790) and Thomas Jefferson (1743–1826), and that still resonates with many of us today.

11.2 Brief Biography

Priestley was born near Leeds, a major city in Yorkshire in the north of England. The family was rather poor but strongly religious, albeit not in conformity with the established Church of England. Priestley showed precociousness from an early age and, as noted by several biographers, at the age of four he could flawlessly recite all 107 questions and answers of the Westminster Shorter Catechism. As an example, question 1 was "What is the chief end of man?" with the answer "Man's chief end is to glorify God, and to enjoy him forever". This cerebral, analytical attitude was to remain with Priestley for the whole of his life.

Priestley studied theology at the Daventry Academy where liberal, enlightened emphases were strong. As a non-conformist, or dissenter from the established Church of England, he was precluded from attending most higher institutions including Oxford and Cambridge. In Daventry he developed his belief in tolerance, abhorrence of dogma, and a passion for the rational analysis of the natural world. Later Priestley moved to the prestigious Warrington Academy where he tutored in languages because of his broad knowledge of French, Italian and German, and his liberal leanings increased. He wrote an English grammar [9] that still makes enjoyable reading with its rigorous analytical style. He also produced several essays on the importance of a liberal education with an emphasis on its development. Another strong interest was electricity, particularly its history, and he published a 700-page book titled "The History and Present State of Electricity" [10] which went through five editions, and translations into French and German. He made several discoveries in this area including the conductivity of charcoal, and he proposed that electrical force was similar to Newton's gravitational force in following an inverse square law.

Priestley then returned to Leeds where he became the minister of the Mill Hill Chapel. There he wrote his book "Institutes of Natural and Revealed Religion" [12] which further emphasized his liberal views and set out his conviction that religious beliefs should be consistent with a scientific view of the world. As a result he challenged a number of basic Christian principles and this evoked a great deal of savage criticism. Priestley questioned the divinity of Christ and the Virgin Birth, and he was one of the founders of the Unitarian Church in both England and America.

Priestley's financial position was never particularly secure and in 1772 he entered the services of Lord Shelburne, who was immensely wealthy and had a large country estate in Wiltshire and a house in London. Shelburne stated that he needed a tutor for his children, a librarian, and an assistant, but in fact the duties were light

Fig. 11.2 Attack on Priestley's house by a riot mob in 1791. (From the Susan Lowndes Marques Collection, by permission)

and some of Priestley's best experimental work on oxygen was carried out at this time. The country estate known as Bowood House is a grand English stately home that can be visited today.

In 1780 Priestley moved to Birmingham where he continued his teaching and research. While he was there he became a member of the famous Lunar Society that included inventors, scientists, manufacturers and others who met together once a month at the time of the full moon to discuss their work. The many members included the engineer James Watt, the physician and philosopher Erasmus Darwin, and the botanist William Withering known for his work on the foxglove and digitalis. However Priestley increasingly faced resistance because of his extreme liberal views. This came to a head in July of 1791 when a dinner was arranged to celebrate the anniversary of the storming of the Bastille in Paris, and this provoked a riot. A mob torched Priestley's house destroying all of his belongings including his scientific equipment (Fig. 11.2). Priestley and his family eventually escaped to London where he lived for a period in the district of Hackney.

The criticism surrounding Priestley became so virulent that in 1794 he sailed with his family to America where they initially lodged in Philadelphia, the capital at that time. Before leaving England he wrote "I cannot refrain from repeating again, that I leave my native country with real regret, never expecting to find anywhere else society so suited to my disposition and habits" [8]. While Priestley was on the high seas in 1794, his contemporary, Lavoisier, was executed by guillotine in Paris. Priestley was offered a professorship at the University of Pennsylvania but declined. The family eventually moved to Northumberland County in Pennsylvania where Priestley's son and others purchased 300,000 acres. Priestley built a new house and laboratory and this is now a national monument. However his scientific work never completely recovered from his rejection in England. Priestley died in 1804 and is buried in Riverview Cemetery in Northumberland.

11.3 First Production of Oxygen

August 1, 1774 is a red letter day for respiratory physiologists because this was when Priestley first produced oxygen. He did this by heating red mercuric oxide (known at the time as *mercurius calcinatus per se*) by focusing the sun's rays using a convex lens of 12 inches diameter. The experiment was not trivial. First in order to produce red mercuric oxide it was necessary to heat mercury exposed to air to near its boiling point for several months. Next was the problem of collecting the gas released from the mercuric oxide when it was heated to a very high temperature. Priestley did this by using glass containers shown in Fig. 11.3. He placed some of the red mercuric oxide in the bottom of the container, filled it with mercury, and then inverted it over a basin of mercury. Stephen Hales (1677–1761) had previously

Fig. 11.3 Apparatus used by Priestley for the production of oxygen and other gases. From [14]

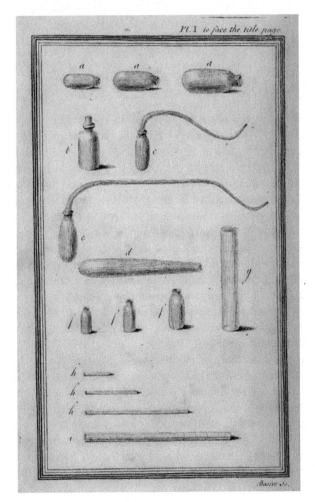

Fig. 11.4 Priestley's account of his discovery of oxygen on August 1, 1774. From [14]

With this apparatus, after a variety of other experiments, an account of which will be found in its proper place, on the 1st of Auguſt, 1774, I endeavoured to extract air from *mercurius calcinatus per ſe* ; and I preſently found that, by means of this lens, air was expelled from it very readily. Having got about three or four times as much as the bulk of my materials, I admitted water to it, and found that it was not imbibed by it. But what ſurprized me more than I can well ex-preſs, was, that a candle burned in this air with a remarkably vigorous flame, very much like that enlarged flame with which a candle burns in nitrous air, expoſed to iron · or liver of ſulphur ; but as I had got nothing like this remarkable appearance from any kind of air beſides this particular modification of ni-trous air, and I knew no nitrous acid was uſed in the preparation of *mercurius calci-natus*, I was utterly at a loſs how to account for it.

used glass containers filled with water inverted over a water bath for collection of gases [5]. When Priestley heated the mercuric oxide by focusing the sun onto it he found that a gas was evolved causing the mercury in the glass vessel to fall.

He then collected a sample of this gas (which he called dephlogisticated air) and subjected it to several tests. First he found that it was insoluble in water. This was important because he had previously produced a gas that had some of the same properties but was water-soluble. This was nitric oxide (NO) which oxidized to nitrogen dioxide (NO_2). Next he placed a candle in the new gas and reported in his immortal statement "but what surprized me more than I can well express, was, that a candle burned in this air with a remarkably vigorous flame ... I was utterly at a loss how to account for it" (Fig. 11.4). He also reported that "a piece of red-hot wood sparkled in it ... and was consumed very fast".

Because it is interesting to see the actual printed version, Fig. 11.4 reproduces the passage in Priestley's chapter from his book "Experiments and Observations on Different Kinds of Air" Volume II published in 1776 [14]. This is frequently quoted as the first record of the discovery. However Fig. 11.5 reproduces the first actual statement in print. The circumstances of this are interesting. Very soon after his first experiment on August 1, 1774, Priestley set out with his patron, Lord Shelburne, for a tour of the European continent. In October they were in Paris and met with members of the Académie des Sciences, and also had dinner with Lavoisier. During this Priestley told Lavoisier about his extraordinary results, and in retrospect it was perhaps strange that he would reveal these to one of his potential competitors before he had published them. However Priestley did not believe in secrecy. In the account of his discovery [14] he wrote "As I never make the least secret of any thing that I

Fig. 11.5 Priestley's first
description of his production
of oxygen in a letter to Sir
John Pringle. From [13]

A

candle burned in this air with an amazing ftrength of
flame; and a bit of red hot wood crackled and burned
with a prodigious rapidity, exhibiting an appearance
fomething like that of iron glowing with a white heat,
and throwing out fparks in all directions.   But to complete the proof of the fuperior quality of this air, I introduced a moufe into it; and in a quantity in which, had
it been in common air, it would have died in about a
quarter of an hour, it lived, at two different times, a
whole hour, and was taken out quite vigorous; and the
remaining air appeared to be ftill, by the teft of nitrous
air, as good as common air.

observe, I mentioned this experiment also, as well as those with the *mercurius cal-cinatus*, and the red precipitate, to all my philosophical acquaintance in Paris, and elsewhere, having no idea at that time, to what these remarkable facts would lead". Conversely Lavoisier put a great deal of importance on secrecy and precedence, and was guilty at times of not citing the work of others whose work he had exploited. In the event Lavoisier was quick to repeat Priestley's experiment before the latter could return to England and continue his work.

Priestley was back in England in November 1774 but did not return to working on the new gas until March 1, 1775. In the meantime he worked on other gases that he had discovered. However on March 8 Priestley decided to test the effects of the new gas on a mouse. He found to his delight that whereas the animal could live only a quarter of an hour in common air, it survived for an hour in his new gas on two different occasions. Furthermore when the mouse was taken out it out was still quite vigorous. At this point Priestley realized the importance of getting the new work into print and on March 15 he wrote a letter to Sir John Pringle, the President of the Royal Society. This was published later in 1775 and an extract is reproduced in Fig. 11.5. Note that Priestley reiterated the effect of the gas on a burning candle and a piece of red-hot wood but he also went on to describe its effect on a mouse.

Priestley then speculated in his typical engaging way that his new gas might be useful for patients of lung disease. He stated "From the greater strength and vivac-ity of the flame of a candle, in this pure air, it may be conjectured, that it might be peculiarly salutary to the lungs in certain morbid cases, when the common air would not be sufficient to carry off the phlogistic putrid effluvium fast enough" (Fig. 11.6). He continued with another wonderful paragraph "My reader will not wonder, that, after having ascertained the superior goodness of dephlogisticated air by mice liv-ing in it, and the other tests above mentioned, I should have the curiosity to take it myself. I have gratified that curiosity, by breathing it, drawing it through a glass siphon, and, by this means, I reduced a large jar full of it to the standard of common air. The feeling of it to my lungs was not sensibly different from that of common air; but I fancied that my breast felt peculiarly light and easy for some time afterwards. Who can tell but that, in time, this pure air may become a fashionable article in luxury. Hitherto only two mice and myself have had the privilege of breathing it" (Fig. 11.6).

Fig. 11.6 Priestley's conjecture on the possible use of
oxygen in patients with lung
disease. From [14]

From the greater ſtrength and vivacity of
the flame of a candle, in this pure air, it may
be conjectured, that it might be peculiarly ſa-
lutary to the lungs in certain morbid caſes,
when the common air would not be ſufficient
to carry off the phlogiſtic putrid effluvium faſt
enough.   But, perhaps, we may alſo infer
from theſe experiments, that though pure de-
phlogiſticated air might be very uſeful as a
medicine, it might not be ſo proper for us in
the uſual healthy ſtate of the body : for, as a
candle burns out much faſter in dephlogiſti-
cated than in common air, ſo we might, as
may be ſaid, *live out too faſt*, and the animal
powers be too ſoon exhauſted in this pure kind
of air. A moraliſt, at leaſt, may ſay, that
the air which nature has provided for us is as
good as we deſerve.

My reader will not wonder, that, after having aſcertained the ſuperior goodneſs of de-
phlogiſticated air by mice living in it, and the
other teſts above mentioned, I ſhould have
the curioſity to taſte it myſelf. I have gratified that curioſity, by breathing it, drawing
it through a glaſs-ſyphon, and, by this means,
I reduced a large jar full of it to the ſtandard
of common air. The feeling of it to my lungs
was not ſenſibly different from that of common
air ; but I fancied that my breaſt felt peculiarly light and eaſy for ſome time afterwards.
Who can tell but that, in time, this pure air
may become a faſhionable article in luxury.
Hitherto only two mice and myſelf have had
the privilege of breathing it.

Whether the air of the atmoſphere was, in
remote times, or will be in future time, better
or worſe than it is at preſent, is a curious ſpe-
culation ; but I have no theory to enable me to
throw any light upon it.

Priestley warned in his disarming fashion "but, perhaps, we may also infer from
these experiments, that though pure dephlogisticated air might be very useful as a
medicine, it might not be so proper for us in the usual healthy state of the body; for,
as a candle burns out much faster in dephlogisticated than in common air, so we
might, as may be said, *live out too fast*, and the animal powers be too soon exhausted in this pure kind of air. A moralist, at least, may say, that the air which nature has
provided for us is as good as we deserve".

He later wondered about possible changes in the atmosphere thus giving his reflections a distinctly modern flavor when he pondered "Whether the air of the atmosphere was, in remote times, or will be in future time, better or worse than it is at

present, is a curious speculation; but I have no theory to enable me to throw any light upon it". A little later he states "Philosophers, in future time, may easily determine, by comparing their observations with mine, whether the air in general preserves the same degree of purity, or whether it becomes more or less fit for respiration in the course of time; and also, whether the changes to which it may be subject are *equable*, or otherwise; and by this means may acquire *data*, by which to judge both the past and future state of the atmosphere. But no observations of this kind having been made, in former times, all that any person could now advance on this subject would be little more than random conjecture. If we might be allowed to form a judgement from the length of human life in different ages, which seems to be the only *datum* that is left to us for this purpose, we may conclude that, in general the air of the atmosphere has, for many ages, preserved the same degree of purity. This *datum*, however, is by no means sufficient for an accurate solution of the problem".

11.4 Oxygen is Produced by Green Plants

The experiments described above with their dramatic findings and colorful descriptions assured Priestley's place in the history of respiration. However he made another remarkable discovery about oxygen, that is that it is produced by green plants. Interestingly this observation took place several years before the work described above.

Back in 1771 Priestley had an interest in "noxious air" or "air infected with animal respiration". For example he tried various ways of improving the air produced in a closed vessel when a candle had burned out. However every intervention such as cooling the air or compressing it had no effect. But in August 1771 he made an extraordinary serendipitous discovery. First he pointed out that everybody knew that animals in a confined space caused the air to become noxious with the result that they died. He assumed that the same would be true for plants but he was in for a big surprise. He wrote "I own I had that expectation, when I first put a sprig of mint into a glass jar, standing inverted in a vessel of water: but when it had continued for growing there for some months, I found that the air would neither extinguish a candle, nor was it at all inconvenient to a mouse, which I put into it". On August 17, 1771 he put a sprig of mint into the noxious air that had been produced when a candle burned out. To his delight, he found that 10 days later, a candle burned in the air perfectly well. He added that he carried out the experiment about ten times during the summer of 1771 and repeated the experiments in the summer of 1772. He found that the best results occurred when spinach was used.

The experiments of 1772 are particularly interesting because of Priestley's distinguished guests. He reported that on June 20 he generated some noxious air by keeping mice in an enclosed space until they died. He then put a sprig of mint into the vessel and found that the air had been restored to such an extent that a candle could burn in it, and a mouse could also live. This experiment was seen by Benjamin Franklin and Sir John Pringle, President of the Royal Society, who were visiting Priestley. Franklin made the droll remark that "I hope this will give some check to the rage of destroying trees that grow near houses". Incidentally sometimes historians claim that Priestley discovered photosynthesis but this is perhaps an unwarranted extrapolation.

11.5   Who Discovered Oxygen?

This question is posed so frequently that it seems appropriate to devote a few lines to it here. In fact it is not a particularly useful question because the answer depends on semantics, for example what is meant by the word "discover". The facts are as follows. Priestley was clearly the first person to describe the production of oxygen. As we have seen, he produced it on August 1, 1774, and described it in his letter to Pringle dated March 15, 1775 which was published later that year. It is true that Priestley called the new gas "dephlogisticated air" because he was wedded to the phlogiston theory expounded by Georg Ernst Stahl (1659–1734), and he interpreted his findings in the light of this erroneous theory. Therefore while he was clearly the first person to produce the new gas, he did not understand its true nature. Scheele produced oxygen as early as 1772, also by heating red mercuric oxide, and called it "fire-air". However although he sent his report to the printer in 1775, it was not published until 1777, that is 2 years after Priestley's report. Scheele also interpreted his finding according to the phlogiston theory.

As we have seen, Lavoisier learned of Priestley's critical experiment in October 1774 and he immediately repeated it. Also Scheele had written to Lavoisier on September 30, 1774 describing his own experiments with heated mercuric oxide although it is likely that the letter reached Lavoisier after Priestley's visit. Incidentally Lavoisier never acknowledged receiving Scheeles' letter and this is consistent with Lavoisier's pattern of not acknowledging the work of his competitors. Both Scheele and Priestley remarked on this failing of Lavoisier.

Lavoisier communicated a memoir to the Académie des Sciences at Easter 1775 titled "On the nature of the principle which combines with metals during calcination and increases their weight", and it was published in May 1775 in a journal edited by Rozier. This is sometimes referred to as the Easter Memoir. He reported on the gas produced by heating red mercuric oxide but mistakenly thought that this was common air [2]. Lavoisier's critical memoir to the Académie des Sciences titled "Experiences sur la respiration des animaux …" in which he clearly described the three respiratory gases, oxygen, carbon dioxide, and nitrogen, was not published until 1777 [6].

Incidentally some historians give the credit for the discovery of oxygen to earlier scientists such as John Mayow (1641–1679), who believed that air had a component that he called "nitro-aerial spirit" and that was used up in a flame. He discovered that both burning lamps and animals expire in a closed space "for want of nitro-aerial particles" but he did not himself produce oxygen.

11.6   Other Gases Discovered by Priestley

Priestley isolated and characterized eight gases in all including oxygen. This record has not been equaled before or since. In 1772 Priestley discovered no less than four new gases. One of these was nitric oxide (NO) although in his terminology this was called "*nitrous air*" which can lead to confusion. He produced the gas by the action of nitric acid (called by him *spirit of nitre*) on brass or other metals. This gas played an important role in his early work on the "goodness" of air. When he added nitrous air to ordinary air in a tube above a water bath there was a reduction in volume of the air by one-fifth. The reason was that the oxygen in the air combined with the nitrous air to form nitrogen dioxide (NO_2) which dissolved in the water. Priestley performed this test with other samples of air, for example the air expired from the lung, and found that the reduction in volume was less. He therefore used the test to show that the "goodness" of the air had been reduced by the lung. In fact it was the use of this test that misled Lavoisier in his Easter 1775 memoir referred to above.

The next gas that Priestley discovered was nitrous oxide (N_2O). He called this "*dephlogisticated nitrous air*" and produced it by heating iron filings with nitric acid. Another discovery was hydrogen chloride (HCl) which he called "*marine acid air*". This was made by heating copper with spirit of salt but he eventually realized that the marine acid air was simply the fumes of the spirit of salt. Finally in 1772 he produced carbon monoxide (CO) which he called "*combined fixed air*". This was done by heating charcoal. For some time this was confused with "*inflammable air*", that is hydrogen, and also with methane since all three gases were flammable.

In 1773 Priestley discovered ammonia (NH_3) which he called "*alkaline air*". This was prepared from the action of hydrogen chloride (spirit of salt) on sal ammoniac (a mineral composed of ammonium chloride (NH_4Cl)). In 1774 Priestley produced oxygen as we have already seen. In the same year he discovered sulfur dioxide (SO_2) which he called "*vitriolic air*". This was done by burning sulfur in a vessel and collecting the effluent gas.

Priestley also worked with two other gases but these had been discovered by others. The first was carbon dioxide (CO_2) which was known as "*fixed air*". This had been discovered by Joseph Black (1728–1799), who worked in Scotland. Priestley collected carbon dioxide from an adjacent brewery where it was evolved during fermentation when he was at the Mill Hill Chapel. He also prepared it by adding oil of vitriol (sulfuric acid H_2SO_4) to chalk. Finally Priestley worked with hydrogen (H_2) which was called "*inflammable air*". This had previously been discovered by Henry Cavendish (1731–1810), who made it by adding diluted oil of vitriol to steel filings. Priestley observed that when a mixture of inflammable and common air was exploded with an electric spark, the glass vessel "became dewy". He told Cavendish about this, who later burned large quantities of the two gases and obtained pure water.

11.7   Two Revolutionaries: Lavoisier and Priestley

The discovery of oxygen and the ensuing overthrow of the phlogiston theory that occurred in the latter part of the eighteenth century is often referred to as a scientific revolution. It certainly freed the confusion surrounding the respiratory gases from the stranglehold previously held by the phlogiston theory, and progress in this area of science was subsequently rapid. Two of the principal players in this revolution were Lavoisier and Priestley, and it is interesting to briefly compare and contrast these two major scientists.

First the similarities. These two men were scientific contemporaries in that, as we have seen, they worked on the discovery and elucidation of oxygen at the same time in the 1770s and 1780s. Indeed it was Priestley's description of his August 1, 1774 experiment when he had dinner with Lavoisier in October of that year that apparently was a crucial turning point in Lavoisier's research. It is true that Priestley was actually 10 years older than Lavoisier but Priestley's interests in theology and electricity delayed his work on oxygen. Of course both Lavoisier and Priestley were outstanding scientists and both were enormously productive. Having said that, both men had major interests outside those of chemistry. Lavoisier was a successful businessman who became wealthy principally as a result of his work with the Ferme Générale, a tax collecting organization. Priestley made major contributions in the area of theology. Sadly it was Lavoisier's links with the Ferme Générale that ultimately resulted in his execution, and it was Priestley's extreme liberal attitudes in theology that forced him to leave his native country. In a sense the final years of both were equally tragic. Lavoisier was guillotined in 1794 at the age of 51, while in the same year Priestley was subjected to such violent persecution in England because of his nonconformist theology that he was forced to leave the country, and his scientific career never recovered.

However there were important differences between the two men. From a scientific point of view it has to be said that while Priestley's discovery of oxygen as recounted above was so dramatic and such a delight to read, Lavoisier's contributions were more important in terms of the advancement of science. Priestley never freed himself from the erroneous phlogiston theory and continued to espouse it until his death. It was Lavoisier who was responsible for overturning the theory and making possible the subsequent advancement of science. Another important difference was the attitudes of the two men to their new discoveries. As we have seen, Priestley showed an almost boyish delight in recounting exactly what he did, and he freely gave his information to Lavoisier to the latter's great advantage. By contrast Lavoisier was secret about his discoveries and placed a great emphasis on precedence. As an example, when he discovered in 1772 that burning phosphorus resulted in a substantial increase in weight, he realized that this was a crucial observation that was inconsistent with the phlogiston theory. The reason was that substances that burnt were thought to release phlogiston and therefore would be expected to lose weight. So in order to assure precedence, Lavoisier secretly deposited this finding as a sealed note in the Académie des Sciences with the date so that subsequently he could prove precedence. Such an action would have been unthinkable to Priestley. If, as sometimes happens in a television show, we were asked whom we would prefer to have dinner with, most of us would choose Priestley although the fare would likely be less lavish.

Lavoisier was a consummate chemist and indeed is often referred to as the father of chemistry. By contrast some historians have argued that Priestley's footing in chemistry was insecure. Indeed at one point Priestley himself stated that he was not "a professed chemist". An amusing anecdote is that in 1777 he wrote to Benjamin Franklin (1706–1790) stating that he "did not quite despair of the philosopher's stone" (an alchemical substance that turned base metals into gold) whereupon Franklin advised him by return, if he found it, "to take care to lose it again" [8].

Finally the social backgrounds of the two scientists were very different. Lavoisier was born into a well-to-do family, had the advantages of an excellent education in some of France's best schools, and became wealthy partly through his work with the Ferme Générale. At his death he owned a substantial amount of property. By contrast, Priestley was born into a poor family, he never showed a particular interest in making money, and indeed had limited means throughout his life. His circumstances were recognized by Lord Shelburne who, as we saw earlier, offered him a post with an associated income. The duties were light and it is generally believed that they were listed by Shelburne to avoid giving the impression that an important objective was to improve the financial situation of Priestley though in fact this was the case.

11.8 Priestley's Contributions to Electricity

As mentioned earlier, Priestley made important contributions to the study of electricity which was a topic of great interest in the scientific world at that time. This is peripheral to the main subjects of this article but deserves a brief mention. His major contribution was a 700-page book on "The History and Present State of Electricity" [10] that had five editions, and apparently is still read by people today for its insights into the development of the subject. Today when we think of electricity we envision topics such as voltage, current and resistance. However in Priestley's time these were unknown. The emphasis was on electrical charges that could be developed by rubbing a material such as glass against a fabric. In fact Priestley invented an electrical machine in which a rotating glass sphere was in contact with leather or soft flannel and thus produced a charge that could be conducted along a wire (Fig. 11.7). Priestley made the important observation that there was no charge inside a hollow cylindrical vessel or cup connected to the wire and he concluded "May we not infer from this experiment, that the attraction of electricity is subject to the same laws with that of gravitation, and is therefore according to the squares of the distances: since it is easily demonstrated, that were the earth in the form of a shell, a body in the inside of it would not be attracted to one side more than another?" Franklin, whose work on electricity particularly that associated with lightning is well known, had suggested this approach [8].

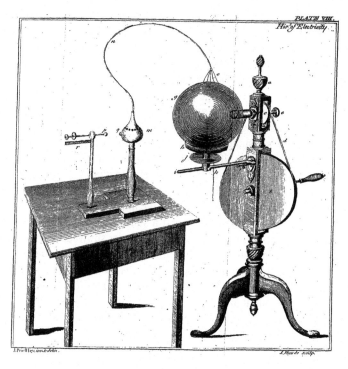

Fig. 11.7 Priestley's machine for generating static electricity (subsequently modified). From [11]

11.9   Priestley and the Enlightenment

The Enlightenment was a major change in attitudes about science, philosophy and religion that occurred mainly in the late seventeenth and eighteenth centuries. Many names are associated with its beginning including those of the Dutchman Baruch Spinoza (1632–1677), the Frenchman Voltaire (1694–1778) and the Englishmen John Locke (1632–1704) and Isaac Newton (1643–1727). It can be thought of as a late product of the Renaissance with its emphasis on the power of human reason, individualism as opposed to institutions, and an opposition to tradition, intolerance and superstition. Priestley was very much a child of the Enlightenment, and is some ways an example of its extremes. This was very evident in his theology. As we have seen, from an early age he was a dissenter from the established Church of England, and his attitudes became more liberal during his lifetime. Although he was a committed clergyman, he gradually estranged himself from his congregations because of his very liberal attitudes.

Priestley argued that all beliefs should be able to stand the scrutiny that he used for scientific investigations. This attitude was consistent with writings from the Warrington Academy where a typical statement was if any idea that is "embraced, shall upon impartial and faithful examination appear to you to be dubious and false, you either suspect or totally reject, such principle or sentiment" [1]. The upshot was not to draw a distinction between the validity of a scientific belief and religious belief. Clearly this attitude would lead to difficulties in traditional Christianity. Priestley therefore found himself moving further and further away from the established church and ultimately allying himself with the emerging Unitarian Church first in England and later in America.

These new attitudes resonated with some people in the new colony in America including Benjamin Franklin and Thomas Jefferson. They also found sympathy with people connected with the revolution in France which had a strong anti-Church element. However, as we have seen, these liberal attitudes met with violent opposition in England and eventually resulted in Priestley being forced to leave for America.

It might be added that many scientists today are aware of tensions between the rigorous attitudes of their profession to evidence on the one hand, and the traditions of organized religion on the other. For a robust contemporary account see Richard Dawkins' book "The God Delusion" [3] while a more conciliatory approach is described by Stephen Jay Gould [4]. Opposing views come from many including McGrath [7].

In conclusion, Priestley will always be remembered as the man who first reported the discovery of oxygen with its remarkable properties of re-igniting an ember of wood and increasing the survival of mice in a closed container. In addition he discovered that the gas he called dephlogisticated air was produced by green plants. Unfortunately he never fully understood the nature of his new gas because he was wedded to the erroneous phlogiston theory. It was left to his contemporary, Lavoisier, to describe the nature of the respiratory gases. Priestley was a dedicated clergyman who developed very liberal views because of the influence of the Enlightenment. As a result his theology was subjected to vitriolic criticism and ultimately his house and equipment were destroyed forcing his emigration to America. While he was on the high seas, Lavoisier was executed by his own countrymen. These two scientific revolutionaries changed the face of respiratory physiology but both had tragic endings. But Priestley's writings continue to give pleasure even after repeated readings. His attitude to work was summed up in his sentence "Human happiness depends chiefly upon having an object to pursue, and upon the vigour with which our faculties are exerted in the pursuit" [10]. We could do worse than have this displayed on our work desk.

References

1. Bright HA. A historical sketch of Warrington Academy. Liverpool: T. Brakell; 1859.
2. Conant JB. The overthrow of the phlogiston theory. Cambridge: Harvard University Press; 1948.
3. Dawkins R. The god delusion. Boston: Houghton Mifflin; 2006.
4. Gould SJ. Rock of ages: science and religion in the fullness of life. New York: Ballantine Pub. Group; 1999.
5. Hales S. Vegetable staticks. London: Print for W. and J. Innys and T. Woodward; 1727.
6. Lavoisier A-L. Expériences sur la respiration des animaux, et sur les changemens qui arrivent à l'air en passant par leur poumon. In: Lavoisier A-L, editor. Oeuvres de Lavoisier, 1777. Paris: Imprimerie Impériale; 1862–1893.

7. McGrath AE. Dawkins' god: genes, memes, and the meaning of life. Malden: Blackwell Publishing; 2005.
8. Partington JR. A history of chemistry Vol 3. London: Macmillan; 1961.
9. Priestley J. The rudiments of English grammar. London: J. Griffiths; 1761.
10. Priestley J. The history and present state of electricity, with original experiments. London: Print for J. Dodsley, J. Johnson and T Cadell; 1767.
11. Priestley J. A familiar introduction to the study of electricity. London: Print for J. Dodsley, in Pall-Mall; T. Cadell, successor to Mr. Millar, in the Strand; and J. Johnson, in Pater-noster-Row; 1768.
12. Priestley J. Institutes of Natural and Revealed Religion. 1–3. Vols. Birmingham: J. Johnson; 1772–1774.
13. Priestley J. An account of further discoveries in air, in letters to Sir John Pringle. J Philos Trans. 1775;65:384–94;.
14. Priestley J. Experiments and observations on different kinds of air. II. Vol. London: J. Johnson; 1776.
15. Schofield RE. The enlightenment of Joseph Priestley. University Park: Pennsylvania State University Press; 1997.
16. Schofield RE. The enlightened Joseph Priestley. University Park: Pennsylvania State University Press; 2004.

Chapter 12
The Collaboration of Antoine and Marie-Anne Lavoisier and the First Measurements of Human Oxygen Consumption

Abstract Antoine Lavoisier (1743–1794) was one of the most eminent scientists of the late eighteenth century. He is often referred to as the father of chemistry, in part because of his book *Elementary Treatise on Chemistry*. In addition he was a major figure in respiratory physiology being the first person to recognize the true nature of oxygen, elucidating the similarities between respiration and combustion, and making the first measurements of human oxygen consumption under various conditions. Less well known are the contributions made by his wife, Marie-Anne Lavoisier. However she was responsible for drawings of the experiments on oxygen consumption when the French revolution was imminent and these are of great interest because written descriptions are not available. Possible interpretations of the experiments are given here. In addition her translations from English to French of papers by Priestley and others were critical in Lavoisier's demolition of the erroneous phlogiston theory. She also provided the engravings for her husband's textbook thus documenting the extensive new equipment that he developed. In addition she undertook editorial work, for example in preparing his posthumous memoirs. The scientific collaboration of this husband-wife team is perhaps unique among the giants of respiratory physiology.

12.1 Introduction

Antoine-Laurent Lavoisier (1743–1794) enjoys a reputation as one of the most eminent scientists of the late eighteenth century (Fig. 12.1). His wife, Marie-Anne Pierrette Paulze Lavoisier (1758–1836) collaborated in many of his studies but her contributions have received relatively little attention. Antoine Lavoisier's interests ranged over a wide area of chemistry. One of his major advances was to clarify the concept of an element as a substance that could not be further broken down by chemical analysis. His extensive classification of known elements was described in his book *Traité élémentaire de chimie (Elementary Treatise on Chemistry)* published in 1789. He was also responsible for recognizing the conservation of mass in chemical reactions, and proposing a nomenclature of various chemicals including acids and salts.

© American Physiological Society 2015
J. B. West, *Essays on the History of Respiratory Physiology,*
Perspectives in Physiology, DOI 10.1007/978-1-4939-2362-5_12

Fig. 12.1 Antoine Laurent Lavoisier (1747–1794) with his wife Marie-Anne (1759–1836). On the *far left* is the drawing board used by Marie-Anne. On the *far right* is a glass jar containing water and resting on a porcelain dish. This jar was presumably used for collecting gas. Next to it is a narrow tube partly filled with mercury. This is a eudiometer for measuring the oxygen concentration of gases. To the *left* of that is a mercury receiver for collecting gas. On the floor there is a large glass vessel that was used for experiments to make water from hydrogen and oxygen. Next to it is a hydrometer for measuring the specific gravity of fluids. This magnificent portrait by Jacques-Louis David (1780) is in the Metropolitan Museum of Art, New York. Reproduced by permission

In the life sciences Lavoisier was the first person to recognize the true nature of oxygen. This had been isolated by Joseph Priestley (1733–1804) and Carl Wilhelm Scheele (1742–1786) but its role in respiration was not understood because of confusion caused by the erroneous phlogiston theory. Antoine with the help of his wife demolished this. Lavoisier also clarified the similarities between respiration and combustion, and he was responsible for the first studies of human oxygen consumption under various conditions.

Because of his eminence, there is a vast literature on Lavoisier's scientific accomplishments. For readers of this article, Holmes [8] is recommended for the physiology, and McKie [17] for the chemistry. Beretta [1, 2] and Prinz [22, 23] are informative about Marie-Anne's drawings. Guerlac [7] is a comprehensive database, and Grimaux [6] is a standard biography in French. The web site Panopticon Lavoisier [20] is exhaustive. The present article concentrates on Lavoisier's main contributions to respiratory physiology, particularly the first measurements of human oxygen consumption, with a major emphasis on the important role of Marie-Anne Lavoisier which has often been overlooked.

12.2 Antoine Lavoisier's Contributions to Respiratory Physiology

1. Identification of the three respiratory gases

Lavoisier was the first person to clearly state the role of oxygen, carbon dioxide and nitrogen in respiration. Here he built on the previous work of other investigators particularly Priestley. This English non-Conformist minister carried out an experiment in August 1774 when, on heating some red mercuric oxide, he found that a remarkable gas was produced. He stated "but what surprized me more than I can yet well express, was that a candle burned in this air with a remarkably vigorous flame …and a piece of red-hot wood sparkled in it" [21]. Furthermore he showed that a mouse was able to survive longer in this gas than in ordinary air, and he famously surmised that it might be useful for people with disease. Thus there was no doubt that Priestley had produced oxygen. However unfortunately he did not understand its nature. Priestley was a follower of the phlogiston theory that had been promoted by Georg Ernst Stahl (1659–1734) and that stated that all combustible materials are composed of ash (calx) and phlogiston (Greek for inflammable), and that during burning, phlogiston escaped leaving the dephlogisticated ash behind. This is in fact the reverse of what happens during combustion when oxygen combines with a combustible material, but the theory was enormously influential in the mid-eighteenth century. Priestley is often credited with "discovering" oxygen just as Columbus "discovered" America but neither Priestley nor Columbus correctly identified what they found.

Carl Wilhelm Scheele (1742–1786) in Sweden had independently produced oxygen even before Priestley although the publication of his findings was delayed [24]. Scheele called the gas "fire-air" but again he was influenced by the phlogiston theory and did not recognize its true nature.

Priestley visited Paris in October 1774 with his patron Lord Shelburne and they had dinner with Lavoisier and some other chemists. At that time Priestley described the experiments that he had carried out with mercuric oxide and naturally these evoked great interest. As a result, Lavoisier repeated the experiments and he was eventually able to understand the chemical processes. For example in 1775 in what became known as the Easter Memoir, Lavoisier stated that "the substance which combines with metals during calcination, thereby increasing their weight, is nothing else than the pure portion of the air which surrounds us and which we breathe". This was the coup de grâce to the phlogiston theory. As we shall see later, Marie-Anne Lavoisier played an important part in this advance because it was she who translated the English texts of Priestley and another proponent of phlogiston, Richard Kirwan (1733–1812), so that Antoine could read them.

In 1777 Lavoisier communicated a memoir to the French Académie des Sciences titled *Expériences sur la respiration des animaux, et sur les changements qui arrivent à l'air en passant par leur poumon* [11]. This included the statement "Eminently respirable air [he later called it *oxygine]* that enters the lung, leaves it in the form of chalky aeriform acids [carbon dioxide] … in almost equal volume…. Respiration acts only on the portion of pure air that is eminently respirable … the excess, that is its mephitic portion [nitrogen], is a purely passive medium which enters and leaves the lung … without change or alteration. The respirable portion of air has the property to combine with blood and its combination results in its red color". This arresting statement was the foundation of all subsequent work on the respiratory gases. Incidentally although a summary like this might suggest that Lavoisier's development of ideas proceeded in a logical sequence, this was far from the case. He was frequently led off on some line of reasoning that turned out to be erroneous, and the whole process was extremely tortured.

2. *The recognition of respiration as combustion*

This advance was also based on the work of earlier scientists, particularly Robert Boyle (1627–1691) and John Mayow (1641–1679). Lavoisier's work was greatly assisted by his development of an ice calorimeter shown in Fig. 12.2. This beautiful engraving was done by Marie-Anne Lavoisier and is another example of her important colloboration. The calorimeter consisted of three compartments, one inside the other. In the center there was a space for a burning flame or a small animal such as a guinea pig. Surrounding that was a compartment containing ice, and the amount of water that was produced when this melted was a measure of the heat that was evolved. Finally there was an outer compartment containing ice that acted as an insulating jacket.

Lavoisier was able to show that the heat evolved per unit of carbon dioxide produced was in the same ratio for both a flame and animals. He summarized his findings in the following striking sentence "La respiration n'est qu'une combustion lente de carbone et d'hydrogène, qui est semblable en tout à celle qui s'opère dans une lampe ou dans une bougie allumée, et que, sous ce point de vue, les animaux qui respirent sont de véritables corps combustibles qui brûlent et se consument". [Respiration is nothing but a slow combustion of carbon and hydrogen, similar in all respects to that of a lamp or a lighted candle, and from this point of view, animals which breathe are really combustible substances burning and consuming themselves].

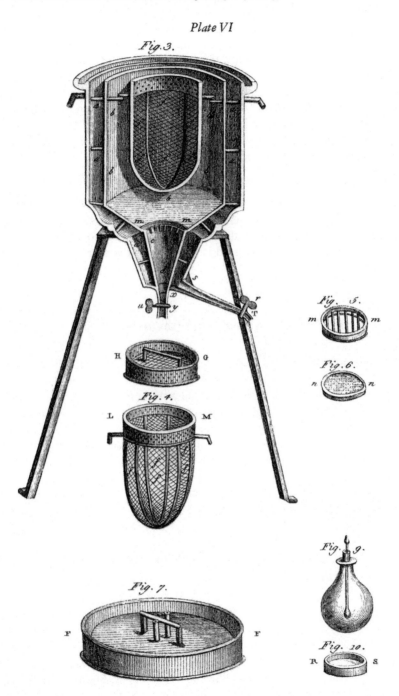

Fig. 12.2 Marie-Anne Lavoisier's engraving of an ice calorimeter used by Lavoisier to demonstrate the similarity of respiration and combustion. See text for details. From [12]

These experiments on combustion were carried out with a junior colleague Pierre-Simon de Laplace (1749–1827) and the two investigators made one of the few errors in Lavoisier's work when they proposed that the combustion actually took place in the lungs. They stated "This combustion is produced within the lungs without evolving perceptible light … the heat developed in this combustion is communicated to the blood which traverses the lungs and is dispersed in the whole animal system" [14]. It is interesting that the site of energy metabolism proved to be very elusive for nearly 100 years after Lavoisier and there were often complicated discussions in textbooks under the heading "animal heat".

3. *Measurements of human oxygen consumption under various conditions*

Lavoisier and his collaborator, Armand Séguin (1767–1835) have the distinction of making the first human measurements of oxygen consumption. These were made during rest and exercise, and with the subject being exposed to different temperatures. These experiments are discussed in detail below together with the drawings by Marie-Anne Lavoisier.

12.3   Contributions of Marie-anne Lavoisier

Some historians of science have argued that Lavoisier's work ranks him along with the other immortals such as Galileo, Newton, Darwin and Einstein. It is interesting to consider his wife's contributions to his work in this context. Marie-Anne is perhaps unique in the sense that her collaborations with her famous husband were so important to his success. Certainly none of the other great scientists listed above had the same support from their spouses. Possible other contenders for this honor among respiratory physiologists in the twentieth century were August (1874–1949) and Marie Krogh (1874–1943).

Marie-Anne Lavoisier's contributions fall into three main areas: translations, graphics, and editorship.

12.3.1   Marie-Anne Lavoisier's Contributions Through Translations

As we shall see later, Marie-Anne found herself exposed to Lavoisier's scientific career at a very early age and with little preparation. The couple were married on December 16, 1771, when Marie-Anne was only 13, about to turn 14. This was shortly after she had emerged from the convent where she received her education. However she was clearly highly intelligent and ambitious. Shortly after the marriage she became interested in the work of her husband and it was not long before she was being tutored in chemistry by one of Lavoisier's colleagues, Jean-Baptiste Bucquet (1746–1780). He was an eminent chemist and physician, a member of the Académie des Sciences, and the author of several books. He had his own private laboratory from which he taught courses in chemistry. It was not surprising that Lavoisier had distinguished colleagues like this. Lavoisier himself had been elected to the Académie at the early age of 25 and through this had extensive contacts with many of the best scientists in France. Another of these was Jean-Hyacinthe de Magellan (1723–1790) who incidentally was a linear descendant of the great Portuguese navigator who discovered the passage to the Pacific Ocean that bears his name. In spite of his foreign-sounding name, Magellan resided in England and was a Fellow of the Royal Society. In a letter to Lavoisier in 1775 he referred to Marie-Anne as a "philosophical wife".

At this time much of the most influential work on the chemistry of the respiratory gases was being done in England. The critical experiments of Priestley who first isolated oxygen have already been referred to. Another active scientist was Joseph Black (1728–1799) who worked in Scotland and clearly described carbon dioxide [3]. A further scientist was Henry Cavendish (1731–1810) who worked on the production of water from hydrogen and oxygen. It was essential for Lavoisier to have access to the results of these influential investigators and Marie-Anne assumed the responsibility. She set herself to learn English which Lavoisier himself never mastered, and her translations were of great importance in enabling Lavoisier to keep up with what was being done in England. Other writings by Marie-Anne in support of her husband's work are also documented [9].

One important book translated by Marie-Anne was by Richard Kirwan titled *Essay on Phlogiston and the Constitution of Acids* and published in 1787 [10]. Kirwan was an influential scientist who was a Fellow of the Royal Society from which he received the prestigious Copley Medal. His book was one of the last and most detailed in the support of the theory of phlogiston. After Lavoisier had read the book he continued to debate the issue of phlogiston with Kirwan and associates for some time, but ultimately Kirwan conceded and acknowledged himself to be converted in 1791.

It was particularly important for Lavoisier to be aware of Priestley's work because this enterprising scientist covered a lot of ground. As an example of his imaginative flair and the elegance of his writing, here is a brief excerpt. "My reader will not wonder, that, after having ascertained the superior goodness of the dephlogisticated air by mice living in it, and the other tests above mentioned, I should have the curiosity to taste it myself. I have gratified that curiosity, by breathing it, drawing it through a glass-syphon, and, by this means, I reduced a large jar to fit to the standard of common air. The feeling of it to my lungs was not sensibly different from that of common air; but I fancied that my breast felt peculiarly light and easy for some time afterwards. Who can tell but that, in time, this pure air may become a fashionable article in luxury. Hitherto only two mice and myself have had the privilege of breathing it". Priestley was clearly a man to be reckoned with.

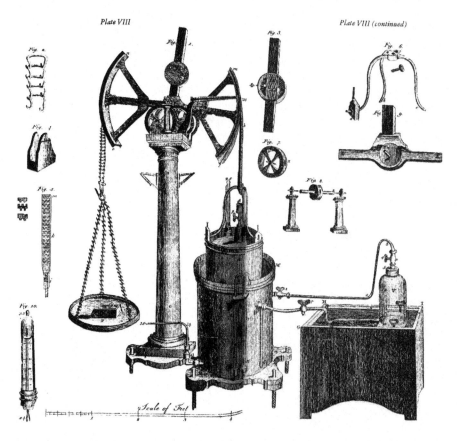

Fig. 12.3 Marie-Anne Lavoisier's engraving of a device for collecting gas and measuring its volume. From [12]

12.3.2   Illustrations Including Engravings

One of Lavoisier's major contributions to chemistry was to develop accurate quantitative procedures. For example he described the law of conservation of mass in chemical reactions by first weighing the reactants of a chemical reaction and subsequently weighing the products. To do this he needed to construct highly accurate balances. He also made careful measurements of the volumes of expired gas using pneumatic troughs or what we now call spirometers. These advances required the development of much new apparatus and Marie-Anne Lavoisier was responsible for making accurate illustrations of the new equipment. In Lavoisier's major work *Traité élémentaire de chimie* of 1789, there are 13 exquisitely engraved plates by Marie-Anne. One example that was described earlier is shown in Fig. 12.2, and equipment for collecting gas is shown in Fig. 12.3. Note the extreme attention to detail including accurate dimensions that would allow other investigators to replicate the instruments. The engravings in *Traité élémentaire de chimie* are all grouped to-

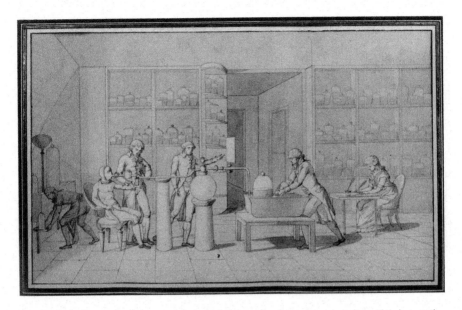

Fig. 12.4 Marie-Anne Lavoisier's drawing of an experiment on oxygen consumption in a resting subject. The expired gas is passed through a glass vessel containing potassium hydroxide in the center and collected on the right by water displacement. Marie-Anne is on the extreme right. See text for details. From [6]

gether at the end of the book and they provide essential information to complement the science described by Lavoisier. Some of Lavoisier's experimental apparatus can be seen today in the Musée des Arts et Métiers in Paris.

Marie-Anne's expertise in engraving is believed to have come from lessons with the famous artist Jacques-Louis David (1748–1825). He was one of the most influential painters in the neoclassical style in France in the late eighteenth century. Figure 12.1 shows one of his best known portraits and there is no equal in the portrayal of a famous scientist with his wife-collaborator. David painted many other famous historical events and his paintings are exhibited in major art galleries around the world.

As discussed below, Marie-Anne also worked on a major book of eight volumes entitled *Mémoires de physique et chimie* which was started by Lavoisier but interrupted by his execution. Surviving documents show that this also was planned to include a number of her engravings but the project was never completed [4].

Of particular interest to us here are two sepia drawings made by Marie-Anne of Lavoisier's experiment on human oxygen consumption. These are shown in Figs. 12.4 and 12.5 which are reproduced from the original illustrations in Grimaux [6]. The drawings have been discussed on several occasions and recently by Holmes [8], Noël [19], Prinz [23], and Beretta [2]. Prinz [22] made a very detailed study of the equipment in the drawing.

Figure 12.4 shows an experiment on a seated subject on the left who is breathing through a close-fitting mask made of copper. He is probably Armand Séguin

Fig. 12.5 Drawing by Marie-Anne Lavoisier of an experiment on oxygen consumption in a subject while exercising. The expired gas is apparently being collected by mercury displacement. The subject's *right foot* is on a pedal that allows him to do mechanical work. Anne-Marie Lavoisier is on the extreme *right*. See text for details. From [6]

who had a major role in these experiments. He is resting and the description below the drawing specifically states "l'homme au repas", that is man at rest. His pulse is being monitored and the man on the far right is thought to be Lavoisier. A valve-box can be seen in the tube near the mask allowing the separation of inspired and expired gas. The expired gas enters a large jar in the center, and the effluent gas from this is being collected in a container on the right using liquid displacement. What appears to be a screen is mounted on the tube between the central jar and the collecting jar. This may contain instructions about the experiment to guide the subject and the man standing with his arm stretched out. Madame Lavoisier on the far right is making notes of the experiment.

Figure 12.5 shows a somewhat similar experiment except that in this case the subject is exercising. The description of the experiment given below the drawing states "l'homme executant au travail", that is man doing work. The subject's right foot can be seen on a pedal which may be attached to wires coming from the table above. This arrangement apparently allows him to exercise. However the details of this are unclear. There is a suggestion of a weight above his foot which perhaps he is being asked to raise. Certainly Lavoisier in some of his descriptions of his experiments on oxygen consumption refers to work done by raising a weight. Just beyond the mouthpiece are two vertical tubes which may be an arrangement for separating the inspired and expired gas. There is another seated man apparently measuring the pulse rate, and there are two other standing men one of whom may be Lavoisier. A laboratory assistant on the far left is carrying supplies. Again Madame Lavoisier on the far right is keeping a record. An electrical machine can be seen in the lower right of the drawing.

A feature of this drawing that is unclear is the jar on the left that is collecting the expired gas. There appears to be an inner container with the tube conveying the expired gas connected at the top. The container is half filled with a dark liquid which is probably mercury. Perhaps this is similar to the gas collecting jar partly filled with mercury on the right of Fig. 12.1. The collecting jar of Fig. 12.5 is enclosed in a much larger glass jar.

These two drawings are fascinating and presumably constitute an authentic depiction of the two experiments. These took place in 1790 which was a volatile period because the Bastille had been stormed on July 14, 1789, and the French revolution was underway. Lavoisier was preoccupied with his many administrative commitments including the tax collecting organization, the Ferme Générale, and perhaps as a result no written records of the experiments have survived. Marie-Anne's drawings are therefore critical in trying to understand the work.

However the interpretation of the drawings is problematic. The central jar in Fig. 12.4 presumably contains potassium lye, that is potassium hydroxide because this substance is referred to in several publications. Lye was easily obtained by leaching wood ash and was well known to investigators of this period. It converted the expired carbon dioxide into potassium bicarbonate and thus removed it from the expired gas. But why this is being done in one experiment and not the other is unclear. Another puzzling feature is why the tube between the valve box and the central jar is so long. Perhaps the valve box did not function perfectly and occasionally some of the inspired gas inadvertently came from the expiratory line. Therefore in order to reduce the chance that any of the toxic potassium lye was inhaled, this was kept far away.

The two drawings depict the first measurements of human oxygen consumption at rest and during exercise. An informative summary of the results of these measurements was set out in a letter from Lavoisier to Joseph Black dated November 13, 1790 [18, 8]. The subject was presumably Séguin who was a coauthor on the reports. Some of the conclusions can be summarized as follows:

1. The quantity of oxygen that a man at rest consumes, or rather converts into fixed acid or carbonic acid during an hour is about 1200 French cubic inches when he is placed in a temperature of 26°. (This is probably about 330 ml.min^{-1} which is a reasonable value.)

2. That quantity increases to 1400 cubic inches under the same circumstances if the person is placed in a temperature of only 12°.

3. The quantity of oxygen consumed or converted into carbonic acid increases during the time of digestion rising to 1800 or 1900 cubic inches.

4. By movement and exercise one reaches as much as 4000 cubic inches per hour, or even more.

5. When by exercise and movement one increases the consumption of oxygen in the lungs, the circulation accelerates. Evidence for this is the increase in pulse rate so that when the person is breathing without hindrance, the quantity of oxygen consumed is proportional to the increase in the number of pulsations multiplied by the number of inspirations.

These dramatic results were also communicated to the Académie des Sciences in a memoir dated November 13, 1790.

A challenging question is exactly how did Lavoisier and Séguin measure oxygen consumption. As noted above Lavoisier's written records are sparse and Marie-Anne's drawings are critical. We know that in some of their work they used a eudiometer, that is a device that measures the reduction in volume of a sample of gas when the oxygen is absorbed by a chemical reaction. Séguin had helped to develop this, and Priestley had previously described a similar device when he was preparing nitric oxide by adding acid to metallic particles. Priestley found that when the gas was exposed to air, a yellow gas (NO_2) was produced and the total volume of gas decreased, and we now know that this was because oxygen was absorbed in the process.

Lavoisier and Séguin used a cylindrical glass tube that was closed at the top with the lower part submerged in a tank of fluid. They found that by igniting a piece of phosphorus in the tube, the reduction in gas volume gave a measure of the amount of oxygen because this was consumed. They were eventually able to make measurements in a tube only 1 inch in diameter and 8 inches in length. The phosphorus was ignited by pressing a glowing piece of charcoal against the outside of the tube. In principle this method could be used to give the concentration of oxygen in inspired and expired gas and so would allow the oxygen consumption to be calculated. The thin cylindrical glass tube on the right of the mercury receiver in Fig. 12.1 is probably one of these eudiometers.

However another method of measuring the oxygen consumption is suggested by Figs. 12.4 and 12.5. In item 1 of Lavoisier's letter to Black, he specifically stated that "the quantity of … oxygen gas that a man at rest … consumes, or rather converts into fixed air or carbonic acid, during an hour …". In other words perhaps Lavoisier sees here a method of measuring oxygen consumption from the carbon dioxide output. Recall that in his 1777 memoir quoted earlier, Lavoisier specifically stated that the amount of oxygen that enters the lung is almost equal in volume to the amount of carbon dioxide that leaves it. So rather than attempting to measure the concentration of oxygen in expired gas, he could measure the volume of carbon dioxide produced by removing it with a caustic alkali such as potassium hydroxide. Indeed in one passage when he was describing the experiments on oxygen consumption, Lavoisier specifically noted that "at each expiration the air is forced to bubble through caustic alkali, where it deposits its carbonic acid" and Séguin also refers to this.

Therefore it seems possible that the measurement of oxygen consumption was made in two stages. First, as Fig. 12.5 shows, the total amount of expired gas over a measured short period was collected. Here Lavoisier was aware that carbon dioxide is soluble in water so he arranged to collect the expired gas over mercury. The experiment was then repeated as shown in Fig. 12.4 but this time the expired gas was passed through a bottle containing caustic alkali. Because the expired carbon dioxide was absorbed, the volume of the expired gas was decreased allowing the amount of expired carbon dioxide to be measured, and from this the oxygen consumption was inferred. In this case, since the treated expired gas did not contain carbon dioxide, it could be collected by water displacement. These two procedures were carried out during both rest and exercise and under other conditions as well.

There is some interesting history about the two drawings shown in Figs. 12.4 and 12.5 [5]. Some writers have suggested that they were made by Marie-Anne Lavoisier after the death of her husband and retouched by David who was her mentor. They were first published in the first edition of the biography by Grimaux in 1896, and the figures reproduced here are from a reprint of the third edition [6]. Graham Lusk (1866–1932), a prominent New York physiologist and nutritionist, recounted an extraordinary event that took place in 1920. He was visiting the apartment of Monsieur de Chazelles, a grand-nephew of Mme Lavoisier, in the Latin Quarter of Paris [15]. There on the wall of a salon was the original portrait shown in Fig. 12.1 which was later acquired by John D. Rockefeller and now hangs in the New York Metropolitan Museum of Art. Then he entered the adjoining study where he saw the two original drawings shown in Figs. 12.4 and 12.5 hanging on the wall. Unhappily the present whereabouts of the two drawings is not known if indeed they are extant.

Lusk made a second trip to France in 1925 [16] and on this occasion he visited the summer home of Madame de Chazelles in the Château de la Canière near Puy de Dôme in central France. Here he saw a collection of instruments used by Lavoisier [26], and with his friend Professor Jean le Goff, he compared the apparatus with the drawings shown in Figs. 12.4 and 12.5. To his astonishment he found a face mask presumably used by Séguin. This was made of copper and fitted with two glass eyes. The edge of the mask had holes so that it could be bound to the head. Séguin himself wrote a brief description of the mask [25].

The Château de la Canière still exists and is now a five-star hotel. Its restaurant is named Le Lavoisier and there are various pictures related to Lavoisier on the wall. However the collection of Lavoisier's instruments has been dispersed. Incidentally Lusk noted that he could trace his tutelage back to Lavoisier. Lusk was a pupil of Carl Voit (1831–1908) in Munich, who was a pupil of Liebig, who was a pupil of Gay-Lussac, who was a pupil of Bertholet and Laplace, who in turn were pupils of Lavoisier.

Recently two additional fascinating pen and ink drawings by Marie-Anne Lavoisier were discovered in the library of the Wellcome Institute in London and these are shown in Figs. 12.6 and 12.7. They further illustrate the collaboration between Marie Anne and her husband. Figure 12.6 shows her taking notes of the experiment, and the similarity of her appearance to that shown in Figs. 12.4 and 12.5 is striking.

Again the design of the experiments is rather obscure. Figure 12.6 shows a subject, presumably Séguin, sitting in a tank of water and exhaling through a tube into a dish that may contain caustic alkali to combine with the exhaled carbon dioxide. His pulse is being measured. The upper part of the man is enclosed in a glass canopy that apparently dips below the surface of the water so that the canopy is gas tight. The result is that as the oxygen content of the canopy gas is gradually reduced because oxygen is consumed by the subject, the canopy gradually falls. Perhaps this experiment was another way of measuring oxygen consumption in a resting subject.

Figure 12.7 is also interesting. The right-hand part shows the same subject as in Fig. 12.6 but this time the tank is shown in cross section so that we can clearly see the glass canopy which presumably is dipping into the water in the tank. An

Fig. 12.6 Another drawing by Anne-Marie Lavoisier of an experiment in which oxygen consumption is apparently being measured. She is seen on the extreme left. See text for details. From the Wellcome Library, by permission

additional feature shown more clearly here than in Fig. 12.6 is that the dish containing the fluid, presumably caustic alkali into which the expired gas is being exhaled, is suspended by a cord with two pulleys with a weight at the other end. Possibly the plan was to measure the weight of the exhaled carbon dioxide from which the oxygen consumption could be inferred. However since a human at rest only exhales about 15 g of carbon dioxide per hour, this hardly seems practicable.

On the left-hand side of Fig. 12.7 we see a figure that is apparently the same subject, presumably Séguin, being carefully weighed. Possibly an attempt was being made to measure the change in weight of the subject as a result of the metabolism that occurred during the breathing period. However, Lavoisier was also interested in the transpiration of water from the skin under various physiological conditions, and this may have been the reason for measuring the weight.

There is no evidence that the experiments shown in Fig. 12.6 and the right-hand part of Fig. 12.7 were ever actually performed. Perhaps these were ideas that never came to fruition. Is it possible that Marie-Anne herself conceived these thought experiments based on her knowledge of physiology to suggest new ways of measuring oxygen uptake and carbon dioxide output?

Incidentally the general appearance of all the subjects in Figs. 12.4, 12.5, 12.6, and 12.7 suggests the same neo-classical attitudes that are shown in Fig. 12.1, and

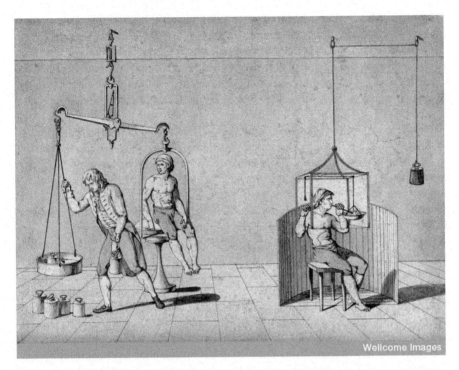

Fig. 12.7 Another drawing by Anne-Marie. The right-hand image shows a cut-away view of the same arrangement as in Fig. 12.6. See text for details. From the Wellcome Library, by permission

which were characteristic of the painter Jacques-Louis David in his other work. These characteristics clearly influenced the drawings of David's pupil, Marie-Anne.

12.3.3   Marie-Anne Lavoisier's Editorial Responsibilities

As alluded to earlier, Marie-Anne was tutored in chemistry shortly after her marriage so that she could involve herself with her husband's work. In addition she collaborated in a number of his experiments by taking notes and preparing the lavish engravings of *Traité élémentaire de chimie*, examples of which are shown in Figs. 12.2 and 12.3. Her attendance as a collaborator in Lavoisier's experiments is documented in Figs. 12.4, 12.5 and 12.6. As a consequence of her knowledge of chemistry and her involvement with the experimental procedures, she was able to make her own contributions on scientific issues.

One example of this is that she added explanatory notes to her translations of scientific articles from English to French. This was particularly the case with Kirwan's book *An Essay on Phlogiston and the Constitution of Acids* which provoked an extensive debate between the author and Lavoisier. As indicated earlier, this ended with the capitulation of Kirwan who then denounced the phlogiston theory in 1791.

However Marie-Anne's editorial responsibilities went beyond this. Late in Antoine's life when the potential dangers of the French Revolution were looming, Lavoisier planned to write a series of memoirs on physics and chemistry (*Mémoires de physique et de chimie*) [13]. However only one or two of these appeared before Lavoisier was imprisoned and subsequently executed. Marie-Anne then assumed the responsibility of publishing what was to be an extensive description of Lavoisier's later work.

Lavoisier had initially planned to publish this series of memoirs with his young collaborator Séguin. He wrote one memoir titled *Mémoire sur la chaleur* (Memoir on heat) that was sent to the Académie des Sciences [14] and in some references to this *Mémoire,* Madame Lavoisier is included as a coauthor. However the publication of these *Mémoires de physique et de chimie* was brought to a halt because of financial difficulties of the printer. Such was the confusion because of the printer's difficulties and the impending revolution that the project was abandoned.

Séguin was chosen to write an introduction to the *Mémoires* but a disagreement developed between him and Marie-Anne apparently because she thought he did not merit the role that he was assuming. As a result of the controversy, Marie-Anne herself wrote an introduction but this was criticized by others. The final upshot was that Marie-Anne was left with a large number of copies of the unbound manuscript and these were never formally published. However the *Mémoires* have now been made available, for example in Panopticon Lavoisier [20].

Finally in a less formal role as a hostess, Marie-Anne must have contributed significantly to Antoine Lavoisier's career. She was described as a charming outgoing woman much given to entertaining [9]. One notable example was presumably the dinner in Paris in 1774 when the guests included Joseph Priestley and his patron, Lord Shelburne. It could be argued that Priestley's description of his experiment in which he heated red mercuric oxide and that as he said "surprized me more than I can yet well express" changed the course of science because it resulted in Lavoisier discovering the true nature of oxygen. In addition, Lavoisier had a wide circle of scientist friends partly through his association with the Académie des Sciences and Marie-Anne's role as a hostess was presumably important in maintaining these valuable contacts.

12.4   Personal Backgrounds of Antoine and Marie-anne Lavoisier

The description of the collaboration between husband and wife given above is better understood if we briefly review their personal backgrounds. As is well known, Antoine Lavoisier's history was tragic in that he was executed by guillotine at a relatively early age. Marie-Anne's personal life was unusually colorful and also makes sad reading.

Antoine Lavoisier had the advantages of a privileged upbringing. He was born in Paris to a wealthy family and had the advantage of a fine education. He initially trained as a lawyer but never practiced in this area. From an early age he was interested in science and he became a member of the French Academy of Sciences at the unusually young age of 25.

Shortly after this he became a member of the Ferme Générale, an organization that, as we might say today, outsourced the collection of taxes for the central government. When he was 28 he married Marie-Anne who was the daughter of Jacques Paulze, a senior member of the Ferme Générale.

Some of his scientific accomplishments were discussed above. However his interests were very wide and, for example, he worked on urban street lighting, the adulteration of cider, the improvement of drinking water in Paris and the preparation of gun powder. As a result of the latter project, his main laboratory was in the Royal Arsenal in Paris for a period.

As the French Revolution accelerated around 1789, Lavoisier as a member of the privileged class was threatened. Ultimately his membership of the Ferme Générale proved to be his undoing and he was guillotined on May 8, 1794 at the age of 51 on the same day as his father-in-law. His body was buried in a nameless grave in the cemetery of the Parc Monceau.

Marie-Anne Pierrette Paulze was born in the province of Loire to an aristocratic family (Fig. 12.8). As mentioned above, her father was a senior member of the Ferme Générale. Marie-Anne's mother died when she was young, and she was schooled in a convent where the education was excellent. However at the age of 13 she returned home because a marriage had been arranged for her to the Count of Amerval aged 50. He was in financial difficulties and a marriage into the wealthy Paulze family was therefore appealing. However, to her credit, Marie-Anne objected vehemently to the proposed match and referred to the Count as "fol d'ailleurs, agrest et dur, une espèce d'ogre" [6] [a fool, an unfeeling rustic, and an ogre]. This placed her father in a difficult position because his superior in the Ferme Générale, a man by the name of Abbe Terray, was strongly in favor of the marriage and Jacques Paulze

Fig. 12.8 Pastel of Marie-Anne Lavoisier as a young woman. From [6]

feared that he would lose his position. He therefore quickly decided to make a more appropriate match for Marie-Anne and approached his 28-year-old colleague, Antoine Lavoisier. The upshot was that the couple were married on December 16, 1771. In spite of this bizarre beginning, the marriage was happy and, as we have seen, scientifically productive although the couple had no children.

After the execution of both her husband and father on the same day, Marie-Anne understandably became severely depressed. She was also impoverished because the revolutionaries seized all of the Lavoisiers' money and possessions including his laboratory equipment and journals. Indeed Marie-Anne was arrested and imprisoned for a short time. However later she was able to recover some of her husband's notebooks and laboratory equipment.

Seven years after her husband's death, Marie-Anne met Count Rumford (1753–1814), an English scientist, also known as Benjamin Thompson. Rumford was a distinguished scientist whose main contributions were in the area of heat conduction and its measurement. He courted her and the couple were married in 1805. However the marriage had difficulties from the beginning. The two personalities were wildly different. Rumford tended to be reclusive while Marie-Anne was very sociable. The result was that the couple were divorced in 1809 but Marie-Anne continued to lead a sociable life in Paris until her death in 1836 at the age of 78.

In conclusion, the story of the two Lavoisier's is stranger than fiction. She was essentially forced into a marriage at an extremely early age but nevertheless made important contributions to science in collaboration with her husband. He was one of the most brilliant scientists in the world in the late eighteenth century and yet was executed by his countrymen. Madame Lavoisier's engravings and drawings and David's magnificent portrait are poignant mementos of an extraordinary couple.

References

1. Beretta M. Introduction to an edition of Antoine Laurent Lavoisier, *Mémoires de physique et de chimie*. 2. Vols. Bristol-Chicago: The University of Chicago Press; 2004.
2. Beretta M. Imaging the experiments on respiration and transpiration of Lavoisier and Séguin: two unknown drawings by Madame Lavoisier. Nuncius. 2012;27:163–91.
3. Black J. Lectures on the elements of Chemistry. Edinburgh: Mundell and Son for Longman and Rees, Edinburgh: London, and W. Creech; 1803.
4. Duveen D. Des illustrations inédites pour les *Mémoires de Chimie*, ouvrage posthume de Lavoisier. Revue d'histoire des sciences et de leurs applications. 1959;12:345–53.
5. Foregger RR. Respiration experiments of Lavoisier. Arch Int Hist Sci. 1960;13:103–6.
6. Grimaux E. Lavoisier, 1743–1794, d'après sa correspondance, ses manuscrits, ses papiers, ses papiers de famille et d'autres documents inédits. 3rd ed. Paris: Alcan; 1899.
7. Guerlac H. Lavoisier Antoine-Laurent. In: Complete dictionary of scientific biography. 8. Vol. Detroit: Charles Scribner's Sons; 2008.
8. Holmes FL. Lavoisier and the chemistry of life. Madison: University of Wisconsin Press; 1985.
9. Kawashima K. Madame Lavoisier: The participation of a salonière in the chemical revolution. In: Beretta M, editor. Lavoisier in Perspective. Munich: Deutsches Museum; 2005.
10. Kirwan R. Essay on Phlogiston and the constitution of acids. London: J. Davis; 1787.

11. Lavoisier A-L. Expériences sur la respiration des animaux, et sur les changemens qui arrivent Ã l'air en passant par leur poumon. In: Lavoisier A-L, editor. Oeuvres de Lavoisier, 1862–1893. Paris: Imprimerie Impériale; 1777.
12. Lavoisier A-L. Traité élémentaire de chimie (Elementary Treatise on Chemistry). Paris: Cuchet; 1789.
13. Lavoisier A-L. Mémoires de physique et de chimie. 2 Vols. Bristol-Chicago: The University of Chicago Press; 2004.
14. Lavoisier A-L, Laplace P-S. Mémoire sur la chaleur. In: Oeuvres de Lavoisier. II. Vol. Paris: Imprimerie Impériale, 1862–1893; 1783. pp. 283–33.
15. Lusk G. Some influences of French science on medicine. JAMA. 1921;76:1–8.
16. Lusk G. Mementoes of Lavoisier. JAMA. 1925;85:1246–7.
17. McKie D. Antoine Lavoisier. London: Gollancz; 1935.
18. Mielo A, editor. Una lettera di A. Lavoisier a J. Black. Archeion. 1943;25:238–9.
19. Noël Y. Commentaire sur les dessins de Madame Lavoisier. In: Lavoisier, editor. Correspondance. 6. Vol. Paris: Académie des Sciences; 1993. pp. 437–8.
20. Panopticon Lavoisier. (1999–2009). http://moro.imss.fi.it/lavoisier/.
21. Priestley J. Experiments and observations on different kinds of air. II. Vol. London: J. Johnson; 1775.
22. Prinz JP. Die experimentelle Methode der ersten Gasstoffwechseluntersuchungen am ruhenden und quantifiziert belasteten Menschen (A.L. Lavoisier und A. Seguin 1790): Versuch einer kritischen Deutung. Sankt Augustin: Academia; 1992.
23. Prinz JP. Lavoisier's experimental method and his research on human respiration. In: Beretta M, editor. Lavoisier in Perspective, Munich: Deutsches Museum; 2005.
24. Scheele CW. Chemische Abhandlung von der Luft und dem Feuer (Chemical Studies on Air and Fire). Upsala; Leipzig, Verlegt von Magn. Swerus Buchhandler, zu finden bey S. L. Crusius; 1777.
25. Séguin A. Mémoire: Sur la salubrité et l'insalubrité de l'air atmosphérique. dans ses divers degrés de pureté. Annales de Chimie. 1814;89:251–72.
26. Truchot M. Les instruments de Lavoisier. Relation d'une visite ã la Caniére (Puy-de-Dome), ou se trouvent réunis les appareils ayant servi ã Lavoisier. Annales de Chimie. 1789;18:289–319.

Chapter 13
Henry Cavendish (1731–1810): Hydrogen, Carbon Dioxide, Water, and Weighing the World

Abstract Henry Cavendish (1731–1810) was an outstanding chemist and physicist. Although he was not a major figure in the history of respiratory physiology he made important discoveries concerning hydrogen, carbon dioxide, atmospheric air, and water. Hydrogen had been prepared earlier by Boyle but its properties had not been recognized and Cavendish described these in detail including the density of the gas. Carbon dioxide had also previously been studied by Black but Cavendish clarified its properties and measured its density. He was the first person to accurately analyze atmospheric air and reported that the oxygen concentration was very close to the currently accepted value. When he removed all the oxygen and nitrogen from an air sample, he found that there was a residual portion of about 0.8 % which he could not characterize. Later this was shown to be argon. He produced large amounts of water by burning hydrogen in oxygen and recognized that these were its only constituents. Cavendish also worked on electricity and heat. However his main contribution outside chemistry was an audacious experiment to measure the density of the earth which he referred to as "weighing the world". This involved determining the gravitational attraction between lead spheres in a specially constructed building. Although this was a simple experiment in principle, there were numerous complexities which he overcame with meticulous attention to experimental details. His result was very close to the modern accepted value. The Cavendish Experiment as it is called assures his place in the history of science.

13.1 Introduction

Henry Cavendish (1731–1810) was a famous chemist and physicist with broad scientific interests, and in particular he was a meticulous experimenter. He was born in Nice, France where his parents were residing at the time. The family was notably aristocratic and could trace its roots back to Norman times. Henry's father was Lord Charles Cavendish who was the son of the 2nd Duke of Devonshire, and Henry's mother was Anne, a daughter of Henry Grey, 1st Duke of Kent. In spite of these illustrious ancestors, the family was not particularly affluent. Nevertheless during his life, Henry became exceedingly wealthy through bequests, and at his death the estate was worth about a million pounds, an enormous sum in those days. The French

© American Physiological Society 2015

J. B. West, *Essays on the History of Respiratory Physiology,*
Perspectives in Physiology, DOI 10.1007/978-1-4939-2362-5_13

scientist, Jean-Baptiste Biot (1774–1862) quipped that he was "le plus riche de tous les savants, et probablement aussi le plus savant de tous les riches" (the richest of all learned men, and probably also the most learned of all the rich) [11].

Henry's mother died when he was aged two and he was brought up by his father. In fact he lived with his father until the latter's death in 1783 when Henry was aged 52. He was first educated at the Hackney Academy in London and this is of interest because it was run by dissenters (people who did not conform to the Church of England) and this reminds us that his contemporary, Joseph Priestley (1733–1799), was a fierce non-conformist. Later Henry entered Peterhouse College in Cambridge although, as was common at the time, he did not graduate, possibly because of religious reservations.

The Cavendish family had a strong tradition of public service including extensive political responsibilities. However these topics did not interest Henry and he devoted his life to science. Here his aristocratic connections enabled him to move in the highest circles. At the age of 29 he became a Fellow of the Royal Society of London and he was elected to its Council when he was 34. He was also a trustee of the British Museum, and he became manager of the recently founded Royal Institution of Great Britain when Humphry Davy was working there. Davy worked on respiratory gases especially nitrous oxide.

As a person, Henry Cavendish had many scientific friends but outside this area he was extremely reclusive, perhaps even pathologically so. He remained a bachelor all his life and had an aversion to women. Various anecdotes are told about this. For example he would order his dinner by leaving a note on the hall table, and he had a separate staircase installed in his house so that he could avoid any female servants. His dress was quaintly old-fashioned and he wore an antiquated wig. He refused to sit for a portrait; the only image that we have of him (Fig. 13.1) was surreptitiously sketched by the artist William Alexander.

Cavendish was a very eminent scientist but not a major figure in physiology. However his work is of considerable interest. He was the first person to prepare and understand the nature of hydrogen. He made this by adding acid to metal filings. The gas had previously been produced by Robert Boyle (1627–1691) but he confused it with ordinary air. Cavendish also worked on carbon dioxide, or "fixed air" as it was called. This had previously been prepared by Joseph Black (1728–1799) but his description was brief and Cavendish elucidated its properties. Another important area of research was the properties of water. Joseph Priestley had previously observed that when a mixture of inflammable and common air was exploded with an electric spark, the container "became dewey". Cavendish went on to study the properties of water in detail. Another interest was the composition of atmospheric air which he was the first to accurately determine. He also worked on nitric acid but all his chemistry research was interpreted in terms of the phlogiston theory. This stated that a fire-like element was released during combustion. Having said this, he was one of the first people outside France to recognize the shortcomings of this theory after the crucial studies by Antoine Lavoisier (1747–1794).

Cavendish worked in other areas including experiments on heat. Here he expanded on the work of Black on latent heat. However the major contribution of

Fig. 13.1 Henry Cavendish (1731–1810). This sketch was made surreptitiously by William Alexander. From [16]

Cavendish to science outside chemistry was in the field of gravitational attraction. His major project now known as "the Cavendish experiment" was a meticulous measurement of the gravitational attraction between lead balls in his laboratory carried out with extreme precision [6]. The results allowed him to calculate the average density of the world which he gave as 5.45 times that of water. He referred to this experiment as "weighing the world". As a result, many physicists credit Cavendish with the first accurate measurement of Newton's gravitational constant.

There are three major biographies of Cavendish. The most recent [9] is over 800 pages long and is exhaustive. This is an extension of an earlier book [8] and contains many letters to and from Cavendish. An earlier biography [1] concentrates on his published papers, and a much earlier volume [16] deals extensively with the controversy between Cavendish and Watt on the discovery of the composition of water. A collection of the Cavendish papers is available [10] and several are reproduced on the Internet. Partington [12] has a good discussion of the chemical studies.

13.2 Hydrogen

One of the first publications by Cavendish was titled "Three papers, containing Experiments on factitious Air" [2] and it appeared in the journal *Philosophical Transactions* in 1766. The purpose of this journal was to "register", that is print,

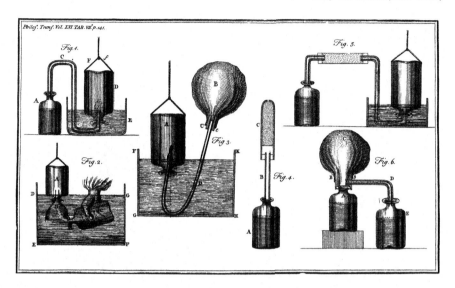

Fig. 13.2 Apparatus used by Cavendish in his studies of hydrogen, carbon dioxide, atmospheric air, and water. From [2]

communications that had been given to the Royal Society. The three papers were presented to the Society on May 29, November 6 and November 13, 1766. *Philosophical Transactions* dealt with all areas of science and the discussions at the Society meetings covered a bewildering array of topics. For example the first paper on inflammable air was preceded by two on a recent solar eclipse, another on the double horns of a rhinoceros, and a third on a very large hernia. Following Cavendish's presentations there was one on men who were 8 ft. tall or more in Patagonia, another on locked jaw apparently cured by electricity, and a third on a swarm of gnats in Oxford. This disconcerting collection emphasizes the lively, effervescent intellectual activity in the Royal Society at the time.

By factitious air Cavendish meant "any kind of air which is contained in other bodies in an unelastic state, and is produced from thence by art" [2]. The term had originally been introduced by Boyle. The first paper was on "Inflammable air" which we know today as hydrogen. Other investigators such as Boyle had previously prepared this gas but had not realized what it was. The detailed studies of Cavendish allowed him to be credited as the first person to recognize its true nature.

Cavendish prepared the gas by the action of acids on various metals. The two acids that he used were spirit of salt (hydrochloric acid) and dilute oil of vitriol (sulfuric acid). He studied three metals, zinc, iron and tin. Various types of apparatus were used in his experiments and these are illustrated in his publication (Fig. 13.2). Cavendish described inflammable air as "permanently elastic" which was a term used to describe all gases at the time. He showed that the volume remained constant in spite of the gas being exposed to water or alkalis. In fact it was so insoluble in water that there was no measurable absorption over a period of several weeks. Finally

Fig. 13.3 Chemical balance used by Cavendish. This was described as "of rude exterior but singular perfection". It is now in the Royal Institution in London. (By permission of the Royal Institution)

he described that it was explosive when it was mixed with common air and exposed to a spark, and he noted that this behavior had been described by others.

An important determination was the density of the new gas. It was easy to measure a given volume by water displacement using the equipment shown in Fig. 13.2. However measuring the weight of a given volume of gas was much more challenging. This was done by placing a known volume of the gas in a bladder as shown in the third image of Fig. 13.2 and weighing it using a sensitive analytical balance. One of Cavendish's balances that still exists today is shown in Fig. 13.3. Other scientists at the time, for example Black, had also developed very accurate balances. Cavendish measured the density of several samples of hydrogen and compared the density with that of common air and water. He reported that the mean density was "8700 times lighter than water". This meant that it was much lighter than air which had "its specific gravity… 7840 times less than that of water". The paper read by Cavendish on May 29 describing his work on inflammable air was warmly received, and the secretary of the Society, Henry Oldenburg, wrote in the Society's Journal Book that "It is impossible to do Justice to the Experiments under the title "Of inflammable air" without citing them wholly" [14].

13.3 Carbon Dioxide

The second of the three papers was read by Cavendish on November 6, 1766 and was about his work on "Fixed air". We now know this as carbon dioxide and it had previously been produced by Joseph Black although his description was brief. Cavendish produced fixed air by the same methods as used by Black, that is by adding acid to alkalis such as calcium carbonate or magnesium carbonate, or heating these substances, that is what is known as calcination. Cavendish explored the properties of this gas using the same techniques as he had for inflammable air. He reported that it was permanently elastic like inflammable air but in other respects it was different. First it was soluble in water and for that reason he collected it over mercury. He also found that it was not flammable and he reported its density as 511 times lighter than water, or 1.57 times heavier than common air.

Cavendish's third paper presented to the Society on November 13 dealt with the gases produced by fermentation and putrefaction. The results showed considerable variability as we might expect and these studies are of less interest.

Cavendish's work on factitious airs earned him the Copley Medal of the Royal Society, its highest distinction. His work was characterized by very careful measurements, and he resisted the temptation to extrapolate beyond exactly what he had found. As an example of his meticulous methods, he kept a sample of fixed air over mercury for "upwards of a year" to see if any change in volume occurred. However his interpretation of his chemical experiments was marred by the fact that he worked within the confines of the phlogiston theory, and therefore the full implications of his discoveries were only seen after Lavoisier had demolished this.

13.4 Composition of Atmospheric Air

In 1783 Cavendish published a paper "An account of a new eudiometer" [3]. This term which had been used by others, particularly Priestley and Lavoisier, is odd. It comes from the Greek εὐδίος (eudios) meaning fine weather, and the instrument was used to measure the "goodness" of air. In effect this was the concentration of oxygen in the air. Previous measurements by Priestley and others had shown considerable variation in the measurement and Cavendish was determined to obtain the correct value. Priestley was the first person to describe the preparation of oxygen.

The method consisted of adding "nitrous gas" (nitric oxide) to air and measuring the reduction of volume. The nitric oxide combined with the oxygen to form nitrogen dioxide (NO_2) which was soluble in water. The nitric oxide was prepared by another method, that is by adding nitric acid to copper. Cavendish experimented by using various proportions of air and nitric oxide to determine the best way of completely removing the oxygen. He also determined the concentration of oxygen by another method, that is by adding inflammable air (hydrogen) to atmospheric air and exploding the mixture with a spark of electricity.

Cavendish's results showed that the concentration of oxygen in air had a mean value of 20.83. This is very close to the value of 20.93 which is accepted today. He also analyzed air collected at high altitude during a balloon ascent by the eccentric American physician and balloonist, John Jeffries. This had essentially the same value.

Cavendish was also able to remove both the oxygen and the nitrogen from an air sample. The nitrogen was removed by adding oxygen and subjecting the mixture to an electric spark which resulted in the formation of nitric oxide. To his surprise he found that a small bubble of gas remained. This was unexplained until about a hundred years later when the gas was shown to be argon by Rayleigh and Ramsey [13]. Cavendish reported that its volume was 1/120 of the total, that is 0.83%. The modern figure for the percentage volume of argon is 0.93.

13.5   Composition of Water

In 1784 Cavendish published a paper showing that burning hydrogen in oxygen produced water [5]. As mentioned earlier, Priestley had previously observed that when inflammable air was exploded with common air by an electric spark the container "became dewey". This experiment was conducted with Priestley's colleague, John Warltire. They also reported that when the explosion was made in an airtight vessel, it was accompanied by a loss of weight. This was attributed by Warltire to the loss of heat which he thought had weight.

Cavendish repeated the experiment in his usual meticulous way and found no loss of weight which did not surprise him because he did not believe that heat had weight. He reported that about one-fifth of the air lost its elasticity (that is disappeared) and that this had condensed into a dew. He went to produce larger amounts of the dewy substance, showed that it had no taste or smell, and concluded "in short, it seemed pure water" [5].

Cavendish reported his findings to Priestley in about March 1783 but the paper was not published until 1784. Meanwhile James Watt (1736–1819) had published a paper on the composition of water [14] and the result was an unpleasant controversy about priority.

13.6   Electricity and Heat

Like many other contemporary scientists, Cavendish was interested in the properties of electricity and heat. It is now recognized that Cavendish made important contributions to the study of electricity but this was not appreciated in his lifetime because he wrote little about these topics. Subsequently James Clerk Maxwell collected and carefully studied Cavendish's notebooks and manuscripts which were subsequently published in 1879 [10].

Cavendish found, but did not publish, the fact that electrical force diminished as the inverse square of the distance just as the case with gravitation. Previously Priestley had suggested this. Cavendish also appreciated the concept of electrical potential which was called the degree of electrification. He measured this using an electrometer containing two gold leaves. When the device was electrified the two leaves diverged and he measured the angle between them. He recognized that in a good conductor the electrical potential was uniform. He also showed that the flow of electricity, which we now refer to as the current, was proportional to the resistance between two points having different potentials. There was no way of measuring current at that time and Cavendish estimated this by holding the electrodes and noting whether the sensation was limited to his fingers, or whether it ascended to his wrists, or his elbows. He also studied the electrical behavior of the torpedo electric fish and compared the fish's electric shock with that from a series of Leyden jars. He stated that electricity was the result of particles that repel each other and we now know that this is a property of electrons.

The work of Cavendish on heat was similarly described in his notebooks and manuscripts and relatively little was published. He believed that the temperature of a substance was related to the degree of motion of its constituents. For example he stated "Heat most likely is the vibrating of the particles of which bodies are composed". He rejected the notion of heat as a fluid with weight which was a popular theory at the time. This was referred to above in his experiments on the production of water. In a paper on the freezing of mercury [4] he used the concept of latent heat which was later developed by Black although Cavendish did not use this term. As in the case of electricity, his treatment of heat was mathematical. Some of his unpublished work on heat was similar to research conducted later by Black and it is said that Cavendish may have delayed publication to give Black an advantage.

13.7   Density of the Earth

The major topics of this paper are Cavendish's studies of hydrogen, carbon dioxide, atmospheric air, and water because of their physiological importance. A brief note about his research on electricity and heat has also been included. However in the later stages of his life, Cavendish carried out a truly extraordinary experiment which he referred to as "weighing the world". In fact the objective was to determine the density of the earth which was the title of the paper [6]. This audacious experiment was so remarkable and gives so much insight into the meticulous experimental methods of Cavendish that it will be briefly described here. History has recognized its significance so that it is generally referred to as "The Cavendish Experiment".

Cavendish was nearly 67 when he embarked on the project and it was the last paper that he published. The experiment was described at length in elegant English in 65 pages in the journal *Philosophical Transactions*. He began by explaining that a friend, the Reverend John Michell, had started to construct an apparatus to measure the density of the earth but he had died before the equipment was completed.

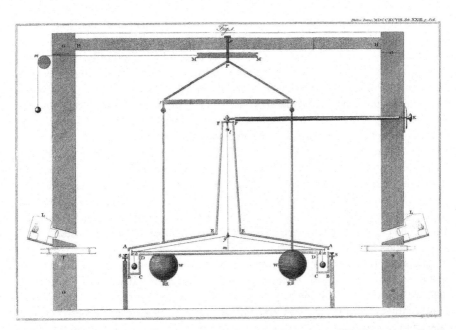

Fig. 13.4 Diagram of the apparatus used by Cavendish to "weigh the world". The apparatus itself is now believed to be in the Royal Institution collection. See text for details. From [6]

Cavendish acquired this and used the same principle but made a number of changes. The basic idea was very simple but carrying it out was extremely challenging. The objective was to measure the gravitational force between two metal balls in the laboratory but the force was vanishingly small being only about 0.02 mg weight (2×10^{-7} N).

Figure 13.4 which is reproduced from Cavendish's report shows the apparatus. The moving part of the experiment was a 6-foot (1.8 m) long wooden rod suspended by a thin wire. At each end of the rod were two lead balls of 2 in. (51 mm) diameter suspended by short wires. This part of the equipment was encased in a narrow wooden box to protect it from air currents. The second part of the apparatus consisted of two massive lead balls each 12 in. (305 mm) in diameter and each weighing about 350 pounds (158 kg). These were suspended from a frame which could be rotated from outside the room. These large lead balls were carefully moved so that they could be placed about 2 in. (5.1 cm) from the small balls. As a result of the gravitational attraction, the rod suspending the small balls rotated slightly and the deflection was measured using a vernier scale and telescopes viewing through the wall of the laboratory room as shown in Fig. 13.4. The room had been specially built of wood away from the house and was 10 ft. (3.05 m) square and 10 ft. high. The walls were said to be 2 ft. (0.61 m) thick. The reason for the special room was to maintain a constant temperature and thus limit convective air currents.

Cavendish found that when the wooden rod was deflected it did not come to rest in a new position but continued to oscillate. He calculated the mean deflection from the extremes of the oscillation. In addition, by timing the period of the oscillations he was able to determine the force required to cause torsion of the wire. There was an elaborate series of calculations that are described in detail in the 65-page paper.

Although the principle of the experiment was simple, there were many complicating factors. These included small variations of temperature that generated convection currents in the air and therefore displaced the wooden rod. Each experiment took several hours to complete. The deflection of the rod caused by the gravitational attraction was only about 0.16 in. (4.1 mm) but Cavendish was able to measure this to an accuracy of better than 1/100 of an inch (0.25 mm) by means of the vernier scale. Interestingly the torsion balance method remains the best way today of measuring the gravitational constant. At about the same time Charles Augustin de Coulomb (1736–1785) also used a torsion balance to measure the electrostatic force of repulsion [7].

The single number that came out of Cavendish's monumental experiment was that the mean density of the earth was 5.448 times that of water. (Oddly enough he actually reported the number as 5.48 because of a simple arithmetic error.) This is close to the modern value. It allows the calculation of the Newtonian gravitational constant G to be 6.74×10^{-11} m^3 kg^{-1} s^{-2} although Cavendish did not do this sum. It differs from the currently accepted value by only 1%. Note that to derive the "weight" of the earth (actually its mass), it is necessary to know its radius. This was first measured by the Greek, Eratosthenes, in about 200 BC.

This prosaic account of the experiment does not do justice to the extreme attention to errors described by Cavendish in his paper, and anyone who is interested should read this. It is available on the Internet. "The Cavendish Experiment" is a fitting denouement to the extraordinary experimental life of this unusual man.

In conclusion, Cavendish was not a major figure in the history of respiratory physiology but he was a scientist of exceptional interest. Although he was a recluse and had many eccentricities, he was a meticulous experimenter. He is best known as the discoverer of hydrogen, and he made important contributions to our knowledge of carbon dioxide and water. His bold experiment to "weigh the world", as he put it, assures his place in the history of science.

References

1. Berry AJ. Henry Cavendish. His life and scientific work. London: Hutchinson; 1960.
2. Cavendish H. Three papers, containing experiments on factitious air. Phil Trans R Soc Lond. 1766;56:141–84.
3. Cavendish H. An account of a new eudiometer. Phil Trans R Soc Lond. 1783;73:106–35.
4. Cavendish H. Observations on Mr. Hutchins's Experiments for determining the degree of cold at which Quicksilver freezes. Phil Trans R Soc Lond. 1783;73:303–28.
5. Cavendish H. Experiments on Air. Phil Trans R Soc Lond. 1784;74:119–53.

6. Cavendish H. Experiments to determine the density of the earth. Phil Trans R Soc Lond. 1798;88:469–526. https://archive.org/details/philtrans07861996.

7. Coulomb C-A. Premier mémoire sur l'électricité et le magnetism. Histoire de l'Académie Royale des Sciences. Paris: l'Imprimerie Royale; 1785a. pp. 569–77.

8. Jungnickel C, McCormmach R. Cavendish. Philadelphia: American Philosophical Society; 1996.

9. Jungnickel C, McCormmach R. Cavendish: the experimental life. Lewisburg: Bucknell; 1999.

10. Maxwell JC. The scientific papers of the Honourable Henry Cavendish. Cambridge UK: Cambridge University Press; 1879.

11. Michaud L-G. In: Cavendish H, editor. Biographie universelle, ancienne et modern. Vol. 7. Paris: Michaud Frères; 1813. pp. 455–7.

12. Partington JR. A history of chemistry. Vol. 3. London: Macmillan; 1961.

13. Rayleigh L, Ramsay W. Argon, a new constituent of the atmosphere. Proc Roy Soc. 1895;57:265–87.

14. Royal Society. Journal books of scientific meetings. Vol. 25, December 8, 1763-December 18, 1766.

15. Watt J. Thoughts on the constituent parts of water and of dephlogisticated air. Phil Trans R Soc Lond. 1784;74:329–55.

16. Wilson G. The life of the Hon. Henry Cavendish. London: Cavendish Society; 1851.

Chapter 14
Humphry Davy, Nitrous Oxide, the Pneumatic Institution, and the Royal Institution

Abstract Humphry Davy (1778–1829) has an interesting place in the history of respiratory gases because the Pneumatic Institution in which he did much of his early work signaled the end of an era of discovery. The previous 40 years had seen essentially all of the important respiratory gases described and the Institution was formed to exploit their possible value in medical treatment. Davy himself is well known for producing nitrous oxide and demonstrating that its inhalation could cause euphoria and heightened imagination. His thinking influenced the poets Samuel Taylor Coleridge and William Wordsworth and perhaps we can claim that our discipline colored the poetry of the Romantic Movement. Davy was also the first person to measure the residual volume of the lung. The Pneumatic Institution was the brainchild of Thomas Beddoes who had trained in Edinburgh under Joseph Black who discovered carbon dioxide. Later Davy moved to the Royal Institution in London that was formed, in part, to diffuse the knowledge of scientific discoveries to the general public. Davy was a brilliant lecturer and developed an enthusiastic following. In addition he exploited the newly described electric battery to discover several new elements. He also invented the safety lamp in response to a series of devastating explosions in coalmines. Ultimately Davy became president of the Royal Society, a remarkable honor for somebody with such humble origins. Another of his important contributions was to introduce Michael Faraday (1791–1867) to science. Faraday became one of the most illustrious British scientists of all time.

14.1 Introduction

Humphry Davy (1778–1829) (Fig. 14.1) was a celebrated chemist who has a special place in the history of respiratory physiology. He was the first person to describe the properties of nitrous oxide which is still extensively used in anesthesia. Shortly after its discovery, the gas was also used as a stimulant resulting in euphoria and heightened imagination. Davy was an early member of the Pneumatic Institution in Bristol UK which is of historical interest because it was one of the first organizations that was formed to exploit the newly discovered respiratory gases in medical practice. In addition, Davy was also one of the first professors at the Royal Institution in London in 1801. This was an important innovation which was developed, in part,

© American Physiological Society 2015

J. B. West, *Essays on the History of Respiratory Physiology,*
Perspectives in Physiology, DOI 10.1007/978-1-4939-2362-5_14

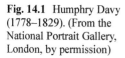

Fig. 14.1 Humphry Davy (1778–1829). (From the National Portrait Gallery, London, by permission)

to inform the general public about scientific research, and it still thrives today. As a research chemist, Davy was remarkably productive. He discovered the elements sodium, potassium, and calcium, and he was the first person to isolate magnesium, boron, and barium. Davy is also well known as the person responsible for developing the miner's safety lamp.

There is an extensive literature on Davy. A readable introduction is Hartley [8]. The biography by Knight [10] is more detailed and contains useful citations to primary sources. Treneer [17] wrote another biography with an emphasis on Davy's relations with other people including his wife and also Faraday. Partington [12] is authoritative on his chemical research. Davy's collected works are available [6].

14.2   Early Years

Davy was born in Penzance, Cornwall near the extreme southwestern tip of England. When he was nine years old, his family moved away and he was put in charge of his godfather, John Tonkin, who was an apothecary-surgeon. Davy did not excel at school and always regarded himself as basically self-educated. After his father's death, Davy was apprenticed to John Borlase, another apothecary-surgeon. This man had a dispensary and Davy liked to tinker there with chemicals. It is said that his friends complained that he might blow them all up.

When Davy was 19 read he Lavoisier's *Traité élémentaire de chimie,* probably in an English translation. The book was considered revolutionary at the time because of its clear classification of the known elements. Davy was particularly interested in Lavoisier's views on heat which was regarded there as a weightless element called "caloric". In a well-known experiment, Davy took two pieces of ice and rubbed them together to produce water by melting. He saw this as disproving Lavoisier's theory because no substance could have been added in the process.

The next chapter in Davy's life begins with the oft-quoted anecdote that Davis Giddy (who later changed his name to Gilbert), a well-connected man with scientific interests, saw Davy swinging on the half-gate of Dr. Borlase's house. Gilbert was impressed by the young boy, allowed him to use his library, and introduced him to a Dr. Edwards who lectured in chemistry at St. Bartholomew's Hospital in London. Edwards encouraged Davy to use the equipment in his laboratory and this was a catalyst in Davy's developing a love of science.

In addition to his scientific interests, Davy enjoyed literature and painting. He wrote several poems at this time and at least three of his paintings still survive. One of his poems, written when he was only seventeen, presages his scientific curiosity. Here is one stanza:

To scan the laws of Nature, to explore
The tranquil reign of mild philosophy;
Or on Newtonian wings sublime to soar
Thro' the bright regions of the starry sky
[14]

Five of his early poems were included in an anthology of Bristol poets [9]. Although he moved on from writing poetry, Davy maintained strong literary connections and had friendships with eminent poets of the Romantic Movement including Samuel Taylor Coleridge, William Wordsworth, and Robert Southey. His influence on Wordsworth has been documented [9]

14.3 The Pneumatic Institution

This interesting organization was formed to exploit the use of the recently discovered respiratory gases for medical practice. Davy joined the Institution in October 1798, and this date emphasizes the very rapid progress in the discovery of the respiratory gases. To recap, Joseph Black had discovered carbon dioxide in 1756 and this was soon followed by the work of Joseph Priestley, Carl Wilhelm Scheele, Antoine-Laurent Lavoisier, and Henry Cavendish, who together not only discovered oxygen, but clearly elucidated the roles of oxygen, carbon dioxide and nitrogen in respiration. So in the space of just 42 years, these critically important gases were discovered and understood, and it was argued that now the time had come to investigate their use in medical practice.

The Pneumatic Institution was the brainchild of Thomas Beddoes (1760–1808) (Fig. 14.2). This colorful man had studied under Joseph Black in Edinburgh, and then continued his work in London and Oxford. As a reader in chemistry at Oxford his lectures were extremely popular. In fact he claimed that the classes were the largest assembled in the University since the thirteenth century! Georgiana, Duchess of Devonshire, visited Beddoes in his laboratory in Bristol where he had started

Fig. 14.2 Thomas Beddoes
(1760–1808). He was a major
figure in the formation of the
Pneumatic Institution. From
[15]

to study the possible medical uses of respiratory gases. She suggested that he should replace the laboratory with a medical pneumatic institution. She tried to persuade the eminent Sir Joseph Banks to give financial support, but he declined partly because of scientific concerns, and partly because he objected to Beddoes' support of the French revolution. Beddoes corresponded with Priestley and the physician Erasmus Darwin (1731–1802), and was also familiar with the Wedgewoods and other members of the Lunar Society in Birmingham. A grateful patient gave him £1500, and Thomas Wedgewood added £1000 with the rather cynical note "that it was worthwhile expending the sum subscribed in order to assure us that elastic fluids would *not* be serviceable as medicine" [15].

James Watt (1738–1819), famous for his work on the development of the steam engine, was also interested because his son had pulmonary tuberculosis and he thought that the new gases might be helpful in his cure. As, a result, Watt built a portable gas chamber for some of the experiments. The Institution was set up in Dowry Square, Bristol (Fig. 14.3) and by April 1800 had several inpatients and some 80 outpatients [8]. Davy moved to Bristol in 1799 as Beddoes' assistant, and soon the Institution was a focus of a number of interesting people including Southey and Coleridge as mentioned earlier. For example Davy was in correspondence with William Wordsworth who asked for Davy's opinion on his poems.

Davy was soon working hard in the laboratory and one of his first interests were the oxides of nitrogen. Here he took considerable risks because one of the gases he inhaled was nitric oxide which, because of its combination with water to form nitrous and then nitric acid, was potentially very dangerous. However he soon began experiments with nitrous oxide (N_2O) which had previously been prepared by Priestley. His initial experiments were done by adding nitric acid to zinc but later he found that he could prepare pure nitrous oxide by heating ammonium nitrate.

In April 1799 he re-breathed four quarts of the gas from a silk bag with his nose closed and reported "a sensation analogous to gentle pressure on all the muscles, attended by a highly pleasurable thrilling, particularly in the chest and extremities... Towards the last inspirations the thrilling increased, the sense of muscular power became greater, and at last an irresistible propensity to action was indulged

Fig. 14.3 Premises of the Pneumatic Institution at 6 Dowry Square, Bristol. (Courtesy of Andy Dingley)

in" [3] (p. 458). In recovering from a period of inhaling the gas he stated "My emotions were enthusiastic and sublime; and for a minute I walked about the room perfectly regardless of what was being said to me…I exclaimed… Nothing exists but thoughts!—the universe is composed of impressions, ideas, pleasures and pains" [3] (pp. 488–489). In another experiment he noted the following "by degrees as the pleasurable sensations increased, I lost all connection with external things, traces of vivid images rapidly passed through my mind and were connected with words in such a manner as to produce perceptions perfectly novel" [3] (p. 488). These reactions have much in common with the poetry of the Romantic Movement.

Davy's reports created considerable interest among his friends and more than 20 of them were keen to try the effects of breathing his new gas. For example, the poet, Robert Southey, reported that breathing the gas "excites all possible mental and muscular energy and induces almost a delirium of pleasurable sensations without any subsequent dejection" [13]. Davy's friend, Samuel Taylor Coleridge, an eminent poet of the Romantic Movement, inhaled the gas and commented on the subsequent euphoria. He reported that while he was inhaling the gas "towards the last, I could not avoid, nor indeed felt any wish to avoid, beating the ground with my feet; and after the mouth-piece was removed, I remained for a few seconds motionless, in great extacy [sic]" [3] (p. 517).

Coleridge and Wordsworth produced a book titled "Lyrical Ballads, with a few other poems", the first poem being Coleridge's "Rime of the Ancient Mariner". In the preface to the book Wordsworth wrote "The first volume of these poems has already been submitted… to ascertain, how far, by fitting to metrical arrangement a selection of the real language of men in a state of vivid sensation, that sort of pleasure and that quality of pleasure may be imparted…" [19]. This suggests that nitrous oxide may have played a part. Certainly Coleridge also had an addiction to opium.

Davy also made another remarkable finding. In reviewing his experiments, he included the following perceptive statement. "As nitrous oxide in its extensive operation appears capable of destroying physical pain, it may probably be used with advantage during surgical operations in which no great effusion of blood takes place" [3] (p. 556). Further experiments were conducted and in one of these while he was suffering from severe toothache he stated that the "uneasiness was for a few minutes swallowed up in pleasure" but the pain returned when he ceased breathing the gas [3] (p. 465). Therefore Davy was close to recognizing that inhaled nitrous oxide could be valuable for anesthesia. However the gas was not used as an anesthetic until 1844 by the American, Horace Wells (1815–1848).

Davy followed up his initial observations on nitrous oxide with an extensive research program. He studied its effects on a series of animals and he measured its absorption by blood in a test tube. Of special interest to us today is that in order to interpret some of the results he obtained by re-breathing the gas from a bag, he realized that he needed to know the volume of the lungs at the end of a forced expiration, that is the residual volume. He therefore measured this by re-breathing hydrogen which he believed was not absorbed by the blood and reported his residual volume as 41 cubic inches. This is equivalent to about 0.72 l and must have been an underestimate since the normal value is at least double that. Nevertheless Davy can be credited with making the first measurement of residual volume. The next measurement was by the French physiologist, Nestor Gréhant (1838–1910). The same principle is used today in the helium dilution method for measuring lung volumes. Davy's results of his work at this time were published in full [3].

As indicated earlier, the reason for setting up the Pneumatic Institution was to determine whether the newly discovered respiratory gases were useful for medical treatment. Unfortunately this was found to be a blind alley. Davy wrote "Pneumatic chemistry in its application to medicine, is an art in infancy, weak, almost useless…". There was an apparent success when a patient with palsy, that is, a disease causing muscular weakness, was treated with nitrous oxide and appeared to recover. However later this was shown to be not the effect of nitrous oxide but simply a delusion resulting from the attentions of Davy and his colleagues.

Towards the end of his period at the Pneumatic Institution, Davy's research took a new turn that later resulted in some of his most important chemical discoveries. In 1818, the Italian physicist, Alessandro Volta (1745–1827), invented the electric battery known initially as a voltaic pile. This consisted of plates of zinc and copper separated by an electrolyte which was either dilute sulfuric acid or salt water brine. The discovery was quickly taken up by other scientists including Davy. He found that using zinc and silver plates with sulfuric acid produced a more powerful battery. However a critical change in Davy's career occurred at this time and this delayed his further experiments with electrochemistry. In January of 1801 he was offered a post at the newly formed Royal Institution in London by Count Rumford and he accepted the offer.

14.4   The Royal Institution

Just as the Pneumatic Institution had been the brainchild of the somewhat eccentric Thomas Beddoes, so the Royal Institution came about as the result of the activities of just as colorful an entrepreneur, Sir Benjamin Thompson also known as Count Rumford (1753–1814) (Fig. 14.4). He was born in Massachusetts and fought on the British side during the American Revolutionary War. At the conclusion of this he moved to London where he had an appointment in the Colonial Office. While there he worked on improvements of guns and was made a Fellow of the Royal Society. He then returned to America for a while and received a knighthood. However in 1784 he moved to Bavaria where he was both a soldier and administrator, and he carried out research on the nature of heat which was to become his principal scientific interest. While he was there he was made a Count of the Holy Roman Empire.

Fig. 14.4 Sir Benjamin Thompson, also known as Count Rumford (1753–1814). He was an important figure in the formation of the Royal Institution. (From the National Portrait Gallery, London, by permission)

Fig. 14.5 The Royal Institution of Great Britain where Davy was one of the first professors. The illustration is from about 1838 and the façade is little changed today. (By Thomas Hosmer Shepherd, c. 1838)

When Rumford returned to London in 1798 he developed proposals for an institution that had two main purposes. The first was to increase public awareness of new scientific discoveries, and the second was to facilitate the application of scientific knowledge to the improvement of manufacturing. The official proposal stated that the organization would be an "institution for diffusing the knowledge, and facilitating the general introduction, of useful mechanical inventions and improvements; and for teaching, by courses of philosophical lectures and experiments, the application of science to the common purposes of life" [1]. This was to be done by setting up an exhibition of improvements in industries, and by lectures at the new institution aimed at both the public and entrepreneurs in industry.

A committee of managers was formed and they held a meeting in the home of Sir Joseph Banks. This is interesting because, as mentioned earlier, Banks declined to support the Pneumatic Institution. Fifty-eight prominent citizens agreed to contribute fifty guineas each, a substantial amount of money at the time. A property in Albemarle Street off Piccadilly was purchased and indeed this remains the home of the Royal Institution today (Fig. 14.5). Incidentally the full name is the Royal Institution of Great Britain.

The first professor at the Institution was a Dr. Garnett but although he had initial success as a lecturer, a replacement was soon sought. Davy was recommended and was appointed in February 1801. His first lecture was an outstanding success and a published report included the following. "The audience were highly gratified and testified their satisfaction by loud applause. Mr. Davy, who appears to be very young, acquitted himself admirably well. From the sparkling intelligence of his eye, his animated manner, and the *tout ensemble*, we have no doubt of his attaining distinguished excellence" [16]. The lectures were so popular that Albemarle Street was made one-way at the time of the lectures, the first street in London to have this distinction.

Davy's first lecture was on electrochemistry or "galvanism", his most recent interest. This was a great success and soon his audience numbered nearly 500 people. Davy was only 23 and handsome, and his popularity was apparently particularly high among the ladies. Two years later he was elected a Fellow of the Royal Society and later he would become its president.

Davy had plans to continue his electrochemical research in the laboratory of the Royal Institution but the managers wanted him to work on more practical issues. They decided that he should give a series of lectures on the chemistry of tanning which was a very important industry at the time. As a result, Davy spent several months visiting tanneries and investigating the chemistry of the process. It seems strange in retrospect that the managers would take this tack when Davy's original lectures were so popular. Indeed after a period, the emphasis of the Institution on the improvement of arts and manufactures declined, and Davy was able to return to his original research interests and continue his brilliant lectures. Incidentally these were critically important for the health of the new Institution which found itself in some financial difficulties. The enormous audiences that Davy attracted put the Institution on a sound financial footing. The Institution today combines outstanding, interesting lectures with serious science, although from time to time some supporters have questioned whether the atmosphere sometime becomes too popular or fashionable.

One of the best known events at the Royal Institution are the Christmas lectures for children. A particularly interesting one was given in 1848 by Michael Faraday, Davy's successor (see below). He described the chemistry of a burning candle and drew an analogy between that and tissue metabolism as had Lavoisier some time before. This lecture influenced the author Charles Dickens. In his novel *Bleak House*, Dickens described an illiterate collector of odds and ends called Krook who was a prodigious drinker of gin, and who died of spontaneous combustion! This provoked an immediate violent objection from the physiologist George Henry Lewes (1813–1878), who complained that the "circumstances are beyond the limits of acceptable fiction and give credence to a scientific impossibility". Dickens countered with other alleged examples of the phenomenon, and the controversy continued for some time [18]. Lewes was a well known physiologist. He wrote a book [11] that influenced Ivan Pavlov (1849–1936) of conditioned reflexes fame, and he was a founding member of the [British] Physiological Society.

The main public lectures at the Royal Institution are the so-called Friday Evening Discourses. These have taken place for over 200 years and the setting is elegant with a string quartet playing during preliminary drinks, and the audience wearing evening dress (black tie). There are some amusing quirks. For example the lecturer is closeted in a locked room for a period before the talk allegedly because one of the speakers became so apprehensive that he absconded. Another oddity is that on the stroke of 8 o'clock, the door of the lecture theatre suddenly opens, the speaker walks to the rostrum, and immediately launches into his topic without any pleasantries such as "Good evening". Finally at the stroke of 9 o'clock the lecturer is supposed to finish abruptly and immediately withdraw from the theatre. The author can talk from first experience because he had the pleasure of giving one of these discourses.

14.5 Further Chemical Researches

This essay concentrates on Davy's work relevant to respiratory physiology, particularly his research on various gases in the Pneumatic Institution, including his studies of nitrous oxide and lung volume. However Davy made many important advances in chemistry including the discovery of several new elements and these will be briefly discussed.

Davy exploited the newly discovered phenomenon of galvanism or electrochemistry with great enthusiasm and success. He believed that electricity was an instrument that could elucidate the composition of various substances. He stated "...substances that combine chemically...exhibit opposite states...they might, according to the principles laid down...attract each other in consequence of their electrical powers" [4]. Early on he found that, using his electrical apparatus, he could decompose water into hydrogen and oxygen and he showed that they were in the correct proportions.

One of his great successes occurred in October 1807 when he applied his new technique to an aqueous solution of caustic potash (potassium hydroxide). First he found that the water was split into hydrogen and oxygen. But then he studied the solid residue and found that small globules of it were attracted to the negative pole of his equipment and burst into flame. After discussion with others he realized that this substance was a metal which he named "potasium" (after potash) and later was called potassium. He had discovered a new element.

He continued using the same techniques to study other solutions. For example when he substituted sodium hydroxide for potassium hydroxide, he was able to produce sodium, another new element. He went on to discover calcium using similar techniques on a mixture of mercuric oxide and lime. Then, following a lead made by others, he applied his electrolysis to a mixture of lime in mercury and succeeded in isolating calcium. In later experiments he was the first to prepare magnesium, boron, and barium. This was an extraordinary list of successes in the early isolation of elements, particularly in the alkaline earths.

Another interesting discovery by Davy was the properties of chlorine. This had previously been discovered by Carl Wilhelm Scheele in 1774, but Davy's contribution was to show that the acid of chlorine (hydrochloric acid) contained no oxygen. This was an important advance because Lavoisier had previously argued that acids had oxygen as an essential component. In fact Lavoisier's word for oxygen was "oxygène" meaning acid-producer. Thus Davy's discovery overturned this central tenet of Lavoisier's system. It was Davy who gave chlorine its name.

Davy's extraordinary successes were soon universally recognized and he was selected as the prestigious Bakerian Lecturer of the Royal Society. His subject was "On the relations of electrical and chemical changes". This and subsequent lectures assured Davy's place as one of the most eminent scientists of the day. He had made extraordinarily rapid progress from the early days of the Pneumatic Institution and the Royal Institution

14.6   Davy Safety Lamp

In 1815 Davy was on holiday in Scotland when he received a letter asking him to investigate the cause of explosions in coal mines. One of the most important mine disasters had been that in the Felling Colliery near Newcastle in the north of England in 1812 as a result of which 92 men and boys lost their lives. On the way back to London, Davy stopped in Newcastle to obtain more information, and when he arrived in London he set about determining the nature of the gas that was responsible for the explosions. This had been thought to be hydrogen by some people but Davy recognized that it was methane. Further research showed that methane in air would only explode at a high temperature so Davy experimented with various designs of lamps where the gases were cooled as they entered. For example he found that if a mixture of methane and air was passed through a metal tube of internal diameter of 1/8th of an inch or less, it would not explode, presumably because of the cooling provided by the wall. His final design was a flame that was contained inside a cylinder of wire gauze, and later this was given the name of the Davy Safety Lamp. Sir Joseph Banks wrote him a highly congratulatory letter which included the words "… I am of the opinion that the solid & effective reputation of that body [the Royal Society] will be more advanced among our contemporaries of all ranks by your present discovery than it has been by all the past" [2].

George Stephenson (1781–1848) also developed a similar lamp and there was an unpleasant dispute about priority. However Davy was awarded the Rumford Medal of the Royal Society. When Banks died in 1820, Davy was the obvious candidate to succeed him as President of the Royal Society. He had risen to the pinnacle of success in British science.

14.7   Later Years and Michael Faraday

Historians generally agree that Davy's reputation declined after 1820. The character of the Royal Society was changing from something of a club for well-connected gentlemen with an interest in science, to a more professional and academic organization. Davy apparently found it difficult to navigate through the various resulting disputes. Banks who was a prominent member of the aristocracy was a difficult man to follow particularly as Davy had humble upbringings. In addition, Davy's health declined and his scientific insight that had initially been so perceptive now showed signs of deterioration.

An important relationship was that between Davy and Michael Faraday (1791–1867). Faraday was born in what is now south London, of a poor family. His father had been a village blacksmith and Michael was largely self-educated. At the age of 14 he became an apprentice to a bookseller and he used this opportunity to educate himself. He developed a particular interest in science with an emphasis on chemistry and electricity.

At the end of his apprenticeship at the age of 20, Faraday was able to attend Davy's lectures at the Royal Society. Tickets were needed to go to these, but a friend of Faraday's made them available. Davy's lectures inspired Faraday to such an extent that the latter wrote a 300-page book based on notes that he had taken at the lectures. He sent this to Davy who was understandably impressed, and the result was that he employed Faraday as a secretary. Later an opening occurred in Davy's laboratory and this allowed Faraday to be appointed as chemical assistant at the Royal Institution. Nevertheless his humble beginnings were always evident at the Institution which was attended mainly by people from high society, and in particular Davy's wife sometimes emphasized these.

Faraday went on to become the most illustrious scientist in Britain at the time and he received numerous awards. His main contributions were in the field of electricity and these included two major discoveries. The first was the property of electricity to cause movement of a conductor in a magnetic field, and this was done in an experiment known as electromagnetic rotation. A pool of mercury had a permanent magnet placed upright in the center, and a wire was dipped into the pool near the magnet. When a current was passed through the wire, it rotated around the magnet. Faraday was ecstatic when he first observed this, and we can still share his excitement when we read about the experiment [7]. This discovery eventually resulted in the development of the electric motor.

The other major discovery was that when a coil of wire was wound around part of an iron ring and a current was passed through it, a current was induced in another coil of wire around the same ring. This was the basis of the electrical transformer that ultimately made possible the distribution of electricity. It could be argued that Faraday was responsible for the industrial use of electricity which until his time had only been seen as an interesting laboratory phenomenon. Another invention well known to electrophysiologists is the Faraday cage. This is a container made of a conducting material which acts as an electromagnetic shield. At one stage Davy is said to have stated that Faraday was his greatest discovery, but later there were charges of plagiarism and the two great scientists fell out. Part of the problem was the tension that so often develops between a teacher and his student when the latter becomes so successful that he outshines his former mentor.

When Davy was 48 years old he had a stroke and subsequently did little scientific work. His writings included a book on fly-fishing which was one of his favorite recreations, and he also wrote his general reflections on science in a book "Consolations in Travel" that was popular for many years [5]. He traveled in Europe on several occasions and died in Geneva where he is buried. There is a memorial tablet in Westminster Abbey.

In conclusion, Davy has a special place in the history of respiratory physiology, particularly that of the respiratory gases. The Pneumatic Institution can be regarded as the coming of age of the respiratory gases in that one of the main reasons for its formation was to determine the value of the newly discovered gases in medical practice. His work on nitrous oxide is exploited daily in general anesthesia. He was a very productive chemist and discovered many elements particularly in the alkaline earth series. He was a brilliant lecturer and as president of the Royal Society rose to the top of his profession. His later years were sad but he had the distinction of promoting Michael Faraday who was one of the greatest British scientists of all time.

References

1. Brown SC, editor. The collected works of Count Rumford. Vol. V: Public Institutions. Cambridge: Harvard University Press; 1970, p. 771.
2. Chambers N, editor. Scientific correspondence of sir Joseph Banks, 1765–1820. Vol. 6. London: Pickering & Chatto; 2007.
3. Davy H. Researches chemical and philosophical, chiefly concerning nitrous oxide, or dephlogisticated nitrous air, and its respiration. London: J. Johnson; 1800.
4. Davy H. The Bakerian lecture: on some chemical agencies of electricity. Phil Trans R Soc Lond. 1807;97:1–56.
5. Davy H. Consolations in travel. London: John Murray; 1830.
6. Davy J, editor. The collected works of sir Humphry Davy. London: Smith, Elder and Co.; 1839.
7. Faraday M. On some new electro-magnetical motions, and on the theory of magnetism. Q J Sci. 1822;12:74–96.
8. Hartley H. Humphry Davy. London: Nelson; 1966.
9. Hindle M. Humphry Davy and William Wordsworth: a mutual influence. Romanticism. 2012;18:16–29.
10. Knight DM. Humphry Davy: science and power. New York: Cambridge University Press; 1996.
11. Lewes GH. Physiology of common life. Vols I and II. London: Blackwood; 1859.
12. Partington J. R. History of chemistry. Vol. 4. London: Macmillan and Co.; 1964. pp. 29–76.
13. Southey CC, editor. Life and correspondence of Robert Southey. Vol. II. London: Longman, Brown, Green, and Longmans; 1850. p. 18.
14. Southey R, Coleridge ST, Lamb C. Annual anthology. Vol. 1. London: T. N. Longman and O. Rees; 1799. pp. 93–99.
15. Stock, JE. Memoirs of the life of Thomas Beddoes, M.D. London: John Murray; 1811.
16. Tilloch A, editor. Philosophical magazine. 1801;9:281–282.
17. Treneer A. The mercurial chemist: a life of Sir Humphry Davy. London: Methuen; 1963.
18. West JB. Spontaneous combustion, Dickens, Lewes, and Lavoisier. News Physiol Sci. 1994;9:276–278.
19. Wordsworth W, Coleridge ST. Lyrical Ballads. London: J. & A. Arch; 1798.

Chapter 15
Denis Jourdanet (1815–1892) and the Early Recognition of the Role of Hypoxia at High Altitude

Abstract Denis Jourdanet (1815–1892) was a French physician who spent many years in Mexico studying the effects of high altitude. He was a major benefactor of Paul Bert (1833–1886) who is often called the father of high altitude physiology because his book *La Pression Barométrique* was the first clear statement that the harmful effects of high altitude are caused by the low partial pressure of oxygen. However Bert's writings make it clear that the first recognition of the critical role of hypoxia at high altitude should be credited to Jourdanet. Jourdanet noted that some of his patients at high altitude had features that are typical of anemia at sea level including rapid pulse, dizziness and occasional fainting spells. These symptoms were correctly attributed to the low oxygen level in the blood and he coined the terms "anoxyhémie" and "anémie barométrique" to draw a parallel between the effects of high altitude on the one hand and anemia at sea level on the other. He also studied the relations between barometric pressure and altitude, and the characteristics of the native populations in Mexico at different altitudes. Jourdanet believed that patients with various diseases including pulmonary tuberculosis were improved if they went to altitudes above 2000 m. This led him to recommend "aérothérapie" where these patients were treated in low-pressure chambers. Little has been written about Jourdanet and his work deserves to be better known.

15.1   Introduction

My dear Colleague:

It is to you that I owe, not only the first idea of this work, but also the material means to execute it, which are so difficult to collect. I have been very happy to see physiological experimentation on one of the most important points of my study confirm entirely the theory which your intelligence had deduced from numerous pathological observations collected on the high Mexican plateaux. For all these reasons I should dedicate this book to you, and I do so with the greater pleasure because you are one of those persons who would make gratitude easy to even the most thankless natures.

Note that this article was authored jointly by John-Paul Richalet, University of Paris, and myself.

© American Physiological Society 2015
J. B. West, *Essays on the History of Respiratory Physiology,*
Perspectives in Physiology, DOI 10.1007/978-1-4939-2362-5_15

[Mon cher Confrère,

C'est à vous que je dois, avec l'idée première de ce travail, les moyens matériels si difficiles à rassembler. J'ai été bien heureux de voir l'expérimentation physiologique, sur un des points les plus importants de ces études, confirmer entièrement la théorie que votre sagacité avait déduite de nombreuses observations pathologiques recueillies sur les hauts plateaux mexicains. A tous ces titres, je devais vous dédier ce livre, et je le fais avec d'autant plus de plaisir que vous êtes de ceux qui rendraient aux natures les plus ingrates la reconnaissance légère à porter.]

This dedication was placed by Paul Bert (1833–1886) at the beginning of his monumental book *La Pression Barométrique* [Barometric Pressure] published in 1878 [3]. Bert is often called the father of high altitude physiology and medicine in large part because this book represents a watershed between the early, largely anecdotal, accounts of the deleterious effects of high altitude on the one hand, and the modern analytical studies on the other. The book put high altitude physiology on a firm scientific foundation and proved that the harmful effects of high altitude were caused by the low partial pressure of oxygen.

This being the case, some readers may be puzzled by Bert's dedication that clearly gives precedence for the recognition of the critical role of hypoxia at high altitude to Denis Jourdanet (Fig. 15.1). In fact Jourdanet's major two-volume publication *Influence de la pression de l'air sur la vie de l'homme: climats d'altitude et climats de montagne* [The influence of air pressure on the life of man: altitude climates

Fig. 15.1 Denis Jourdanet (1815–1892). By permission of the Wellcome Trust

and mountain climates] was published in 1875 3 years before Bert's magnum opus. Furthermore as early as 1862 Jourdanet's book *L'air raréfié dans ses rapports avec l'homme sain et avec l'homme malade* [Rarefied air in relation to the healthy man and the sick man] discussed the effects of the hypoxia of high altitude on man.

It is therefore remarkable that in spite of Jourdanet's groundbreaking idea as acknowledged by Bert (Fig. 15.2), very little about Jourdanet has been published. Apparently there is no full description of the man and his work in English. There are two or three biographical articles in French or Spanish but these do not discuss his scientific contributions. The aim of this paper is to redress this omission and show

Fig. 15.2 Paul Bert (1833–1886). From [26]

that Jourdanet was an innovative scientist as well as a successful physician, and that his contributions should be better known.

15.2   Brief Biography

This section is based on the account by Duffau [8–10] with additions from Auvinet [1] and Auvinet and Briulet [2]. Denis Jourdanet (1815–1892) (Fig. 15.1) was born in Juillan near Tarbes in the Hautes-Pyrénées, France, on May 1. As a young boy he was influenced by his uncle who was a village priest and he became very interested in the study of Latin. At the age of 13 he entered the Petit Séminaire de Saint Pé in Bigorre in the Hautes-Pyrénées where he remained from 1828 to 1833 and demonstrated a lively intellect. After a period in the college d'Aire he moved to Paris with his cousin Antoine where he began to study medicine. He enjoyed the contact with patients but not the formal lectures. Furthermore according to his biographer, Father Duffau, he had a preference for the easy life of the salons. In 1842 at the age of 26 before he had finished his medical studies, he decided to seek his fortune in America and on February 15 he embarked on the ship Arago at Le Havre for Mexico. He reached Veracruz on April 10.

Apparently his plans were vague and he was short of money, but fortunately he was helped by a representative of Yucatan in Veracruz by the name of Don José Dolores Castro. As a result he went to Campeche in the Yucatan Peninsula where his knowledge of medicine attracted many patients and he rather prematurely set up a medical practice in spite of the fact that this was not authorized. However an operation for cataract surgery (sic) impressed the French consul, A. Laisné of Villévêque, and as a result he was given the necessary legal authorization to practice medicine.

Jourdanet was subsequently accepted into the society of wealthy people, and his situation suddenly changed for the better when he married Señorita Rita Estrada, a daughter of Jose Maria Gutierrez de Estrada. This family enjoyed a special status in Campeche. Gutierrez Estrada was a prominent diplomat who had been a Secretary for Foreign Affairs of Mexico. Jourdanet therefore happily found himself to be the son-in-law in a family of considerable wealth.

In 1846 Jourdanet returned to Paris with his new wife where he was able to complete his formal study of medicine at the Sorbonne. He was influenced by the famous French toxicologist, Mathieu Orfila and his thesis dealt with the prevention and treatment of tetanus. His wife who had been affected by pulmonary tuberculosis for 12 years noted a definite improvement in her disease during her stay in Paris. Her brother had died of the same disease in 1847.

In 1848 the couple returned to Mexico but they decided not to live in Campeche where the tropical climate worsened the health of his wife. They therefore moved to Puebla, altitude 2200 m, southeast of Mexico City. Here Jourdanet began to develop what became one of his primary interests, that is the beneficial effects of altitude on some diseases. He made a series of measurements of barometric pressure at various increasing altitudes including the summit of the volcano La Malinche, altitude

4461 m. He came to the conclusion that the increased altitude of the high plateau improved his wife's health and this led him later to write extensively on the value of a reduced barometric pressure in the treatment of disease.

In 1851 after 2 years in Puebla, Jourdanet moved to Mexico City, also at an altitude of about 2200 m. There he became a member of the Faculty of Medicine of Mexico and earned his second title of Doctor of Medicine. He also founded a Franco-Mexican college in an imposing colonial building known as the House of the Masks. This enabled French teachers to educate the children of French families who had settled in Mexico. There were occasional serious outbreaks of cholera in Mexico City and Jourdanet's reputation increased as he dealt with these. His patients included several celebrities, and one was the successful businessman, Manuel Escandón, who generously rewarded Jourdanet for his services.

Unfortunately Jourdanet's wife died in 1859 and he decided to return to France. In 1860 in Paris he had an opportunity to discuss his new findings at a meeting of the French Academy of Medicine. However he had a mixed reception. He argued that residence at an increased altitude could contribute to health and his first article on the topic was published in 1861 [14]. He also discussed his concept of what he called *anoxyhémie* where he drew a comparison between the physiological effects of anemia at sea level on the one hand, and reduced oxygen levels at high altitude on the other. This led him to promote the use of low-pressure chambers in the treatment of some illnesses. These topics are discussed below.

At this time Jourdanet became involved in politics. There was considerable tension between Mexico and its creditors, chiefly France, Britain, Spain and the United States, because interest payments had been suspended by the Mexican president. The problem was discussed at the Convention of London of 1861 and ultimately led to a French campaign beginning in 1862. Jourdanet was summoned to the Tuileries Palace by the Emperor Napoleon III to discuss the situation, and Jourdanet's father-in-law headed a Mexican commission to Archduke Maximilian of Austria who was subsequently crowned as the Emperor of Mexico. Maximilian invited Jourdanet to be his personal physician but Jourdanet declined.

Jourdanet returned to Mexico in 1864 and continued his clinical work. His 1861 book [14] was renamed and was consulted by many officers of the Franco-Mexican troops on the effects of altitude on health. As a result of this service to the armed forces, Jourdanet was awarded the Cross of the Legion of Honor.

Jourdanet remarried in 1865, again into a wealthy family. His new wife was Señorita Juana Beistegui y García, daughter of a prominent owner of silver mines. Her sister, Loreto, subsequently married Alphonse Dano who was the Minister Plenipotentiary of France in Mexico. The two families continued to maintain close links. The Jourdanets left for France in 1867 after the defeat of the French troops in Mexico, and the execution of Maximilian. They returned to Paris and took an apartment on the Champs Elysées.

At this point Jourdanet decided to leave the practice of medicine and concentrate on science. At the Sorbonne he met Paul Bert who was 18 years his junior and the two developed a collaboration that resulted in the dedication of Paul Bert's magnum opus as cited earlier. Jourdanet who was now wealthy financed Bert's extensive laboratory in the Sorbonne where most of his work on the effects of high altitude was carried out. He also subsidized the publication of Bert's book [3].

In 1875 Jourdanet found himself associated with the tragic flight of the French balloon "Zenith" with the three aeronauts, Tissandier, Croce-Spinelli and Sivel. The balloon reached an altitude of about 8600 m resulting in the deaths of the last two from hypoxia. Bert had previously learned that the balloonists had insufficient oxygen, and he tried to warn them about this but the warning letter arrived too late. The disaster caused a sensation in France. Jourdanet had witnessed the rapid ascent of the balloon and reported that he felt serious misgivings about the flight.

Jourdanet retained his love of Mexico and subsequently published French translations of two books in Spanish about the history of the country [5, 28]. He died in Paris on May 6, 1892 and was buried in the vault of the Dano family in the Passy Cemetery.

15.3   Jourdanet's High Altitude Studies

Jourdanet's high altitude studies were reported by him in several articles and books [14–20, 22]. However his most important publication was *Influence de la pression de l'air sur la vie de l'homme* in two volumes published by Masson in Paris, 1875 (Fig. 15.3). This is a very handsome production with many beautiful engravings and a number of colored maps. Each volume is divided into five parts each consisting of several chapters which themselves are comprised of several numbered "articles". There is also an appendix and some supplementary notes. A second edition was issued in 1876. The entire first edition is available on the Internet. This digitized copy is from the Countway Library in Cambridge, MA, and includes a note at the beginning from Jourdanet with his signature.

The book begins with a long historical introduction on the physics of the atmosphere starting with Aristotle who speculated on the weight of the air. The contributions of Galileo, Torricelli, von Guericke and several others are then described but, perhaps betraying its French origin, the only engraving besides Galileo is a fine one of Blaise Pascal. There is also a section on the Loi de Mariotte which many of us know as Boyle's Law. Entertainingly, Pascal estimated the total weight of the atmosphere at about 8×10^{18} livres (about the same in English pounds). There is a discussion of the properties of the atmosphere including the relations between temperature and altitude and the amount of water vapor. This is followed by a short, rather speculative chapter on possible changes in the atmosphere over geological time.

There is a long section on geography including altitudes of many places in the world. Facing page 85 there is a beautiful engraving showing what is purported to be Mt. Everest based on an aquarelle made by Hermann Schlagintweit in 1855 (Fig. 15.4). The altitude of the summit is given as 8816 m which is puzzling because Schlagintweit gave a value of 29,000 ft corresponding to 8839 m. The modern accepted value is close to this being 8848 m. However it is notable that in Appendix 1 of Bert's *La Pression Barométrique*, Bert gives a table of the heights of mountains

Fig. 15.3 Title page of
Jourdanet's major publication
"Influence de la pression de
l'air sur la vie de l'Homme"
[22]

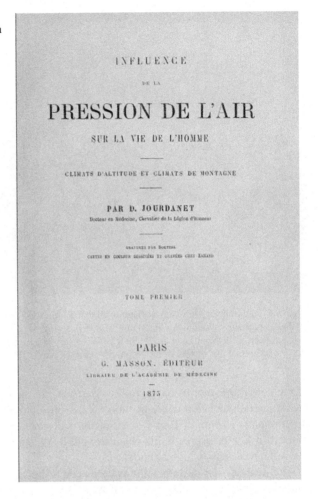

INFLUENCE

DE LA

PRESSION DE L'AIR

SUR LA VIE DE L'HOMME

CLIMATS D'ALTITUDE ET CLIMATS DE MONTAGNE

PAR D. JOURDANET

Docteur en Médecine, Chevalier de la Légion d'honneur

GRAVURES PAR BOETZEL
CARTES EN COULEUR DESSINÉES ET GRAVÉES CHEZ ERHARD

TOME PREMIER

PARIS

G. MASSON, ÉDITEUR

LIBRAIRE DE L'ACADÉMIE DE MÉDECINE

1875

and the associated barometric pressures which he says is based on Jourdanet's data. Mt. Everest is shown to have an altitude of 8840 m with a barometric pressure of 248 mm Hg. This is very close to the accepted value of about 250 mm Hg. Actual measurements on the summit by climbers in good weather when the pressure tends to be high are 253 mm Hg [33].

There is a very long section with extensive tables on the altitudes of many high places in the world. This is followed by an extended section on Mexico including its geography and the number of inhabitants in various regions.

Jourdanet then turns to the medical and physiological effects of going to high altitude. First he describes in some detail the results obtained by Paul Bert in his low-pressure chambers. By exposing various animals to air at low barometric pressures on the one hand (hypobaric hypoxia) and low concentrations of oxygen at normal pressure (normobaric hypoxia) on the other, Bert was able to show the critical role

Fig. 15.4 Aquarelle by Hermann Schlagintweit dated June 1855 of what was thought to be Everest but is probably Makalu. An engraving of this is facing page 85 in Jourdanet's major book. Schlagintweit gave the altitude as 29,000 ft which is very close to the actual value. This is one of the first images of this section of the Himalayan range. From [22]

of the partial pressure of oxygen. Also Bert was the first physiologist to measure the oxygen dissociation curve of blood under physiological conditions. There are descriptions of some of Bert's experiments where he exposed humans including himself to low pressures in his chambers although Jourdanet points out that Bert himself never went to high altitude.

The next section describes the experiences of humans ascending to high altitudes. One of the first descriptions of Acute Mountain Sickness was provided by the Spanish priest Joseph de Acosta in 1590 and his account is given in full albeit in French. Other famous ascents for example by Horace Bénédict de Saussure and Alexander von Humboldt are also described. The features of Acute Mountain Sickness are set out including its effects on respiration, circulation, and the gastrointestinal and central nervous systems. There is also a description of the famous balloon ascent by Glaisher and Coxwell in 1862 when Coxwell became paralyzed and had to use his teeth to seize the cord to open the valve of the balloon. Much of this historical section is similar to that in Bert's book [3] published 3 years later.

The next section deals with the physiological differences of inhabitants of high altitude compared with sea level dwellers. There is a brief discussion of the enlargement of the chest in high altitude natives. Special emphasis is given to the movement of populations in Mexico between different altitudes including the problems of the indigenous population living on the slopes of the volcano Popocatépetl.

In the second volume of this large book Jourdanet discusses the pathophysiology of high altitude, a topic that he had previously dealt with in less detail in "De l'anémie des altitudes et de l'anémie en général dans ses rapports avec la pression de l'atmosphère" published in 1863 [17]. A major emphasis is the analogy that he draws between the effects of simple anemia, that is a reduced concentration of red blood cells at sea level, and the effects of a reduced oxygen level in the blood at high altitude. For the latter he introduces the term "l'anoxyhémie" and he uses this and the term "anémie barométrique" interchangeably. This is confusing for us now because of course we use a strict definition of the term anemia. However Jourdanet was clearly referring to the overall effects of going to high altitude rather than the concentration of red cells in the blood. For example he states "l'anoxyhémie des altitudes a donc son analogue dans l'anémie hypo-globulaire du niveau de la mer". [The anoxemia of altitudes is analogous to the anemia caused by a reduced number of red cells at sea level]. He goes on to say "l'oxygène étant l'agent vital par excellence, sa diminution par défaut de globules fait la faiblesse des anémiques; sa diminution dans le sang par défaut de pression doit produire les mêmes résultats". [Oxygen is absolutely the vital agent, its reduction by a lack of red blood cells causes the weakness of anemia; its reduction in the blood by the low pressure causes the same results]. Incidentally he uses the two spellings "anoxyhémie" and "anoxyémie" interchangeably.

It is interesting that Jourdanet was drawn to consider changes in the blood at high altitude by his experience as a surgeon. He noticed that in high altitude dwellers the blood tended to be thick and of dark color, and it flowed slowly. But in spite of the thick blood, patients at high altitude exhibited some of the same features seen in anemia at sea level. For example there was a rapid pulse, sometimes dizziness, and occasional fainting spells. However the definitive studies of the polycythemia of high altitude had to wait for the work of François Viault [31] following the inspired suggestion of Bert (3, p. 1000 in the English translation).

In discussing the increase in altitude which is necessary for anoxyhémie Jourdanet drew from his own experience. As we have seen he initially lived in Campeche at sea level on the Gulf of Mexico and then moved to the inland plateau at an altitude of about 2200 m where he clearly saw the effects of altitude. He therefore concluded that the critical increase in altitude was about 2000 m. Jourdanet also published two other articles, one on statistics of Mexico and another on syphilis [21; 23].

In his book La Pression Barométrique, Bert mentions the scientific debt that he owes to Jourdanet in several places. We have already noted the dedication of the book to Jourdanet which is reproduced at the beginning of this article. In addition there is an interesting section on page 274 of the English translation where Bert refers to a criticism of Jourdanet's work by a M. Leroy de Méricourt. This author refers to the following statement by Jourdanet "An ascent above 3000 m is a barometric disoxygenation of the blood, just as a bleeding in a corpuscular disoxygenation" in pejorative terms calling it "strange". Bert retorts "I was very desirous of reporting this opinion because it shows well what the sentiment of the most learned and the best authorities was in 1866. We must, in fact, wait for the theory expressed by M. Jourdanet, the accuracy of which I have demonstrated experimentally, to be considered soon as a thing so simple and evident that everyone will claim to be its originator, or at least will refuse it any merit of originality".

Finally, near the end of the long chapter in which Bert describes various theories and experiments on the effects of going to high altitude, he summarizes the contributions made by Jourdanet as follows. "… it is to M. Jourdanet… that we shall give the credit for having found the true explanation of the symptoms of decompression, as he already has the credit for having so clearly defined and described them by the name of anoxemia.

However, we must note here again that the basis of the theory rested only on reasoning and deductions, very well connected, to be sure, but not sufficient to establish complete proof to minds accustomed to the precision of scientific methods. It was necessary to make experimental proof of anoxemia and its effect upon the production of the symptoms which appear in rarefied air. … M. Jourdanet himself permitted me to subject to experimental test both his own theory and all those which have deserved to be examined thus. The account of the experiments which I made with the help of the apparatuses which I had secured, thanks to him, will form the second part of this work." (Page 351 in the English translation.)

15.4 Aerotherapy

Jourdanet goes on to discuss the beneficial effects of high altitude on some diseases. He cites evidence that pulmonary tuberculosis can be improved by going to high altitude and the same is true of yellow fever and typhus. As a result of these studies he came to the conclusion that subjecting patients to a low barometric pressure in a chamber would be valuable in some instances. He therefore embarked on a form of treatment known as aerotherapy. This is discussed in more detail in his book *Aérothérapie. Application artificielle de l'air des montagnes au traitement curatif des maladies chroniques* [Aerotherapy. The artificial use of mountain air in the *curative treatment of chronic diseases*] [18].

Metal chambers which allowed humans to be exposed to either an increased or decreased air pressure had been introduced a number of years before Jourdanet constructed his own [13]. For example in 1834 Junod [24] constructed a chamber that could be used to both increase and decrease the pressure in the treatment of some types of lung disease. This early chamber was a copper sphere measuring about 1.5 m in diameter and it was claimed that the circulation to the internal organs was increased and that the patient felt an increased sense of well-being because of an increase in cerebral blood flow. Pravaz [27] built a large chamber that could accommodate twelve patients at one time and the therapeutic regime was named "le bain de comprimé" [the bath of compressed air]. Hyperbaric chambers had also been used for the treatment of decompression sickness in workers constructing caissons underwater. One of the first was a caisson employed during the excavation of the bed of the Loire River. This chamber was built by Triger [30].

Jourdanet's low-pressure chamber had a volume of six cubic meters and is shown in Fig. 15.5. Its use is described in the last chapter of Jourdanet's book [22] and a number of clinical examples are given claiming its effectiveness although of course by modern standards these anecdotal records are of limited value.

On the other hand Jourdanet's colleague, Paul Bert, made extensive use of low-pressure chambers in *La Pression Barométrique* [3]. Figure 15.6 shows low-pressure chambers in his laboratory in the Sorbonne, and on the extreme left we can see part of Jourdanet's chamber. As mentioned earlier, the wealthy Jourdanet was Bert's patron and was responsible for the costs of setting up the elaborate laboratory. All Bert's high altitude experiments were carried out with low-pressure chambers and in one experiment he decompressed himself to the pressure at the summit of Mt. Everest although of course he could only remain conscious by inhaling an enriched oxygen mixture. During this experiment Bert reported that the flame of a candle in the chamber became very blue, and a sparrow vomited and seemed extremely sick. However Bert himself stated that it "seemed to me as if I could have gone lower yet,

Fig. 15.5 Jourdanet's low-pressure chamber that was constructed for aerotherapy, that is exposure of patients with lung and other diseases to low pressures. From [13]

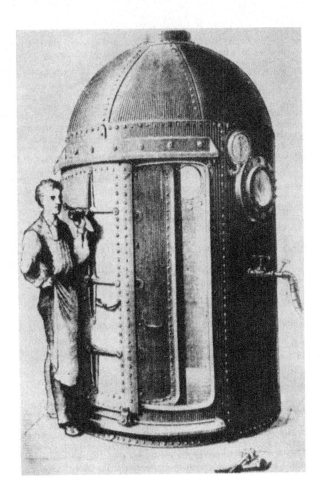

Fig. 15.6 Bert's laboratory at the Sorbonne showing his double human hypobaric chambers at the far right, an air pump in the center, and Jourdanet's low-pressure chamber on the far left. From [32]

with no inconvenience, and I was quite ready to do so, had not my esteemed pumps, weary with work, refused to continue exhausting the air of the cylinders" (3, pages 706–708 in the English translation). Incidentally a few years later the Italian physiologist Angelo Mosso (1846–1910) decompressed himself to the extraordinarily low barometric of 192 mm Hg although of course the chamber was enriched with oxygen. This pressure corresponds to an altitude of about 10,800 m [6].

15.5 Criticism of the Claim by Jourdanet and Bert that Hypoxia is the Critical Factor in the Physiological Responses to High Altitude

Today we accept that the physiological responses to hypoxia are determined by the low partial pressure of oxygen and we recognize that this was first proved by Paul Bert and his predecessor Denis Jourdanet. However for a number of years after the publication of *La Pression Barométrique* [3] there was heated criticism of the role of hypoxia and indeed the issue was disputed for a further 30 years. The interesting history of this debate has been well documented by Kellogg [25].

One of the first scientists to query Bert's finding was the eminent Russian physiologist Setschenow [29]. He is often referred to as the father of Russian physiology and he worked in Vienna with Carl Ludwig where he constructed a blood-gas pump that was used to liberate the gases from a blood sample. He argued that the affinity of oxygen for hemoglobin was so great that even at a PO_2 of 20–30 mm Hg, the blood should be well-oxygenated. Criticism also came from the German physiologists Fraenkel and Geppert [11]. Again they questioned the accuracy of the oxygen dissociation curve used by Bert. Here they were influenced by the publication of a paper by Hüfner [12] showing that the hemoglobin remained essentially completely saturated down to a PO_2 of about 20 mm Hg. Their conclusion was that the arterial blood maintained its high oxygen concentration even at the altitude of Mont Blanc (4807 m) and therefore that mountain sickness at moderate altitudes could not be caused by hypoxemia.

Another opponent of Bert's theory was Angelo Mosso, professor of physiology at Turin, whose work was briefly referred to earlier. He carried out low-pressure chamber experiments in which he measured the concentrations of oxygen and carbon dioxide in the expired gas samples and concluded that the deleterious effects of the low barometric pressure were more closely associated with a low PCO_2 than a low PO_2. He coined the term "acapnia" to refer to the low PCO_2 and argued that this was the primary cause of mountain sickness. As additional evidence he pointed out that voluntary hyperventilation at sea level which reduced the PCO_2 but increased the PO_2 in the blood can cause cerebral symptoms similar to those of high altitude. Mosso argued that exposure to a low barometric pressure was similar to the situation existing in a blood gas analyzer where the vacuum extracted CO_2 from the blood. The British physiologist Joseph Barcroft criticized this notion by pointing out that humans are exposed to an essentially zero partial pressure of carbon dioxide even at sea level. However Mosso clung to his acapnia theory until his death in 1910.

By the end of the first decade of the 20th century most of the objections to Bert's theory were muted and for example, when Douglas et al [7] wrote up their account of the Pikes Peak expedition in 1913, they clearly attributed the effects of high altitude to hypoxia. However even today most physiologists credit Paul Bert with introducing the hypoxia theory of high altitude and it is salutary to remember that he gave the credit to his patron Denis Jourdanet.

15.6 Relations between Jourdanet and Bert in Their Later Years

It is sad to report that relations between Jourdanet and Bert became increasingly strained in their later years. First we should remember that Jourdanet was 18 years older than Bert and also that Jourdanet, partly through his marriages, had developed strong links with very successful and highly placed families. His first marriage had been into the family of Gutierrez Estrada who had been a Secretary for Foreign Affairs in Mexico. His second marriage was into the family of a prominent wealthy owner of mines in Mexico. Moreover his second wife's sister married the Minister Plenipotentiary of France in Mexico. The families of the two sisters remained close and in fact Jourdanet and his wife were buried in the imposing vault of the Dano family in the Passy Cemetery in Paris although interestingly neither his nor his wife's name are on the plaque of the vault. It is easy to understand that with these connections Jourdanet had many friends in high places and that his politics were rather conservative. He was also a man of substantial wealth in his later years.

By contrast the younger Bert was a strident liberal with some anti-establishment sympathies. He was politically active and, for example, strongly campaigned against the role of the church in teaching in French public schools. One of his books [4] was a scathing criticism of the Jesuits. Bert's appointment at the Sorbonne was opposed by church authorities on the grounds of his very liberal and anti-clerical ideas and apparently it was only with the support of Louis Pasteur that Bert's appointment was successful. The contrast between the two men is suggested by Figs. 15.1 and 15.2. Jourdanet looks like a conventional pillar of the establishment whereas Bert suggests a certain insouciance.

Presumably relations between the two men were cordial enough when they were discussing scientific matters. However according to Jourdanet's biographer, Father Duffau, Jourdanet complained "Je suis obligé, souvent, de subir, dans les séances de laboratoire, des digressions radicales ou anticléricales. Je désapprouve de telles idées, et Paul Bert sera peut-être peu satisfait d'en voir les preuves dans mon livre." [I am often obliged during laboratory sessions to take part in radical or anti-clerical discussions. I disapprove of such ideas, and Paul Bert may be unhappy to see the evidence in my book]. Jourdanet went on "Paul Bert s'est détourné de la science et lancé dans la politique où il s'est fait remarquer par son orgueil et sa passion de sectaire." [Paul Bert has turned away from science and become involved in politics where he is noted for his pride and sectarian passion]. According to Duffau, Bert then broke off all relations with his collaborator and even sought to belittle the importance of Jourdanet's work. Do not forget, Bert said, that we have a serious quarrel about the origin of man following the publication of the "Origin of Species" by Charles Darwin. The critics of Darwin have published vile objections to the theory of "l'Homme qui descend du singe" [man is descended from apes] and the denial of God's role in the creation of the world.

To finish on a happier note, Jourdanet later founded the Chair of Medical Geography in the Society of Anthropology in Paris. This was a forum for continuing discussions and educational activities on the effects of different climates and altitudes on human life which were important legacies of Jourdanet's work.

References

1. Auvinet G. El Dr Denis Jourdanet, un médico francés en México (1842–1867). In: Leticia Gamboa, Guadalupe Rodriguez, Estela Munguia, Coords, editors. From colonial to contemporary Mexico. Editorial ICSyH/BUAP. ISBN/ISSN 978-607-487-345-0. Mexico; 2012.
2. Auvinet G, Briulet B. El doctor Denis Jourdanet; su vida y su obra. Gac Méd Méx. 2004;140:426–9.
3. Bert, P. La pression barométrique. Paris: G. Masson, 1878. English translation by M.A. Hitchcock and F.A. Hitchcock. Columbus: College Book Company; 1943.
4. Bert P. The Doctrine of the Jesuits. Boston: B.F. Bradbury; 1880.

5. Díaz del Castillo B. L'Histoire véridique de la Conquête de la Nouvelle-Espagne. Traducción del español al francés de D. Jourdanet. 1876;2(8).
6. Di Giulio C, West JB. Historical vignette: Angelo Mosso's experiments at very low barometric pressures. High Alt Med Biol. 2013;14:78–79.
7. Douglas CG, Haldane JS, Henderson Y, Schneider EC. Physiological observations made on Pike's Peak, Colorado, with special reference to adaptation to low barometric pressures. Philos Trans Roy Soc Lon B. 1913;203:185–318.
8. Duffau F. Le Docteur Jourdanet, sa vie, ses ouvrages (1815–1892), Annuaire du Petit Séminaire de Saint-Pé, Tarbes: E. Croharé; 1893. pp. 270–328.
9. Duffau F. Docteur Jourdanet 1815–1883. Notice biographique. Les contemporains #197, 19 July 1896a.
10. Duffau F. Le Docteur Jourdanet, sa vie, ses ouvrages (1815–1892), Annuaire du Petit Séminaire de Saint-Pé, Tarbes: E. Croharé; 1896b. pp. 410–54.
11. Fraenkel A, Geppert J. Ueber die Wirkungen der verdfinnten Luft auf den Organismus. Eine Experimental-Untersuchung. Berlin: August Hirschwald; 1883.
12. Hüfner, G. Untersuchungen zur physikalischen Chemie des Blutes. Z Physiol Chem. 1882;6:94–111.
13. Jacobson JH, Morsch JH, Rendell-Baker L. Clinical experience and implications of hyperbaric oxygenation. the historical perspective of hyperbaric therapy. Ann N Y Acad Sci. 1965;117:651–670.
14. Jourdanet D. Les altitudes de l'Amérique tropicale comparées au niveau des mers, au point de vue de la constitution médicale. In-8, Baillère éd. Paris; 1861.
15. Jourdanet D. L'air raréfié dans ses rapports avec l'homme sain et l'homme malade. Paris: J.-B. Baillière et Fils; 1862a.
16. Jourdanet D. Prophylaxie de la fièvre jaune par des fièvres d'autre nature. L'Union médicale 1862b;15:531–537.
17. Jourdanet D. De l'anémie des altitudes et de l'anémie en général, dans ses rapports avec la pression de l'atmosphère. Paris; 1863a.
18. Jourdanet D. Aérothérapie. Application artificielle de l'air des montagnes au traitement curatif des maladies chroniques. Paris: J.-B. Baillière, 1863b.
19. Jourdanet D. Précis d'aérologie médicale et d'aérothérapie. 1864a.
20. Jourdanet D. Le Mexique et l'Amérique tropicale. Climat, hygiène et maladies. Paris: Baillère et fils; 1864b.
21. Jourdanet D. La Statistique du Mexique, in-8. Bulletin de la Société mexicaine de géographie et de statistique; 1865.
22. Jourdanet D. Influence de la pression de l'air sur la vie de L'homme. Paris: Masson; 1875.
23. Jourdanet D. Les syphilitiques de campagne de Fernand Cortez. Médicales Etudes sur la chronique of Bernal Diaz del Castillo, compagnon d'armes of Fernand Cortez. Paris: G. Masson; 1878
24. Junod VT. Recherches physiologiques et thérapeutiques sur les effets de la compression et de la raréfaction de l'air, tant sur le corps que sur les membres isolés. Rev Med Franc Etrange. 1834;3:350.
25. Kellogg RH. La Pression Barométrique: Paul Bert's hypoxia theory and its critics. Respir Physiol. 1978;34:1–28.
26. Olmsted JMD. Father of aviation medicine. Sci Am. 1952;186:66–72.
27. Pravaz. 837–1838. Mémoire sur l'application du bain in d'air comprimé au traitement des affections tuberculeuses, des hémorragies capillaires et des surdités catarrhales. Bull Acad Natl Med (Paris). 1837–1838;2:985.
28. Sahagún B. Histoire générale des choses de la Nouvelle Espagne. Traducción y anotaciones de Denis Jourdanet y Rémi Siméon. Paris: G. Masson; 1886.
29. Setschenow J. Zur Frage über die Atmung in verdünnter Luft. Pflügers Arch Gesamte Physiol. 1880;22:252–61.
30. Triger. Mémoire sur un appareil Ã air comprimé, pour le percement des puits de mines et autres travaux, sous eaux et dans les sables submergés. Compt Rend. 1841;13:884.

31. Viault F. Sur la quantité d'oxygène contenue dans le sang des animaux des hauts plateaux de l'Amérique de Sud. C R Acad Sci. 1891;112: 295–298.
32. West JB. High life: a history of high-altitude physiology and medicine. New York: Oxford University Press; 1998.
33. West JB. Barometric pressures on Mt. Everest: new data and physiological significance. J Appl Physiol. 1999;86:1062–6.

Chapter 16
Centenary of the Anglo-American High Altitude Expedition to Pikes Peak

Abstract 2011 marks the 100th anniversary of one of the most important early research expeditions to high altitude. The principal participants were J.S. Haldane and C.G. Douglas from Oxford, Y. Henderson from Yale, and E.C. Schneider from Colorado College. Pikes Peak just outside Colorado Springs proved to be an excellent venue because of its substantial altitude of 4300 m, convenient access via a cog railway, and comfortable living accommodation. The expedition had a classical design with measurements made first at sea level, then on the summit for five weeks, and then at sea level again. The extensive scientific program included descriptions of acute mountain sickness, many measurements of partial pressures in alveolar gas and arterial blood, changes in ventilation including periodic breathing, exercise measurements, and a large number of blood studies. One error was the conclusion that the arterial PO_2 could considerably exceed the alveolar value implying oxygen secretion by the lung, but this should not detract from the other important advances. Mabel FitzGerald was invited to be a member of the expedition but did not join the men on the summit. Instead she visited various mining camps in Colorado at somewhat lower altitudes and carried out classical studies of alveolar gas partial pressures and hemoglobin values. Indeed her results of alveolar PO_2 and PCO_2 are frequently cited today. The Pikes Peak expedition was a model and it had an extensive influence on later studies.

The centenary of one of the most important early expeditions to high altitude provides an opportunity to look at the state of high altitude and respiratory physiology 100 years ago, and the important events leading up to that period. A key figure was the French physiologist Paul Bert (1833–1886). He is often referred to as the father of high altitude physiology because of the publication of his book *La Pression Barometrique* in 1878 (Bert, 1878a). In fact the book deals with high pressure as well as low barometric pressure and so the title of father of environmental physiology is also appropriate. The book is now easily available because it was digitized by Google, but if your French is a bit rusty, there is an excellent English translation by the two Hitchcocks [2]. The book makes very good reading, particularly if you are interested in the history of high altitude physiology, because the first chapter of some one hundred pages is devoted to this. Bert wrote with an urbane, droll, witty style which is a pleasure to read.

© American Physiological Society 2015

J. B. West, *Essays on the History of Respiratory Physiology,*
Perspectives in Physiology, DOI 10.1007/978-1-4939-2362-5_16

Prior to Bert's work there were numerous theories about the reasons for the deleterious effects of high altitude and some of these were very bizarre. For example one was that weakening of the joint between the femur and pelvis because of the low barometric pressure meant that the hip muscles had to work very hard causing fatigue. Another frequent error was that the low barometric pressure resulted in rupture of blood vessels because of the reduced pressure around them. This is a misconception because all the pressures in the body decrease together, but this error occasionally surfaces in students today when they first start thinking about high altitude physiology.

Bert showed conclusively that the deleterious effects of high altitude were caused by the low partial pressure of oxygen, that is, the oxygen concentration multiplied by the ambient barometric pressure. He did this by exposing animals in a large jar to gas mixtures containing low concentrations of oxygen at normal barometric pressure on the one hand, and a reduced pressure of air on the other. The experiments described in detail in his book are very convincing but it is interesting that his explanation was not accepted for many years by a number of prominent physiologists.

All Bert's experiments were carried out using low-pressure chambers in the Sorbonne in Paris and some of the accounts are harrowing. In fact on one occasion Bert exposed himself to the barometric pressure at the summit of Mt. Everest (about 250 mmHg) but showed that he could survive by breathing oxygen. However Bert never studied subjects at high altitude although he advised on several balloon ascents, particularly the disastrous flight of the balloon Zenith in 1874 when two of the balloonists died of acute hypoxia. The event caused a sensation in France (Bert [2], translation, pp. 963–973).

The first person to install a scientific laboratory at high altitude was the Frenchman Joseph Vallot (1854–1925). He was not a physiologist but a botanist and geologist and was responsible for the construction of the Observatoire Vallot in 1890 at an altitude of about 4350 m on Mont Blanc. Although the hut was located in a very remote place and access to it was very challenging, it was built with typical French panache including a well-appointed kitchen and a room decorated with Chinese hangings and elegant Persian rugs. A number of physiological studies were done in the Observatoire Vallot including some of the first observations of periodic breathing at high altitude. One of the most dramatic events was the death of Dr. Jacottet in 1891 of what was almost certainly high altitude pulmonary edema (West [25], pp. 141–143). A serious disadvantage of the Observatoire Vallot was the difficult terrain on Mont Blanc and Fig. 16.1 shows Professor Jules Janssen en route to the Grands Mulets during a reconnaissance. The arrangements suggest a possible pattern for a senior physiologist who can count on the cooperation of graduate students and post-doctoral fellows.

Only about three years after the installation of the Observatoire Vallot, Angelo Mosso (1846–1910) who was professor of physiology in Turin arranged for the construction of the Capanna Margherita on one of the peaks of the Monte Rosa. This was extraordinarily high at 4559 m and again access was very difficult involving a long climb over glaciers in a region with hidden crevasses. The hut owes its name to Queen Margherita who was a keen alpinist, and she actually visited the hut soon

Fig. 16.1 Professor Jules Janssen en route to the Grands Mulets (3050 m) on Mont Blanc during a reconnaissance. The difficulties of the route can be seen. From Vivian [24]

after its construction and spent the night there. The Capanna Margherita was the site of extensive physiological studies carried out by the burgeoning international high altitude physiology community. Interestingly, Mosso who made many important contributions did not accept Bert's conclusion that hypoxia was the culprit, but instead Mosso blamed the low carbon dioxide level in the body, a condition that he called acapnia. Another important high altitude physiologist in the early part of the century was Nathan Zuntz (1847–1920) who was a professor of animal physiology in Berlin. In 1910 he organized an important expedition to the Alta Vista Hut (altitude 3350 m), on Tenerife, Canary Islands.

So this was the situation in 1910 when J.S. Haldane (1860–1936) and Yandell Henderson (1873–1944) discussed the possibility of another physiological expedition to high altitude. Haldane (Fig. 16.2) came from a distinguished old Scots family and he attended Edinburgh University where his initial interest was philosophy. Later he turned to medicine but interestingly his first major paper was actually on philosophy [12]. Following graduation he was an instructor at University College, Dundee where he carried out a project on the foul air of the slums of Dundee and reported abnormally high concentrations of carbon dioxide. In 1887 he became a lecturer in physiology in Oxford and in 1905 published with Priestley a landmark paper on the importance of carbon dioxide on the control of ventilation [15]. Haldane's contributions to physiology were very extensive and went far beyond his

Fig. 16.2 John Scott Haldane (1860–1936). From Cunningham and Lloyd [3]

interests in high altitude [14]. They included important studies on mine air and decompression sickness and in fact he was a key figure in industrial physiology.

The origin of the Pikes Peak Expedition makes entertaining reading. Haldane met Yandell Henderson, who was a physiologist at Yale University, at the International Congress of Physiology in Vienna in 1910 and the conversation was recounted by Henderson [17]. Haldane remarked that what was needed was "a nice, comfortable mountain". This sounds rather whimsical but Haldane was referring to the difficult access and the spartan conditions of high altitude stations such as the Observatoire Vallot and the Capanna Margherita. According to Henderson, Haldane stated that to reach the Capanna Margherita "one must climb several thousand feet over snow and ice… the climate is arctic even in midsummer… worst of all the investigator must cook for himself". He continued, it is "open to question whether the effects observed were due to the barometric pressure… or bad cooking". In fact Joseph Barcroft (1872–1947), a distinguished high altitude physiologist from Cambridge, had made a similar remark about the Capanna Margherita when he stated "the difficulty of transport greatly restricts the possibilities both of research and gastronomy".

By contrast Henderson pointed out that Pikes Peak, a mountain outside Colorado Springs in the Rockies, had a number of advantages including easy access via a cog railway, a substantial altitude of 4300 m, several rooms at the hotel on the summit that could be set aside for the expedition, and the possibility of a warm laboratory. Haldane later remarked on the "excellent cuisine". In other words this venue meant that the effects of high altitude, that is, the sustained hypoxia, could be studied in the absence of complicating factors such as difficult access, limited laboratory instrumentation, severe cold, and poor food that had to be prepared by the investigators. Incidentally the same principles were enunciated by Griffith Pugh (1909–1994) when he planned the Silver Hut Expedition 50 years later in 1960 [21]. In this case a group of physiologists lived at an altitude of 5800 m for several months in a warm environment with ample food.

The design of the Pikes Peak expedition was classical. Physiological measurements were first made at sea level in Oxford or New Haven. Then there was a short period at Colorado Springs altitude 1830 m, where equipment was readied, following which the four investigators Douglas, Haldane, Henderson and Schneider ascended together on the cog railway (Fig. 16.3). They spent five uninterrupted weeks at high altitude studying the changes during acclimatization, and then descended via Colorado Springs back to sea level where further studies were carried out. C.G. Douglas (1882–1963) was a collaborator of Haldane's in Oxford. E.C. Schneider (1874–1954) had previously been at Yale but at the time of the expedition was on the faculty of Colorado College in Colorado Springs. This was convenient because it meant that supplies or equipment could be brought up from nearby if necessary. An extensive series of projects were carried out on the summit and these will be summarized here. The official report [5] which is easily available from JSTOR is 133 pages long and written in a relaxed leisurely style that modern editors would not condone.

Fig. 16.3 Members of the expedition. From left to right, Henderson taking samples of alveolar gas, Schneider sitting and recording his respiration, Haldane standing, Douglas wearing a "Douglas bag" to collect expired gas for the determination of oxygen consumption during climbing. From Henderson [18]

16.1 Acute Mountain Sickness

All four investigators developed acute mountain sickness shortly after arriving at the summit with headache that was severe at times, loss of appetite, nausea, intestinal disturbances, hyperventilation, cyanosis and periodic breathing. However after 2 or 3 days, most of the symptoms disappeared although the hyperpnea and dyspnea remained as did the periodic breathing in some subjects. Many visitors arrived at the summit by train and some of these showed extreme acute mountain sickness with nausea, vomiting, and fainting. The official description [5] includes "… the scene in the restaurant and on the platform outside can only be likened to that on the deck or in the cabin of a cross-channel steamer during rough weather".

16.2 Alveolar Gases During Acclimatization

Extensive measurements of the PO_2 in alveolar gas and arterial blood were made. Figure 16.4 shows alveolar gas partial pressures from Douglas during the whole period. This is a reproduction of the top panel of Fig. 16.6 in the official report. The results in the other three subjects were similar. Note that after 3 or 4 days in Colorado Springs, the PCO_2 fell from the sea level value of about 40 to about 32 mmHg and continued its decline during most of the five weeks ending at a value

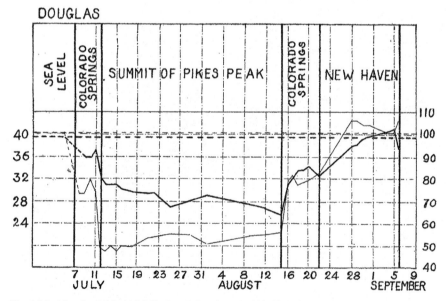

Fig. 16.4 Alveolar PCO_2 (thick line) and alveolar PO_2 (thin line) for Douglas. The other members of the expedition showed similar changes. From Douglas et al [5]

of about 25 mmHg. Consistent with this the alveolar PO$_2$ fell precipitously to about 50 mmHg on arrival at the summit and then gradually increased over the five week period ending at about 56 mmHg. On descent to Colorado Springs both the PCO$_2$ and PO$_2$ increased, but it is interesting that over the three week period after the expedition when the subjects were in New Haven at sea level, the PCO$_2$ only gradually returned to its normal value of 40. This was a particularly useful finding because there are relatively few data on the deacclimatization process on return to sea level, and in fact these are some of the best data in the literature albeit 100 years old!

Of course the investigators recognized that the cause of the reduced PCO$_2$ shown in Fig. 16.3 on ascent to altitude was hyperventilation. However they were at a loss to explain the mechanism and there is a somewhat tortuous description of possible reasons such as lactic acid from the hypoxia stimulating the respiratory center, or resetting of the exciting threshold for CO$_2$. Of course they were unaware of the critical role of the peripheral chemoreceptors that were only discovered 15 years later by the two Heymans [19].

16.3 Arterial PO$_2$ and Comparisons with the Alveolar Values

Extensive measurements of the arterial PO$_2$ were made by an indirect technique involving the inhalation of carbon monoxide and then deriving the arterial PO$_2$ from the relation between the hemoglobin bound to oxygen and carbon monoxide. Haldane had developed this method because of his interest in the toxic effects of carbon monoxide in mines. However in the event it was unfortunate that this indirect technique was used because the measurements were erroneous for reasons that are still not fully understood. Another method for measuring the arterial PO$_2$, that is by using an aerotonometer, had been described by the Kroghs [20] and in retrospect this would have been preferable.

The result of the comparisons of alveolar and arterial PO$_2$ on the summit gave an arterial value that averaged about 36 mmHg above the alveolar value and this was interpreted as evidence for oxygen secretion by the lung. Of course this was an error but it should not detract from many other important observations made during the expedition. Haldane actually believed in oxygen secretion throughout his life and indeed a whole chapter of the second edition of his book *Respiration* [16] published a year before his death was devoted to oxygen secretion. One of the reasons why he took this view was that he was aware that the swim bladder of the fish often contained gas with a much higher PO$_2$ than the surrounding water. He also knew that the swim bladder was a diverticulum of the gastrointestinal tract as was the lung, and so he argued that if the fish could do it, humans could too. In fact Haldane was something of a vitalist and in the introduction to the first edition of *Respiration* (Haldane, 1921) we read "the mechanistic theory of life is now outworn and must soon take its place in history as a passing phase in the development of biology". Haldane's attitudes to vitalism have been discussed elsewhere [22].

16.4   Respiratory Gas Exchange During Rest and Exercise

Many measurements were made during the exercise of walking uphill along the railway track. In particular, the extreme hyperventilation that this elicited was measured. The investigators concluded that the oxygen cost of a given work level appeared to be the same as at sea level, a finding that has been confirmed many times since. There was a surprising amount of attention devoted to the respiratory exchange ratio (CO_2 output divided by O_2 uptake). The value at rest was shown to have a mean of 0.83 but on exercise it increased to nearly 1 or even exceeded this which the investigators recognized was an unsteady state. One of the reasons for the interest in this ratio at rest may have been that this was used to calculate the alveolar PO_2 from the PCO_2 when analyzing alveolar gases as for example in the studies by Mabel FitzGerald described later.

16.5   Periodic Breathing

Extensive observations of periodic breathing were made although the phenomenon had been described previously in both the Observatoire Vallot and the Capanna Margherita. However the investigators were able to show that the administration of oxygen by mask abolished the periodic breathing. This result was in contrast to measurements made by Mosso in the Capanna Margherita who thought that the periodic breathing was caused by the low PCO_2. The report states that the different result was probably because Mosso did not deliver a high enough concentration of oxygen. It is interesting that Douglas apparently needed a period of voluntary hyperventilation to elicit periodic breathing.

16.6   Estimates of Blood Circulation Rate

Attempts were made to measure the output of the heart in response to the hypoxia but only indirect methods were available. These included measurements of heart rate, pulse pressure, and a crude form of ballistocardiograph. Nevertheless the conclusion was that the blood circulation rate in acclimatized subjects at rest was similar to that at sea level and in the event this agreed with later measurements using modern techniques.

16.7   Blood Studies

Extensive measurements of the hemoglobin concentration, oxygen capacity and blood volume were made. Fig. 16.5 shows the results from Douglas. The other subjects showed similar findings. It was shown that the hemoglobin concentration and oxygen capacity of the blood increased over the period of five weeks at high altitude. An interesting observation was the rapid increase in hemoglobin concentration that occurred within the first few days, and this was correctly attributed to a reduction in blood volume. The decrease in blood volume has been confirmed in many studies but the cause is not fully understood. Insensible fluid loss because of hyperventilation is one factor, and diuresis may be another.

16.8   Principal Factors in the Process of Acclimatization

These were summarized at the end of the report [5]. The first factor was the increased secretory activity of oxygen by the lung. As we have seen this was an error resulting from the method of measuring the arterial PO_2. Second was the increase in ventilation which was attributed to the increased exciting threshold of alveolar carbon dioxide. Here the true cause was stimulation of the peripheral chemoreceptors which were not discovered until several years later. The third factor was the increased percentage hemoglobin of the blood. The mechanism of this was obscure at the time. The report correctly concluded that the process of acclimatization was rapid during the first few days but took several weeks to become complete. It was also pointed out that rapid ascent to high altitude without the advantages of acclimatization, for example in a balloon, would be life-threatening.

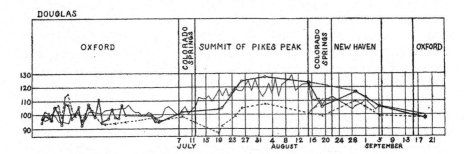

Fig. 16.5 Changes in the blood for Douglas on Pikes Peak. The pre-expedition values are shown as 100. The continuous thick line shows the oxygen capacity of the blood, or the total amount of hemoglobin. The continuous thin line shows the percentage of hemoglobin. The interrupted line shows the blood volume. The findings from the other members of the expedition were similar. From Douglas et al [5]

Fig. 16.6 Mabel Purefoy
FitzGerald (1872–1973).
From Torrance [23]

16.9   Contributions of Mabel Purefoy Fitzgerald (1872–1973)

Whether Mabel FitzGerald (Fig. 16.6) was actually a member of the Pikes Peak expedition is arguable. She offered her services and was invited to join the expedition but in fact did not accompany the four men to the summit. As a result she is not mentioned in the official report [5] and her extensive contributions were written up under her own name [8, 9].

She was born near Basingstoke in 1872 and by 1897 was studying chemistry and biology at Oxford, first with Francis Gotch and Gustav Mann, and later with Haldane. She did well in her examinations and her studies resulted in several publications [6, 7, 10]. Nevertheless she was not allowed to take a degree because this was not possible for women at that time. In 1907 she obtained a scholarship to move to the Rockefeller Institute in New York, and in 1911 she learned of the upcoming Pikes Peak expedition and offered her services. Why she did not join the four men on the summit is not entirely clear. It is said that this was because she was unchaperoned but more likely her presence there would have complicated the living conditions. In the event Haldane suggested that she visit various mining camps in Colorado at somewhat lower altitudes and measure alveolar gas concentrations and hemoglobin levels in blood. She did this accompanied only by a mule and indeed it would be very difficult to think of a more hazardous situation for a single woman than that.

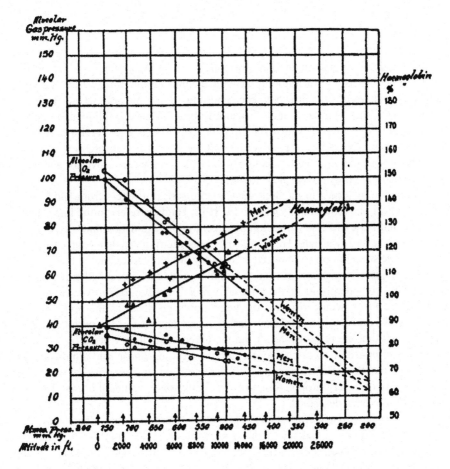

Fig. 16.7 FitzGerald's graphs of alveolar PO$_2$ and PCO$_2$, and hemoglobin concentration plotted against barometric pressure. The points for PO$_2$ are calculated from the PCO$_2$ values assuming an respiratory exchange ratio of 0.83. This graph is a composite of two published by FitzGerald [9]. The composite was prepared by Haldane for his Silliman lectures at Yale in 1916. From Haldane [13]

FitzGerald made very extensive measurements of alveolar PCO$_2$ at altitudes between 1524 and 4300 m, and remarkably these are still cited today. She used a simplified Haldane apparatus which allowed her to measure the alveolar PCO$_2$ in the last expired gas within a few minutes. This is a technique that Haldane had developed for measuring the PCO$_2$ in foul air in slums and mines [11]. She also measured the hemoglobin concentration of blood using a calorimetric method. The results that she obtained in 1911 were subsequently supplemented in 1913 when FitzGerald went to the Appalachians in North Carolina, and where she was able to obtain data at somewhat lower altitudes. Her results are shown in Fig. 16.7. It can be seen that both the measured PCO$_2$ and the calculated PO$_2$ fell linearly with barometric pressure. She also noted the fact that the PCO$_2$ in women was less than in men although she was not aware of the mechanism. Figure 16.7 also shows the linear increase in hemoglobin concentration as barometric pressure is reduced. These results are standardized in that the value for men at sea level is given as 100. Again the lower values in women are shown although there are relatively few data.

A fascinating feature of Mabel FitzGerald's contributions is that there is a very extensive archive of her correspondence and much other material in the Bodleian Library in Oxford. This came about because Robert Torrance, a fellow and tutor of St. John's College, Oxford, was invited to look at her effects after her death in 1973. He arranged for the papers to be sent to the Bodleian where there are now some 40 large boxes of material that is uncatalogued. This is a treasure trove for someone who would be willing to write her biography.

Among the materials in the archive is a large series of letters from Haldane to FitzGerald. In one letter dated October 14, 1913, Haldane refers to the linear plots of PCO_2 and PO_2, and also hemoglobin concentration, against barometric pressure reported in her 1913 article. He states "It seems pretty clear that your published curves come straight down to normal just as they are represented in the Phil. Trans. [her 1913 paper]. I can't imagine that there can be a sudden kink at 760 mm pressure but we can't go underground to the infernal regions!" He goes on to wonder whether the alveolar PCO_2 would rise and the hemoglobin go down as the pressure was increased. As noted earlier, Haldane was of course unaware of the peripheral chemoreceptors that are responsible for the increased ventilation caused by hypoxia. Since the response of these chemoreceptors is very non-linear, and almost absent above a normal arterial PO_2, it seems unlikely that Haldane's supposition is correct. The response of the erythropoietin-generating cells to PO_2 values above normal has apparently not been described.

In another letter dated August 17, 1914, written just a few days after Britain declared war on Germany, Haldane wrote "As you may imagine, this war has come on everybody and everything here like a deluge, and of course upset all plans. All are doing what they can to help and the country is behind the government absolutely. We mean to do our best to crush Prussian militarism, without, if possible, crushing the real Germany". Haldane had a high regard for German physiology in 1914 and had worked with the physiological chemist, Ernst Salkowski, in Berlin in 1886.

Torrance studied the FitzGerald archive and wrote an article titled "Mabel's normalcy: Mabel Purefoy FitzGerald and the study of man at altitude" [23]. The use of the term normalcy seems curious to me. Apparently Torrance was alluding to high altitude acclimatization which in a sense returns the organism to its normal state. However Mabel seems quite exceptional to me.

There is a curious coincidence that is worth mentioning. When FitzGerald and her sisters moved to Oxford in 1896 they took the house at 12 Crick Road in an attractive part of north Oxford. Strangely enough, Haldane was living next door in 11 Crick Road. The houses remain the same today except that there is now a blue plaque on Haldane's house. Curiously Haldane and Mabel FitzGerald apparently did not meet at this time in spite of the fact that they were next-door neighbors. Haldane moved to another house in Oxford a few years later and of course got to know FitzGerald well when she worked with him. In fact one of the projects that they did together was to develop normal values for alveolar PCO_2 in men, women

Fig. 16.8 Mabel FitzGerald after receiving her MA
degree from Oxford University at the age of 100. From
Dill [4]

and children, and one of the subjects was Haldane's son, J.B.S. Haldane, who was
then 12 years old [10].

FitzGerald was lost sight of after this for some time until Daniel Cunningham of
Oxford was organizing the Haldane Centenary Symposium that was held in Oxford
in 1961. He looked in the Oxford telephone directory and was surprised to find that
she was living in Crick Road near the University Laboratory of Physiology. She
was invited to the Symposium and the resulting book [3] shows a photograph of
her with Douglas and J.B.S. Haldane, the son of J.S. Haldane. After she resurfaced,
the Oxford authorities were apparently embarrassed at her former treatment and she
was given an honorary MA degree in 1972 at the age of 100 (Fig. 16.8).

16.10   Conclusion

The Anglo-American Pikes Peak expedition of 1911 was one of the most important early physiological expeditions to high altitude and resulted in many important
advances. The design of the expedition was classical in that measurements were
first made at sea level then for a continuous period of five weeks at an altitude of
4300 m, and then at sea level again to study the deacclimatization process. The alveolar PCO_2 fell abruptly on ascent to the summit and then decreased more slowly
over the five week period to a value of about 25 mmHg. Following the return to
sea level the PCO_2 returned gradually over some three weeks to its pre-ascent level.
The cause of the hyperventilation was attributed to the resetting of the set point for

PCO_2 but of course the investigators were not aware of the role of the peripheral chemoreceptors which were only discovered some 15 years later. Arterial PO_2 was determined by an indirect technique which unfortunately was erroneous. The conclusion that the lung secreted oxygen should not detract from the other important advances that were made. Exercise studies showed that the oxygen cost of a given work level was the same as at sea level. Periodic breathing was shown to be common and abolished by oxygen administration. Extensive blood studies showed an increase in hemoglobin concentration over the five weeks and it was recognized that the rapid initial increase was caused by a reduction of blood volume. The three principal factors in the acclimatization process were reported as secretion of oxygen by the lung, a large increase in ventilation, and an increase in hemoglobin concentration of the blood. Mabel FitzGerald was invited to join the expedition but did not accompany the four men to the summit. Instead she made extensive studies of alveolar gas and hemoglobin concentrations in mines at lower altitudes and her results are still cited today. The Pikes Peak expedition had an enormous influence on later high altitude studies.

Footnote Paton Lecture given at the Physiological Society meeting in Oxford, July 2011.

References

1. Bert P. La pression barometrique. Paris: Masson; 1878a.
2. Bert P. La pression barometrique. Paris: Masson; 1878b. (Hitchcock MA & Hitchcock FA (1943), English translation. College Book Co., Columbus, OH. Reprint by the Undersea Med. Soc., 1978).
3. Cunningham DJC, Lloyd BB, editors. The regulation of human respiration: The proceedings of the J.S. Haldane Centenary Symposium. Blackwell; 1963.
4. Dill DB. Mabel Purefoy FitzGerald—our second centenarian. Physiologist. 1973;16:247–8.
5. Douglas CG, Haldane JS, Henderson Y, Schneider EC. Physiological observations made on Pike's Peak, Colorado, with special reference to adaptation to low barometric pressures. Philos T R Soc Lon B. 1913;203:185–318.
6. FitzGerald MP. An investigation into the structure of the lumbo-sacral-coccygeal cord of the Macaque monkey (Macacus sinicus). Proc Roy Soc Lond Ser B. 1906;78:88–144.
7. FitzGerald MP. The alveolar carbonic acid pressure in diseases of the respiratory and circulatory systems. J Pathol Bacteriol. 1909–1910;14:328–43.
8. FitzGerald MP. The changes in the breathing and the blood at various altitudes. Trans R Soc Lond Ser B. 1913;203:351–71.
9. FitzGerald MP. Further observations on the changes in the breathing and the blood at various high altitudes. Proc Roy Soc Lond Ser B. 1914;88:248–58.
10. FitzGerald MP, Haldane JS. The normal alveolar carbonic acid pressure in man. J Physiol (Lond.). 1905;32:486–94.
11. Foster CL, Haldane JS, editors. The investigation of mine air. Philadelphia: Lippincott; 1905.
12. Haldane JS. Life and mechanism. Mind. 1884;9(33):27–47
13. Haldane JS. Oxygen and environment. Silliman lectures of 1916. New Haven: Yale University Press; 1917.
14. Haldane JS. Respiration. New Haven: Yale University Press; 1922.

15. Haldane JS, Priestley JG. The regulation of the lung-ventilation. J Physiol. 1905;32(3–4):225–66.
16. Haldane JS, Priestley JG. Oxygen secretion in the lungs. In: Respiration, 2nd edn. New Haven: Yale University Press; 1935. pp. 250–96.
17. Henderson Y. Life at great altitudes. Yale Rev. 1914;3:759–73.
18. Henderson Y. Adventures in respiration. Baltimore: Williams & Wilkins; 1938.
19. Heymans J-F, Heymans C. Sur les modifications directes et sur la régulation réflexe de l'activité du centre respiratoire de la tête isolée du chien. Arch Int Pharmacodyn Ther. 1927;33:272–370.
20. Krogh A, Krogh M. On the tensions of gases in the arterial blood. Skand Archiv Physiol. 1910;23:179–92.
21. Pugh LGCE. Physiological and medical aspects of the Himalayan Scientific and Mountaineering Expedition, 1960–61. BMJ. 1962;2:621–33.
22. Sturdy, SW. Biology as a social theory: John Scott Haldane and physiological regulation. Brit J Hist Science. 1998;21:315–340.
23. Torrance RW. Mabel's normalcy: Mabel Purefoy FitzGerald and the study of man at altitude. J Med Biogr. 1999;7(3):151–65.
24. Vivian R. L'Epopée Vallot au Mont Blanc. Paris: Denoel; 1986.
25. West JB. High life: A history of high-altitude physiology and medicine. New York: Oxford University Press; 1998.

Chapter 17
Alexander M. Kellas and the Physiological Challenge of Mt. Everest

Abstract Alexander M. Kellas (1868–1921) was a British physiologist who made pioneering contributions to the exploration of Everest and to the early physiology of extreme altitudes, but his physiological contributions have been almost completely overlooked. Although he had a full-time faculty position at the Middlesex Hospital Medical School in London, he was able to make eight expeditions to the Himalayas in the first two decades of the century, and by 1919 when the first official expedition to Everest was being planned, he probably knew more about the approaches than anybody else. But his most interesting contributions were made in an unpublished manuscript written in 1920 and entitled "A consideration of the possibility of ascending Mount Everest." In this he discussed the physiology of acclimatization and most of the important variables including the summit altitude and barometric pressure, and the alveolar Po_2, arterial oxygen saturation, maximal oxygen consumption, and maximal ascent rate near the summit. On the basis of this extensive analysis, he concluded that "Mount Everest could be ascended by a man of excellent physical and mental constitution in first-rate training, without adventitious aids [supplementary oxygen] if the physical difficulties of the mountain are not too great." Kellas was one of the first physiologists to study extreme altitude, and he deserves to be better known.

At the beginning of this century it was customary for explorers and geographers to cite three unattained goals: reaching the North and South Poles and ascending Mt. Everest. The first two were readily accomplished within the first decade, but the last took more than half the century.

When the Royal Geographical Society met in London on May 18, 1916, Alexander M. Kellas, Lecturer in Chemistry at the Middlesex Hospital Medical School, delivered a paper entitled "A consideration of the possibility of ascending the loftier Himalaya." The President of the Society, Sir Thomas Holdich, introduced Kellas with the following words: "The poles having been reached, it is obvious that the next object of importance on the earth's surface to be attacked by adventurers is the highest mountain in the world. There are, perhaps I should say not unfortunately, a good many difficulties in the way of reaching it. In the first place, you have to deal with a Government which has up to the present time forbidden you to approach within 100 miles of the mountain's base. In the next place, the mountain itself is

© American Physiological Society 2015

J. B. West, *Essays on the History of Respiratory Physiology,*
Perspectives in Physiology, DOI 10.1007/978-1-4939-2362-5_17

probably—though of this we have no sufficient evidence—of considerable difficulty; and there is thirdly the main obstacle, the effect of the rarity of the air at great heights on the human frame" [15].

Kellas had been invited by A. R. Hinks, Secretary of the Royal Geographical Society, to prepare this lecture. Hinks wrote to him on April 14, 1916, as follows[1]: "If you could give us a paper with some general title like 'The possibilities of climbing above 25,000 ft' it would be the subject of first-rate interest," and he went on, "especially since no-one perhaps in the world combines your enterprise as a mountaineer and your knowledge of physiology." Hinks was remarkably perceptive. Indeed, Kellas was unique in his knowledge of the physical problems of approaching Mt. Everest and the physiological challenge presented by its enormous altitude.

The story of the ten major expeditions to Mt. Everest beginning with the reconnaissance in 1921 and culminating with the first successful ascent in 1953 is one of the sagas of the twentieth century. Most historians have, not surprisingly, concentrated on the problems of finding the best route and the technical difficulties of reaching the summit. However, of equal importance was the gradual understanding of the physiological factors that made it possible for humans to reach these great heights.

It is remarkable that one man, Alexander M. Kellas, made major contributions both to the geography of the approaches to Mt. Everest and to the physiology of human performance at extreme altitude. In fact, when the first official reconnaissance was being planned in 1919, Kellas probably knew more about ways of getting to the mountain than anybody else. But in addition, Kellas had given more thought to the physiological problems of reaching the summit and had climbed more often to altitudes above 20,000 ft (6100 m) than anyone. His paper "A consideration of the possibility of ascending the loftier Himalaya" [15] and particularly his unpublished manuscript "A consideration of the possibility of ascending Mount Everest" were landmarks in the early physiology of extreme altitude. Despite this, his name is almost unknown in high-altitude physiology or indeed in physiology at all.

Alexander Mitchell Kellas (Fig. 17.1) was born in Aberdeen, Scotland, on June 21, 1868. His father, James Fowler Kellas, was Secretary and Superintendent of the Mercantile Marine Company in Aberdeen and married Mary Boyd Mitchell. They had nine children (six sons and three daughters) of whom Alexander was the second oldest. One of Alexander's brothers, Henry, was an advocate (attorney) in Aberdeen and looked after Alexander's affairs when he was abroad. His letters are a valuable source of information about Alexander.[1] Alexander never married, but Henry had four children, three of whom are still alive. Interestingly, one became British Ambassador to Nepal; another became a Doctor of Science and lectured in histology; and the third is a retired general practitioner in Swindon, Wiltshire, UK.[2]

Alexander Kellas was educated at Aberdeen Grammar School and in 1889 went to Edinburgh to study for two years at the University and the Heriot-Watt College

---

[1] Archives of the Royal Geographical Society, 1 Kensington Gore, London SW7 2AR, UK; Kellas file.

[2] The last, Henry R. Kellas, died after this manuscript was completed.

Fig. 17.1 Alexander M. Kellas
(1868–1921). Archives of the
Royal Geographical Society,
by permission

there. He then moved to University College, London, where he obtained his BSc in 1892. He already had a stepbrother in London, James N. F. Kellas; Alexander used to visit the family on Sunday evenings and later brought them small presents from Tibet and India.[3]

For a time Alexander Kellas was a Research Assistant at University College, London, with Sir William Ramsey, who worked on the chemistry of the noble gases. In 1895 Kellas went to Heidelberg University in Germany to study for his DPhil degree, which he received in 1897. He was appointed Lecturer in Chemistry at the Middlesex Hospital Medical School in London in 1900 and held that appointment until 1919. The University of London awarded him the degree of DSc in 1918.

Kellas developed an early love for hill walking. His brother Henry wrote to the Secretary of the Royal Geographical Society after Alexander's death as follows[4]: "As a boy, the hills seemed to have a fascination for him, and he was ardently devoted to walking and climbing, first among the Grampians, and later on in Wales and on the continent. When he was only 14, he and a younger companion slept under the shelter stone in the Cairngorms, ascending Ben Macdhui and other mountains." His first expedition to the Himalayas was in 1907, when he spent the months from August to October exploring mountains in Kashmir and Sikkim, the region of India north of Darjeeling and east of Mt. Everest. This was followed by further visits in the summers of 1909, 1911, 1912, 1913, 1914, and 1920 and the early spring of 1921.

It was impossible for Kellas to visit India between 1914 and 1918 because of the First World War. However, we know that he had planned expeditions to Sikkim to take place in both 1915 and 1916 because a typewritten proposal for these still exists. The proposal sets out an ambitious plan including making a large-scale map

---

[3] Mary Elsie Kellas, daughter of James N. F. Kellas, wrote a family history that was never published but contains useful information about the family.

[4] Kellas wrote an 18-typewritten page account of this adventure, which is still extant. Actually, the 3-day trek took place in August 1885, so Kellas was just 17 year old at the time.

of the north and west approaches to Everest and obtaining a series of photographs showing the best prospects for climbing the mountain.

Kellas particularly explored the mountains of north Sikkim and thus became familiar with the approaches to the Everest region from the east and north [7]. He also spent some time in southern Tibet especially near the village of Kampa Dzong. Kellas was not a prolific writer but accounts of some of these expeditions were published in the Alpine Journal and the Geographical Journal [11–14, 16, 18, 19].

These explorations to extremely remote regions were carried out during the summer vacations, and it is remarkable that Kellas was able to find the time. The expeditions themselves lasted 3 or 4 months and at that time it would take some 3 weeks to sail from London to India and an additional 3 weeks to return. In spite of this, Kellas was described as a conscientious faculty member who was commended for his teaching [26]. However, there are indications in one of his letters of tensions between the medical school administration and himself over the amount of extra leave he requested for these expeditions.[5]

On most of these expeditions, Kellas went alone accompanied only by some native porters. Occasionally he had an English companion; for example, Henry Morshead accompanied him on his expedition to Kamet in 1920. Kellas was apparently the first Himalayan explorer to recognize the immense value of the Sherpas, the people of Tibetan origin who live near Mt. Everest, to exploration in these regions [27]. Since his early contact with the Sherpas, they have played critical roles in expeditions to Everest and other Himalayan mountains.

From a physical point of view, Kellas must have been remarkably tough to withstand the rigors of these small expeditions to remote areas at great altitudes. It is believed that he probably made more ascents over 20,000 ft than anyone else [26]. However, his appearance belied his athletic abilities. The well-known British climber, George Leigh Mallory, described him in a letter to his wife during the early stages of the 1921 Everest reconnaissance expedition thus: "Kellas I love already. He is beyond description Scotch and uncouth in his speech—altogether uncouth. He arrived at the great dinner party 10 min after we had sat down, and very dishevelled, having walked in from Grom, a little place four miles away. His appearance would form an admirable model to the stage for a farcical representation of an alchemist. He is very slight in build, short, thin, stooping, and narrow-chested; his head… made grotesque by veritable gig-lamps of spectacles and a long-pointed moustache. He is an absolutely devoted and disinterested person" [30].

The first two decades of this century were an exciting time in the physiology of extreme altitude. In the late 1900's many people were of the opinion that ~21,500 ft (6,500 m) represented the maximum height attainable by humans. Kellas quotes

---

[5] Letter from Kellas to Hinks, September 29, 1917: "When my assistant comes back at the end of the war, there should be a possibility of getting leave of absence provided they [medical school] rescind a minute of the School Council, which I believe prevents me from asking for extra leave. If they do not, their attitude to my Himalayan aspirations has been so unsympathetic, that I would be forced to look for another post if I wished to do more work in connection with high altitudes…. A scientific man is not necessarily merely a scientific factotum." Royal Geographical Society Archives; Kellas file.

the President of the Alpine Club, T. W. Hinchliff, who wrote in 1876 after visiting Santiago, Chile, as follows[6] "I could not repress a strange feeling as I looked at Tupungato (21,550 ft) and Aconcagua (23,080 ft) and reflected that endless successions of men must in all probability be forever debarred from their lofty crests.... Those who, like Major Godwin-Austen, have had all the advantages of experience and acclimatization to aid them in attacks upon the higher Himalaya agree that 21,500 ft is near the limit at which man ceases to be capable of the slightest further exertion" [9].

However in 1909 the Duke of the Abruzzi led an expedition to the Karakorum that was designed "to contribute to the solution of the problem as to the greatest height to which man may attain in mountain climbing," as the Duke's biographer put it [4]. His party reached 7500 m (24,600 ft) without supplementary oxygen, far higher than anyone had been before. This feat astonished climbers and physiologists alike. The English physiologists Douglas, Haldane, and their co-workers [5] estimated from the reported barometric pressure of 312 Torr that the alveolar P_{O_2} was only 30 Torr, and they concluded that adequate oxygenation of the blood would be impossible under these conditions without active secretion in the lung. However, this conclusion was disputed by Marie Krogh [22], who had recently developed a technique for measuring the diffusion characteristics of the lung using carbon monoxide. She argued that Douglas and his colleagues had markedly underestimated the pulmonary diffusing capacity. This was just one of the exchanges between the two camps arguing for and against active secretion of oxygen by the lung in a controversy that lasted into the mid-1930's.

Kellas had trained in chemistry, and he taught this subject to medical students at Middlesex Hospital Medical School for some 18 years. At this time, many students would have received relatively little chemistry before coming to medical school, and Kellas lectured in both inorganic and organic chemistry. His obituary in the *Mddlesex Hospital Journal* [25] refers to his great interest in teaching, especially to the weaker students, and he wrote three short textbooks to assist with his laboratory classes in chemistry.

Kellas' early research was on various aspects of organic and inorganic chemistry. His DPhil thesis was devoted to the esterification of benzoic acids, and after his period with Sir William Ramsey, he studied the distribution of argon in human expired gas and in various vegetable and animal substances. He also carried out an extensive investigation on "The molecular complexity of liquid sulphur," which was published in 1891.

Kellas gradually became increasingly interested in the physiology of high altitude, which was not surprising given his scientific background and his immense personal experience at extreme altitude. He became an authority on acute mountain sickness and contributed two short articles on this subject [20]. At one stage he even began studying to take a medical degree at the Middlesex Hospital Medical School[7].

---

[6] Unpublished manuscript. "A consideration of the possibility of ascending Mount Everest." Royal Geographical Society Archives; Kellas file.

[7] Letter to A. R. Hinks dated October 9, 1917: "At present I am reading Anatomy and Physiology in my very limited spare time, and might manage with hard work to take a medical degree in 3

In 1918 he collaborated with J. S. Haldane, one of the most eminent British physiologists of the day, on a study of acclimatization obtained by repeated exposures to low pressure in a chamber at the Lister Institute in London. A description of this work was published in the *Journal of Physiology (London)* and is referred to again below [8].

Although Kellas' early Himalayan expeditions were primarily exploratory in nature, he became progressively more interested in using them to study the physiological problems of extreme altitude. At one stage he wrote a tentative proposal for a medical scientific expedition to remain for a period of several months at an altitude of 20,000 ft to study the physiology of acclimatization. In a letter to Hinks dated October 9, 1917[1], he suggested possible locations for a long-term camp, for example, the summit of Kanchenjhau (22,700 ft), and added "it might even be possible to drag up in small parts the framework of a small wooden hut." He planned to carry out the types of experiments that Haldane and his colleagues had made on Pikes Peak. It is remarkable that such an expedition was not organized until over 40 years later, and even then the altitude was a more modest 19,000 ft. [28].

The major published contribution in the area of physiology by Kellas was his paper entitled "A consideration of the possibility of ascending the loftier Himalaya," which was read at the meeting of the Royal Geographical Society on May 18, 1916, and subsequently published in the *Geographical Journal* [15]. The fact that the article was published in a geographical rather than physiological journal might indicate that Kellas thought of himself more as an explorer than a physiologist, although another factor may well have been that he received some financial support for his Himalayan expeditions from the Royal Geographical Society and presumably wanted to report his findings directly to them. The paper contains many features of interest that throw light on the state of high-altitude physiology at the time. For example, Kellas correctly predicts the barometric pressure on the summit of Mt. Everest to be 251 Torr (for a mean air temperature of 0 °C) based on the work of FitzGerald [6]. This correct value contrasts with later estimates made in the 1940's, when the pressure was thought to be much lower based on the inappropriate use of the standard atmosphere.

The article in the *Geographical Journal* created a great deal of interest, particularly among climbers, and it was extensively reviewed in a subsequent issue of the *Alpine Journal* [15]. However, it is probable that Kellas' choice of the *Geographical Journal* resulted in relatively few physiologists seeing the article, and this no doubt contributed to the fact that Kellas' work is not well known.

Kellas' most interesting study was entitled "A consideration of the possibility of ascending Mount Everest". Although two complete manuscripts survive, one in the archives of the Royal Geographical Society and the other in the archives of the Alpine Club library,[8] the paper was never published. One manuscript was sent to A. R. Hinks on April 24, 1920, at a time when Kellas was very busy preparing for what

---

years or thereby." Royal Geographical Society Archives; Kellas file.

[8] The two manuscripts are very similar though not identical; the one in the archives of the Alpine Club appears to be a slightly later version with more notes and corrections.

were to be his last two Himalayan expeditions. It is not clear why the manuscript was never published, although it may simply be that because he died abroad and never returned to England, no one pressed for publication. A French translation of the manuscript was published in a very obscure place, the proceedings of the Congres de l'Alpinisme held in Monaco in 1920 [17]. Kellas apparently presented the paper at this meeting when he was en route to India.[9]

Kellas begins by stating his main question: "Is it possible for man to reach the summit of Mount Everest without adventitious aids [by which he meant supplementary oxygen], and if not, does an ascent with oxygen appear to be feasible?" He divides the problems to be overcome into two groups: "I. Physical difficulties" and "II. Physiological difficulties," and he considers each of these in considerable detail.

The introduction to the paper is interesting because he cites the altitude of Everest as 29,141 ft (8882 m). The official altitude at that time was 29,002 ft, having been determined by Sir Andrew Waugh, Surveyor-General of India, in 1852. Waugh succeeded Sir George Everest, the Surveyor-General after whom the mountain was named; this was when Everest was first identified as the highest mountain in the world. However, in 1909 Sir Sidney Burrard, a later Surveyor-General, recomputed the altitude from the original sightings but using different values for the refractive errors caused by the bending of light in the lower atmosphere, and he came up with the figure of 29,141 ft. It is not clear why Kellas preferred the higher altitude when most people accepted the official lower value. It was not until the 1950's that D. L. Gulatee used new sightings to establish the now-accepted altitude of 29,028 ft (8848 m).

The first physical difficulty discussed by Kellas is that of access to the mountain. He points out that "The mountain has so far never been visited by white men, and it is unlikely that any mortal has reached an altitude of even 20,000 ft (6096 m) upon it." However, with his unrivaled experience of the geography of the region, Kellas was able to suggest three possible routes from Darjeeling depending on whether the authorities of Tibet or Nepal gave permission. He believed the former to be more likely, which indeed proved to be true. Kellas' knowledge of the area was supplemented by explorations made by Captain J. B. Noel, who at one stage disguised himself as a Tibetan to reconnoiter the Himalayan range east of Everest [24].

Kellas then goes on to discuss the best time of the year for an ascent. He correctly reasoned that the monsoon period would be impossible because of the heavy snow-

---

[9] Whether Kellas actually attended the meeting is unclear. His name is not included in the list of "adherents present au congres" or acknowledged in the opening or closing sessions. The dates of the congress were May 1–10, 1920. Hinks stated in a letter[1] to Morshead of May 3, 1920: "he [Kellas] sailed from Liverpool on 30 April," and Kellas wrote to Hinks[1] on July 12: "On arrival in Darjeeling in the end of May I at once started for Jongri and Alukthang." It is conceivable that Kellas left the boat in a Mediterranean port, e.g., Marseilles and presented the paper at the end of the congress (it is the last paper listed in the proceedings). It is also possible that someone else presented his paper, but there seems to have been no one at the congress who knew Kellas well and who would have been interested in giving such a technical paper. The published article contains many typographical errors suggesting that Kellas never saw the proofs.

fall and recommended the months immediately preceding or succeeding the summer monsoon. These have proved to be the most suitable periods up to the present time. Kellas also considered possible routes on Everest itself. Here he was at a very considerable disadvantage because no climber had been anywhere near the mountain and available photographs were taken from many miles away. The two routes to the summit selected by Kellas were the east face and the northeast ridge. The latter was a good choice and was the route attempted by all the early expeditions from the north side, but the former proved to be exceptionally difficult and indeed was not climbed until 1983.[10] Kellas also made the point that the climbing on Mt. Everest above an altitude of 26,500 ft looked easier than on some other Himalayan peaks such as Kanchenjunga, K2, and Nanga Parbat. In this he was correct.

Kellas' most interesting studies from our point of view are in the section headed "Physiological difficulties." He divided this section into four parts: (1) information from balloon ascents, (2) studies in low pressure chambers, (3) observations up to altitudes of 20,000 ft, and (4) the physiology of acclimatization to high altitude.

Under balloon ascents, Kellas briefly describes the experiences of the Englishmen Glaisher and Coxwell and the three Frenchmen led by Tissandier, who both ascended to altitudes near the summit of Mt. Everest. All five men lost consciousness, and in the ill-fated French ascent only Tissandier survived, the other two men dying from acute hypoxia. However, Kellas was quite clear that these dramatic consequences would not be expected during a climb up Mt. Everest because of the advantages coming from acclimatization. He also noted that even higher ascents in more recent times had been made safe by the balloonists inhaling oxygen.

The air chamber experiments fall into two categories. The acute exposure experiments gave results similar to those found in ballooning in that they also showed that humans could not survive the low barometric pressure on Mt. Everest without losing consciousness. Kellas quoted E. H. Starling's influential textbook [31] which stated that "the lowest limit at which life is possible corresponds to an oxygen tension in the alveoli of 27–30 mm, which is distinctly above that calculated for Mt. Everest." Parenthetically, recent work shows that these estimates are too low [34].

Kellas then briefly describes a very interesting set of experiments in which two subjects (J. S. Haldane and himself) spent 4 consecutive days in a low-pressure chamber at altitudes equivalent to 11,600, 16,000, 21,000, and 25,000 ft. This was the study referred to earlier [8] that was carried out at the Lister Institute in London and was apparently the only time that Kellas had an opportunity to collaborate with J. S. Haldane, one of the leading British high-altitude physiologists. Both subjects spent several hours in the chamber on each of the 4 days, and many interesting physiological observations were made. On the 4th day, when the pressure was only 312 Torr (corresponding to an altitude of 25,000 ft), Haldane's alveolar P_{CO2} was 19.8 and P_{O2} 30.1 Torr. Nevertheless he was able to cycle on the ergometer and do

---

[10] Kellas subsequently revised his ideas on the best route in a letter to Hinks dated January 26, 1921. He wrote: "So far as my knowledge of the mountain goes, the SE arete may probably be the best route, but the NE arete seems practicable, although of extraordinary length.[1] In any event, the SE ridge was the one selected for the first successful ascent in 1953 and again for the first 'oxygenless' ascent in 1978."

3300 foot-pounds of work for 4 min but stopped because "he was exhausted and vision was becoming blurred." This work rate corresponds to 456 kg.m.min^{-1} which is not a great deal less than the 600 kg.m.min^{-1} measured as the maximal work rate for subjects at an equivalent altitude after several weeks of acclimatization [33]. Administering oxygen at a low flow rate resulted in a dramatic improvement. Haldane stated: "The light seemed to increase and there was a short apnea. At the same time the lips and face became bright red."

This remarkable series of experiments showed that as little as 3 days of acclimatization apparently increased tolerance to an altitude of 25,000 ft. The full description published in *the Journal of Physiology (London)* makes good reading and contains a number of vignettes disdained by modern-day editors. For example, on the last day in the chamber, which was the climax of the experiment (simulated altitude 25,000 ft), Haldane apparently cut the experiment short and "come out about 4 PM, as it was necessary to catch a train." Presumably this was the last train to Oxford. J. S. Haldane's son J. B. S. Haldane, who later became an eminent biologist, also took part in some of the acute experiments. On one occasion when the pressure was 330 Torr, outside observers noted that he could not stand properly and looked very blue and shaky. The paper records: "The emergency tap was therefore opened so as to raise the pressure. There is a correspondingly indignant and just legible note 'some bastard has turned tap,' after which the notes become quite legible again as the pressure rose."

An interesting sidelight on these experiments is that Kellas apparently tolerated the very low pressures much better than J. S. Haldane. In connection with this the paper states: "In the experience of one of us (A. M. K.) ordinary mountain sickness has never been experienced, either in himself or among the native carriers accompanying him, at heights up to 23,180 ft in the Himalayas." However, paradoxically, Kellas showed more cyanosis than Haldane. The article states: "A further point which appeared quite definitely was that the symptoms of anoxaemia did not coincide in the different subjects with the degree of blueness of the face. The contrast in this respect between A. M. K. and J. S. H. was very marked. The former was always much bluer than the latter, but was otherwise much less affected, and retained his faculties when the latter was helpless." This is surprising. The more intense cyanosis of Kellas suggests that he did not increase his ventilation as much as Haldane, and one would therefore have expected less tolerance to low pressure. Unfortunately, the data given in the paper on alveolar P_{CO_2} are rather sparse; they do not show obvious differences between the two subjects.

The next section of Kellas' manuscript is devoted to observations on physiologists and mountaineers up to altitudes of 20,000 ft Some 14 foolscap pages of manuscript are devoted to mountain sickness, including the symptoms and signs, variability between subjects, and etiology. This section shows that Kellas had a very good understanding of this condition, and most of what he wrote remains true today. He rightly attributes the condition to hypoxia, as first shown by Paul Bert in 1878, although he wonders whether respiratory alkalosis may also play a role. Kellas clearly recognized that adequate acclimatization was the key to tolerating the extreme oxygen lack on the summit of Mt. Everest.

The last section of the manuscript under the heading "The process of acclimatization to altitude" is of great interest. Here Kellas considers in turn the various features of physiological adaptation to extreme altitude. Although there are many factual errors because he had almost no data on which to base his work, Kellas' insight in asking all the right questions was outstanding.

The central issue as Kellas saw it was: Can sufficient physiological adaptation occur to allow a climber to ascend from a camp at ~25,500 ft to the summit (29,141 ft) in 1 day? He starts by reviewing the oxygen dissociation curve including the striking difference between hemoglobin solution and blood and the effects of carbon dioxide and lactic acid on the position of the curve. This part of the manuscript has some overlap with his earlier study that was read to the Royal Geographical Society in 1916 (several of the figures are the same). Note that this was not long after the factors affecting the position of the oxygen dissociation curve were first described; the influence of lactic acid was first understood in 1910 [2].

Kellas then goes on to ask a crucial question: What is the alveolar P_{O_2} on the summit of Mt. Everest? He correctly argues that the hyperventilation caused by hypoxia will reduce the alveolar P_{CO_2} and correspondingly increase the alveolar P_{O_2}. He concluded that for an altitude of 29,141 ft (8,882 m) the P_{O_2} would be 23.6 Torr (Table 17.1). The details of this calculation are not given, but it is clear that Kellas was strongly influenced by the measurements of Mabel P. FitzGerald, who made an extensive survey of the alveolar P_{CO_2} at various altitudes in the Rocky Mountains during the Anglo-American Expedition to Pikes Peak in 1911 [6]. From her measurements (made at much lower altitudes) she made a linear extrapolation that gave an alveolar P_{CO_2} of 19 Torr at the altitude of the summit of Mt. Everest. She also assumed a respiratory exchange ratio (R) of 0.83 at rest. It appears that Kellas used the same equation as FitzGerald

$$P_{A_{O_2}} = 0.2093\,(P_B - 47) - P_{A_{O_2}}/R$$

Table 17.1 Comparisons of Kellas' values and prediction with currently accepted values on physiology at extreme altitudes on Mt. Everest

	Kellas Value	Current Value	Comments
Summit altitude, ft	29,141	29,028	Kellas used Burrard's altitude rather
m	8,882	8,848	than official value in 1920 of 29,002 ft (8,840 m)
Summit barometric pressure, Torr	251	250–253	Kellas calculated this value for a mean air temperature of 0°C. His result was considerably more accurate than later estimates based on the Standard Atmosphere
Alveolar P_{O_2}, Torr	23.6	35	Kellas' values were much too low because he underestimated the degree of hyperventilation and assumed a normal arterial pH
Arterial S_{O_2}, %	42	70	
Arterial pH	7.4	>7.7	
$\dot{V}_{O_2\,max}$ near summit, ml/min	970	~1070	Remarkably accurate prediction
Maximum climbing rate near summit, ft/h	300–350	<330	"The last 100 m took us more than an hour to climb" Messner (22)
Recommended altitude of highest camp, ft	25,500	26,200	Messner and Habeler ascent in 1978 (22)
Recommended route (Kellas' letter dated 1921)	SE Ridge	SE Ridge	
Highest human habitation, ft (not on Everest)	20,000	19,500	Aucanquilcha mine caretakers in Chile (31)

S_{O_2}, O_2 saturation; $\dot{V}_{O_2\,max}$, maximum O_2 uptake.

where PA_{O2} and PA_{CO2} are alveolar P_{O2} and P_{CO2}, respectively, and P_B is barometric pressure. Kellas apparently assumed a P_{CO2}, of 18.6 and R of 0.83. For a barometric pressure of 267, which is the value for an assumed air temperature of 15 °C (see later), this would give his alveolar P_{O2} of 23.6 mmHg.

Parenthetically, we now know that this value of P_{CO2} is much too high and therefore the calculated P_{O2} is much too low. Actual measurements of the alveolar P_{CO2} and P_{O2} on the Everest summit gave values of ~7.5 and 35 Torr respectively [34].

Kellas pointed out that the fall in P_{CO2} at high altitude would displace the oxygen dissociation curve to the left and increase the oxygen saturation of the blood, which would be advantageous. However, he also recognized a disadvantage of this leftward shift in a footnote, namely that the dissociation of oxygen from hemoglobin in peripheral tissues would be impaired. Having said this, he argued from Barcroft's measurements on the peak of Tenerife that the position of the oxygen dissociation curve is normal at high alkalinity of the blood [1]. He therefore calculated an arterial oxygen saturation of 42%. We now know that this is far too low because of the marked respiratory alkalosis that occurs near the Everest summit [34].

Kellas then discusses the increase in erythrocytes at high altitude. Curiously, some studies he quoted had not shown this to occur, but he concluded that there would be a substantial increase, perhaps up to $8 \times 10^6/mm^3$, a 60% rise over the normal value of 5×10^6. This is actually considerably higher than is seen in acclimatized climbers.

There is a brief section on the possible secretion of oxygen by the alveolar epithelium at high altitude. Here Kellas was on difficult ground. His recent eminent collaborator, J. S. Haldane, staunchly promoted oxygen secretion, whereas other physiologists such as Krogh and Barcroft equally strongly argued for passive diffusion. Kellas tactfully concluded that the issue was not settled.

There is a short section on evidence for the more rapid circulation of blood at high altitudes during moderate exercise. Kellas points out that the pulse rate during moderate exercise is always increased at high altitude (compared with sea level), though it may be nearly normal at rest. He concludes that the higher cardiac output would enhance oxygen delivery to the tissues. Current evidence is that, after acclimatization, cardiac output returns to the sea-level values for a given work level [29].

Kellas then deals with the relationships between barometric pressure and altitude. He tabulates and plots the pressure against altitude for two values of the "mean temperature of the air column," namely 15 and 0 °C. These give barometric pressures on the Everest summit of 267 and 251 Torr, respectively. Again the details of these calculations are not given, but he probably used the Zuntz et al. formula [35] as had been done by FitzGerald [6]. Incidentally the value of 251 is very close to that of 248 Torr quoted by Paul Bert in his book *La pression barométrique* published in 1878 [3].

The next section is entitled "Limits of permanent acclimatization to high altitudes." Kellas writes that "the experiences of most mountaineers, who have climbed above 20,000 ft seemed to indicate that there is a distinct depreciation of strength above that altitude...." He therefore chose this altitude as the limit of permanent acclimatization, although he recognized that there were differences of opinion and indeed many present-day physiologists would choose a lower altitude. For example, Keys and his colleagues on the 1935 International High Altitude Expedition described a group of miners who lived at an altitude of 17,500 ft and who preferred to climb every day to the mine at 19,000 ft rather than live there [21]. Largely based on this observation, it has often been said that 17,500 ft is the highest altitude for permanent habitation.

Interestingly enough, very recent reports indicate that a small group of men now live indefinitely at an altitude of 19,500 ft [32]. They are caretakers of the Aucanquilcha mine in north Chile, and the longest period of residence has been ~2 year. These remarkable men apparently vindicate Kellas' value.

The next section of the manuscript deals with the key issues of the maximum rate of climbing and the maximum oxygen consumption at extreme altitudes. Again, the importance of this discussion is not so much that Kellas obtained nearly the right answers; this was partly by chance, since he had so few data. But Kellas recognized that the questions were crucial, and he approached them in two ways.

First, he assumed that maximum climbing rate is proportional to the oxygen concentration of the arterial blood. He stated that experience shows a maximum rate of ascent of ~1,000 ft/h at an altitude of 16,000 ft, where he calculated the arterial oxygen saturation to be 80%. He also had evidence for a rate of ascent of ~600 ft/h at an altitude of 23,000 ft, where he believed the saturation to be 60%. Since the 20% fall in oxygen saturation resulted in a 40% loss of climbing rate, he calculated that the climbing rate near the Everest summit, where he believed the saturation to be ~40%, would be 300–350 ft/h. This value of 5–6 ft/min of vertical ascent is remarkably close to present-day estimates of the maximum climbing rate near the summit. For example, Rheinhold Messner, the first man (with Peter Habeler) to reach the Everest summit without supplementary oxygen reported that "the last 100 m took us more than an hour to climb" [23]. Kellas also made the calculation a different way using the relationship between alveolar P_{O_2} and altitude. Again the results were similar.

Consistent with these predictions of maximum climbing rates, Kellas also calculated maximum oxygen uptakes at various altitudes partly based on measurements made by Haldane and his colleagues on Pikes Peak [5]. He concluded that near the summit of Mt. Everest maximum oxygen uptake would be ~970 ml/min, a value remarkably close to that of 1070 ml/min measured on well-acclimatized subjects with the same inspired P_{O_2} as the Everest summit [33]. He also pointed out that some 400 ml/min of that oxygen uptake would be required for "vital processes," that is, metabolism not related to climbing.

Kellas noted in the final section of the manuscript that this calculated maximum climbing rate would require the highest camp to be at an altitude of at least 25,500 ft and that a very early morning start would be necessary. Subsequent history bore out this prediction too. When Rheinhold Messner and Peter Habeler made their historic first ascent of Everest without supplementary oxygen in 1978 they started their final climb from the South Col at an altitude of 26,200 ft (7986 m) at 5:30 a.m.[11]

---

[11] Messner and Habeler's ascent seems to bear out Kellas'eprediction. However, in fairness it should be added that 25,500 ft was actually much too low for the final camp in the early days of

The last few lines of the manuscript are under the heading "General conclusion": "Mt. Everest could be ascended by a man of excellent physical and mental constitution in first rate training, without adventitious aids if the physical difficulties of the mountain are not too great, and with the use of oxygen even if the mountain can be classed as difficult from the climbing point of view." It took 58 years for this assertion to be proved true!

Six days after Kellas mailed a copy of this manuscript to the secretary of the Royal Geographical Society, he left for India for another Himalayan expedition and never returned to England. The circumstances of his death are worthy of a Greek tragedy.

Kellas was apparently under a great deal of strain during his last years as a lecturer at the Middlesex Hospital Medical School. There are several references in his letters to the very demanding work that he was required to do during the years of World War I, 1914–1918, when the Medical School was understaffed, and he made the decision to resign from the faculty of the School in 1919. In a letter dated October 21, 1919, addressed to A. R. Hinks[1] he referred to a "peculiar and continuous annoyance... a disturbance which medical men tell me is due to overwork and which takes the form of malevolent aural communications, including threats of murder." There is no other reference in his correspondence to this bizarre condition that suggests an incipient paranoid psychosis.

Nevertheless Kellas apparently worked hard to prepare for the forthcoming expedition to Kamet, on which he was proposing to study the effects of breathing oxygen at extreme altitudes. His list of projects including measuring the alveolar P_{O2} and P_{CO2} using the gas analysis techniques described by Haldane, determining changes in hemoglobin concentration and erythrocyte count, and assessing the beneficial effects of breathing oxygen. This work was supported by the Oxygen Research Committee of the British Admiralty in association with the Royal Geographical Society. Kellas had earlier proposed the use of the chemical sodium peroxide in a breathing circuit to prepare oxygen at high altitude, but on this occasion he planned to test the value of oxygen from compressed gas cylinders.

The expedition to Kamet was not completely successful. In a letter from Darjeeling dated November 17, 1920, to Hinks,[1] Kellas wrote that the "lateness of arrival of scientific apparatus and oxygen cylinders caused a retreat from Kamet after reaching only about 23,600 ft.... Although I managed to get all the essential experiments carried out, I am so dissatisfied with certain factors that I intend to arrange a small expedition for next year, and complete the ascent."

During the winter, Kellas remained in India at Darjeeling because he wanted to obtain additional information about the best approaches to Everest. While he was there he received an invitation to take part in the first official reconnaissance expedition to take place in the spring of 1921. After many years of unsuccessful

---

climbing Everest. J. N. Collie, president of the Alpine Club, stated this when he first saw Kellas' manuscript in 1920. Many modern climbers believe that one reason for the failures of the expeditions of the 1920's and 1930's to reach the summit was that too much climbing was left to the last day. When John Hunt planned the successful first ascent in 1953, he was adamant that the last camp should be well about the South Col, and in fact it was located at 27,900 ft.

attempts, permission had finally been obtained from the Dalai Lama to approach Everest through Tibet. Naturally Kellas was enormously elated, since much of the previous 15 years of his life had been a preparation for this event.

However, in early 1921 he was back in the Himalayan ranges climbing alone with a few Sherpas. In a letter to Hinks of May 18, 1921 he wrote[1]: "my main object in ascending Kabru (besides training coolies for the Mt. Everest expedition) was to obtain for your use a photograph of Mt. Everest and all the peaks to the N.W.... part of the [Everest reconnaissance] expedition starts today, and Raeburn and I get off tomorrow... after return from Mt. Everest I intend to try Kabru again." Kellas took his photographic equipment up to ~21,000 ft on Kabru to obtain the best possible photographs. He returned to Darjeeling from Kabru on May 10, which meant that he had only 9 days of rest before starting on the momentous Everest reconnaissance expedition.

The trek from Darjeeling followed the route that Kellas had earlier recommended and knew well. They went north through the humid jungle of Sikkim and crossed into Tibet over the Jelep La pass. Several members of the expedition suffered from diarrhea, a very common condition in lowland treks because of the poor hygiene. Kellas, who was almost 53 years old, was badly affected. He grew so weak that he had to be carried on an improvised stretcher. Mallory, who was one of the expedition members, wrote: "Can you imagine anything less like a mountaineering party? It was an arrangement which made me very unhappy and which appalls me now in the light of what has happened... and yet it was a difficult position. The old gentleman (such he seemed) was obliged to retire a number of times en route and could not bear to be seen in his distress and so insisted that everyone should be in front of him" [30].

Tragically, Kellas died as the expedition was approaching the Tibetan village of Kampa Dzong. It was here that the expedition members had their first view of Everest. Mallory described it thus: "It was a perfect morning as we plodded up the barren slopes above our camp ... we had mounted perhaps a thousand feet when we stayed and turned, and saw what we came to see. There was no mistaking the two great peaks in the west: that to the left must be Makalu, gray, severe, and yet distinctly graceful, and the other away to the right—who could doubt its identity? It was a prodigious white fang excrescent from the jaw of the world" [10].

Kellas was buried on a hillside south of the village, in a place which looks out across the arid Tibetan plain to the distant snows of the Himalayas where there rose the three peaks of Pauhunri, Kangchenjhau, and Chomiomo, which Kellas alone had climbed. Mallory described the scene: "It was an extraordinarily affecting little ceremony, burying Kellas on a stony hillside—a place on the edge of a great plain and looking across it to the three great snow peaks of his conquest. I shan't easily forget the four boys, his own trained mountain men, children of nature, seated in wonder on a great stone near the grave while Bury read out the passage from I Corinthians" [30].

Kellas' death was ascribed to heart failure, but that means little, since it was a catchall for various conditions in those days. Perhaps he developed high-altitude pulmonary edema; the expedition had just come over a 17,500 ft pass, although, of course, Kellas had been much higher many times before. Or perhaps the combined effects of severe diarrhea with the tremendously demanding climbing program that he had completed over the last 6 month was too much for him.

So Kellas, who had spent much of the last 15 year of his life studying the physical and physiological problems of climbing Mt. Everest and who probably knew more about these subjects than anyone else alive, died just as the first official reconnaissance expedition had its first view of the mountain they came to climb. It would be difficult to imagine a more dramatic end to the life of this remarkable man.

Acknowledgements I am indebted to Christine Kelly, archivist, and other members of the staff of the library of the Royal Geographical Society and also to the library of the Alpine Club, Senate House Library of London University, and the British Library, British Museum. Assistance is also acknowledged from the staff of the Biomedical Library, University of California, San Diego.

References

1. Barcroft J. The respiratory function of the blood. 1. Lessons from high altitudes. Cambridge: Cambridge University Press; 1925.
2. Barcroft J, Orbeli L. The influence of lactic acid upon the dissociation curve of blood. J Physiol Lond. 1910;41:355–67.
3. Bert P. La pression barométrique. Paris: Masson; 1878. p. 1156.
4. De Filippi F. Karakoran and Western Himalaya. London: Constable; 1912. p. xv.
5. Douglas CG, Haldane JS, Henderson Y, Schneider EC. Physiological observations made on Pikes Peak, Colorado, with special reference to adaptation to low barometric pressures. Philos Trans R Soc Land B Bio Sci. 1913;203:185–318.
6. FitzGerald MP. The changes in the breathing and the blood at various high altitudes. Philos Trans R Soc Lond B Biol Sci. 1913;203:351–71.
7. Geissler P, Kellas AM. Ein Pioneer des Himalaja. Dtsch Alpenztg. 1935;30:103–10.
8. Haldane JS, Kellas AM, Kennaway EL. Experiments on acclimatization to reduce atmospheric pressure. J Physiol Lond. 1919–1920;53:181–206.
9. Hinchliff TW. Over the hills and far away. London: Longmans Green; 1876. pp. 90–1.
10. Howard-Bury CK. Mount Everest: the reconnaissance, 1921. London: Arnold; 1922. p. 184.
11. Kellas AM. Mountaineering in Sikkim and Garwal. Alpine J. 1912a;26:52–4.
12. Kellas AM. The mountains of Northern Sikkim and Garwal. Alpine J. 1912b;26:113–42. [Also published under the same title. Geogr J. 1912;40(241–263):1].
13. Kellas AM. The late Dr. Kellas' early expeditions to the Himalaya. Alpine J. 1912c/1922;34:408–44.
14. Kellas AM. A fourth visit to the Sikkim Himalaya, with ascent of the Kangchenjhau. Alpine J. 1913;27:125–53.
15. Kellas AM. A consideration of the possibility of ascending the loftier Himalaya. Geogr J. 1917;49:26–47. [A summary was published in the Alpine J. 31: 134-138, 1917.].
16. Kellas AM. The approaches to Mount Everest. Geogr J. 1919;53:305–6.
17. Kellas AM. Sur les possibilites de faire l'ascension du Mount Everest. Congres de l'Alpinisme, Monaco. C. R. Seances Paris. 1920;1:451–521. 1921.
18. Kellas AM, Morshead HT. Dr. Kellas' expedition to Kamet in 1920. Alpine J. 1920/1921;33:312–19.
19. Kellas AM, Morshead HT. Expedition to Kamet. Geogr J. 1921;57:124–30, 213–9.
20. Kellas AM, Shockley WH. Mountain sickness. Geogr J. 1912;40:654–5 (1913;41:76–7).
21. Keys A. The physiology of life at high altitudes. Sci Monthly. 1936;43:280–312.

22. Krogh M. The diffusion of gases through the lungs of man. J Physiol Lond. 1915;49:271–96.
23. Messner R. Caption to front-end paper. Everest. London: Kaye & Ward; 1979.
24. Noel JBL. Through Tibet to Everest. London: Arnold; 1927.
25. Obituary. Middlesex Hosp J. 1921;22:66–9.
26. Obituary. Nature London 1921;107:560–1.
27. Obituary. Geogr J. 1921;58:73–5.
28. Pugh LGCE. Physiological and medical aspects of the Himalayan scientific and mountain-eering expedition,1960–1961. Br Med J. 1962;2:621–27.
29. Pugh LGCE. Cardiac output in muscular exercise at 5800 m (19,000 ft). J Appl Physiol. 1964;19:441–7.
30. Robertson D. George Mallory. London: Faber & Faber; 1969. p. 151, 155.
31. Starling EH. Principles of human physiology. London: Churchill; 1912. p. 1226.
32. West JB. Highest inhabitants of the world. Nature (Lmd). 1986;324:517.
33. West JB, Boyer SJ, Graber DJ, Hackett PH, Maret KH, Milledge JS, Peters RM Jr, Pizzo CJ, Samaja M, Sarnquist FH, Schoene RB, Winslow RM. Maximal exercise at extreme altitudes on Mount Everest. J Appl Physiol. 1983a;55:688–98.
34. West JB, Hackett PH, Maret KH, Milledge JS, Peters RM Jr, Pizzo CJ, Winslow RM. Pulmonary gas exchange on the summit of Mount Everest. J Appl Physiol. 1983b;55:678–87.
35. Zuntz N, Loewy A, Muller F, Caspari W. Atmospheric pressure at high altitudes. In: Hohenklima and Bergwanderungen in ihrer Wirkung auf den Menschen. Berlin: Bong; 1906. pp. 37–39 [English translation of relevant pages in High Altitude Physiology, edited by J. B. West. Stroudsburg, PA: Hutchinson, 1981.].

Chapter 18
T. H. Ravenhill and His Contributions to Mountain Sickness

Abstract Thomas Holmes Ravenhill (1881–1952) was an important pioneer in high-altitude medicine but almost nothing has been published about him. He wrote a landmark paper in 1913 that included the classification of high-altitude sickness that is still in use, and it also contained the first accurate descriptions of high-altitude pulmonary edema and high-altitude cerebral edema, although he used different terms. The work was done while he was medical officer at the Collahuasi and Poderosa mines in northern Chile at altitudes he gave as 4690–4940 m. Remarkably, the paper was then forgotten until it was rediscovered over 50 year later, but it is now cited in any comprehensive study of high-altitude illness. Ravenhill graduated in medicine from the University of Birmingham, Birmingham, UK, in 1905 and 4 year later went to the mines where he spent 2 year. Subsequently, he served in the Royal Army Medical Corps in the 1914–1918 war and was awarded the Military Cross. He returned to general practice but after a few years gave up medicine altogether. He then made important contributions to archeology and spent the last third of his life in London as a painter, mainly in watercolors. It is unclear to what extent his war experiences brought about his dramatic career change.

In 1913, a remarkable paper appeared in the Journal of Tropical Medicine and Hygiene [36]. It was entitled "Some experiences of mountain sickness in the Andes," and the author was T. H. Ravenhill, M.B., B.C. (sic), (This is a typographical error in the *Journal*). The degree is M.B., B.Ch. (Bachelor of Medicine, Bachelor of Chirurgery). Late Surgeon to the Poderosa Mining Co., Ltd., Chile, and to La Compania Minera de Collahuasi, Chile. These mines are in a remote part of north Chile close to the Bolivian border, and the altitudes were given as 15,400–16,200 ft (4690–4940 m). The paper contains vivid, accurate clinical descriptions of what we now know as acute mountain sickness, high-altitude pulmonary edema, and high-altitude cerebral edema. Indeed, Ravenhill's classification of high-altitude illnesses is still in use today, although we now use different terms. Ravenhill's work was years ahead of its time and was essentially forgotten until the paper was rediscovered more than 50 years later.

Almost nothing has been published about Ravenhill, except for a one-paragraph obituary in the British Medical Journal [10]. However, Ravenhill was a fascinating man with several unusual aspects to his character, and he deserves to be better known. For example, he made significant contributions to archeology and later became a

© American Physiological Society 2015

J. B. West, *Essays on the History of Respiratory Physiology,*
Perspectives in Physiology, DOI 10.1007/978-1-4939-2362-5_18

productive painter. In addition, the story of how his paper on high-altitude illness dropped out of sight while high-altitude pulmonary edema was independently rediscovered in Peru [6–9, 24, 25, 28], not far to the north, has some unexpected twists.

18.1 Family and Early Years

Thomas Holmes Ravenhill (THR) was born in Bordesley, an inner suburb of Birmingham, England, on September 17, 1881 (Fig. 18.1). His parents lived at 113 High Street, Bordesley, a rather seedy part of town in those days, which has now been redeveloped. The street, which is now a major thoroughfare, has been renamed Digbeth Street, and number 113 is now a police station that was built in 1911. Ravenhill's father, also named Thomas Holmes Ravenhill, was born in 1847 and was a surgeon. His degrees were M.R.C.S.Eng. (Member, Royal College of Surgeons, England) 1868 and L.S.A. (Licentiate, Society of Apothecaries) 1869 (Queen's College, Birmingham), and he was Honorary Surgeon at the Birmingham Lying-In Charity Hospital. THR's father's father was a clergyman, and was also named Thomas Holmes Ravenhill (see the family tree in Table 18.1). Incidentally, it is unusual in England for father and son to have all three names the same.

THR attended King Edward's School in Birmingham from 1890 to 1900, and an extensive record of his school career exists.[1] The School has an enviable reputa-

Fig. 18.1 Thomas Holmes Ravenhill as a captain in Royal Army Medical Corps. This photograph was taken 6 years after his period in the mines. (Courtesy of Marjorie Rosenthal)

---

[1] There were extensive links between the Ravenhill family and the Schools of King Edward the Sixth in Birmingham. THR spent 10 years at King Edward's School, and his brother 11 years. All four sisters had connections with the High School for Girls. Grace taught languages there from 1906 until her marriage in 1925. Annie was initially secretary to the headmistress and was later a teacher. Mary was high school secretary, and Margaret (daughter of Adelaide) also attended the School.

Table 18.1 Ravenhill/Rosenthal family tree

		Thomas Holmes RAVENHILL m. Mary Frances Vincent BARTLETT			
Thomas Holmes RAVENHILL (1847–1907) m. Adelaide Ann CUTLER (1855?–1923)	Edmund Burton	William Henry	Florence Mary	John Oldfield	Catherine Elizabeth Bartlett
Mary Holmes (1878–1925)	Annie Sibyl (1879–1939)	**Thomas Holmes RAVENHILL (1881–1952)**	Edmund Lawrence Bartlett (1883–1959) m. Edith Frances WANSBOROUGH (?–1847)	Grace Stephanie Frances (1884–1967) m. Sir Geoffrey Ingram TAYLOR (1886–1975)	Adelaide Elizabeth (1886–?) m. Rev. George David ROSENTHAL (1880–1938)
			Margaret Mary (1915–)		Michael David Holmes ROSENTHAL (1919–1981) m. Marjorie (Molly) Mary WILSON (1922–)
			David Stuart Holmes (1948–)		Mark Geoffrey Thomas ROSENTHAL (1952–) m. Monica SMYTHE (1951–)
			Rowan Madoc (1975–)		Amber Dilys (1977–)

tion, and traces its origin back to 1547. THR was only an average student, and he showed no particular aptitude for science or art. However, he excelled in sports. He was a member of the School's First Rugby XV from 1897 and captained the team in 1899–1900. He was also the School's cross-country running champion in 1898 and 1899 and a fine cricketer. After he left school, he continued his rugby by captaining the Old Edwardian's rugby team, and he also played for the Midland Counties in 1902–1904.

THR entered the Medical School of the University of Birmingham in 1900 and graduated M.B.Ch.B. in June 1905. A copy of his university record exists, showing the subjects that he studied, with the dates. After his graduation, he became House Surgeon at the Queen's Hospital in Birmingham in 1906. His activities for the next 2 years are unclear, but his address in the Medical Directory was given as 113 High Street, Bordesley, and he probably helped with his father's practice. His father died in 1907.

THR had a younger brother, E. L. B. Ravenhill,[2] who also attended King Edward's School and then went to farm in Canada where he worked with horses. He returned to England in 1914 to become Riding Master to the Brigade of Guards. There were also four sisters, one of whom married Sir Geoffrey Taylor,[3] an eminent

---

[2] Obituary in the *Old Edwardian's Assoc. Gazette* June 1959.

[3] The best account of G. I. Taylor is in the *Biographical Memoirs of Fellows of the Royal Society* 22: 565–633, 1976. Taylor made important contributions to fluid and solid mechanics and their application in meteorology, aeronautics, and engineering. His papers were collected by G. K. Batchelor and published in four volumes as *The Scientific Papers of Sir Geoffrey Ingram Taylor,* Cambridge, UK: Cambridge Univ. Press, 1963. Particularly interesting are no. 4 "Pressure distribution over the wing of an aircraft in flight," which reports one of the first measurements of the pressures around an airfoil, and number 54. "The formation of a blast wave by a very intense explosion. II. The atomic explosion of 1945," where he calculated the yield of the Trinity site explosion from a motion picture record of the rate of increase of the ball of fire. Taylor dispersion, which refers to

Cambridge fluid dynamicist. Another sister married the Rev. G. D. Rosenthal, and they had a daughter who is still living and a son whose widow is alive (Table 18.1). THR never married. His brother married but had no children, so the Ravenhill line of this branch of the family died out. However, there are many other Ravenhills alive from other parts of the family.

18.2   High-Altitude Studies

In 1909, Ravenhill went to the Poderosa and Collahuasi mines in north Chile for 2 years as medical officer, or "Surgeon," as the position was then called. The evidence for these dates is that, in his article, he states that on January 15, 1912, he had been away from the "altitudes" for 5 months, and he also states that he was there for 2 years. Thus reasonable dates of his stay in Chile would be from August 1909 to August 1911.

As a result of his period there, Ravenhill wrote his landmark paper "Some experiences of mountain sickness in the Andes" [36]. It gives excellent descriptions of acute mountain sickness, high-altitude pulmonary edema, and high-altitude cerebral edema. His choice of the *Journal of Tropical Medicine and Hygiene* for the paper was in the tradition of reports by British physicians on diseases in colonies and other remote areas, although, of course, the subject had little to do with either tropical medicine or hygiene.

Ravenhill begins his paper by noting that mountain sickness is known as "puna" in Bolivia and "soroche" in Peru. He adds that the term "puna" is employed loosely by the inhabitants, who use the word not only to describe the illness that most people suffer on arrival at high altitude but also the dyspnea, which affects everybody who lives there. Ravenhill correctly restricts the term to the first meaning. He also points out that the location of the mines is a good place to study puna, because most newcomers come from Antofagasta on the coast by train and therefore avoid complicating factors such as fatigue and insufficient food. He includes a brief but interesting description of the rail journey.

Ravenhill then describes "puna of a normal type," which we now call acute mountain sickness. The clinical description could hardly be bettered. He wrote

---

diffusion across a moving front of fluid, is well-known to physiologists interested in gas flow in the lung. Taylor was a keen sailor, and the *Biographical Memoirs...* article has a fine photograph of him taken in 1927 at midnight with his wife Grace Stephanie at the Lofoton Islands of Norway, above the Arctic Circle.

There is an interesting connection with high altitude here. G. I. Taylor's grandfather was George Boole, an eminent mathematician, who married Mary Everest, niece of Sir George Everest, Surveyor General in India after whom Mt. Everest was named. George Boole was responsible for the Boolean logic used extensively in modern computers.

It is a curious fact that the symptoms of puna do not usually evince themselves at once. The majority of newcomers have expressed themselves as being quite well on first arrival. As a rule, towards the evening the patient begins to feel rather slack and disinclined for exertion. He goes to bed, but has a restless and troubled night, and wakes up next morning with a severe frontal headache. There may be vomiting, frequently there is a sense of oppression in the chest, but there is rarely any respiratory distress or alteration in the normal rate of breathing so long as the patient is at rest. The patient may feel slightly giddy on rising from bed, and any attempt at exertion increases the headache, which is nearly always confined to the frontal region.

Ravenhill then notes that the pulse is nearly always high and "there is at times re-duplication of the pulmonary second sound." This last is a remarkable prescience of the pulmonary hypertension associated with alveolar hypoxia, which was not described in animals until 1946 by von Euler and Liljestrand [49] and in humans at high altitude until about 1956 [43].

Ravenhill goes on to point out that the patient usually sleeps better on the second night and by the fourth day of arrival is probably very much better. He specifically notes that there is no "epistaxis or other hemorrhages, dyspnea or extreme vertigo" in puna of the normal type, "which is not a very serious condition at the altitude in question." This is an interesting observation, because some earlier descriptions of acute mountain sickness gave prominence to bleeding from the nose and other mucous membranes, which has always been puzzling because these are not features of the modern condition.

Of course, there had been many descriptions of acute mountain sickness before Ravenhill's, and it is interesting that he does not cite any previous references. Indeed, his paper has no references at all. Traditionally, the first extended account of acute mountain sickness is attributed to the Jesuit priest Joseph de Acosta [2], although the description is not typical and sounds more like acute gastroenteritis. Here is a section from the English translation of 1604 [3].

There is in Peru, a high mountaine which they call Pariacaca... when I came to mount the degrees, as they call them, which is the top of this mountaine, I was suddenly surprised with so mortall and strange a pang, that I was ready to fall from the top to the ground.... I was surprised with such pangs of straining and casting, as I thought to cast up my heart too; for having cast up meate, fleugme, & choller, both yellow and greene; in the end I cast up blood, with the straining of my stomacke.

Following de Acosta, there were many descriptions of high-altitude illness in the seventeenth-nineteenth centuries. The first book devoted to mountain sickness was probably that written by Meyer-Ahrens in 1854 [30]. The best compendium of early descriptions is Chap. 1 of Paul Bert's monumental "La Pression Barométrique" [12], where he devotes 170 pages to the topic. However, none of the descriptions there rings as true as Ravenhill's.

Following the section on puna of a normal type, Ravenhill goes on to describe two "divergent types of the disease... [1]. Those in which cardiac symptoms, and [2] those in which nervous symptoms predominate." He describes three cases of puna of a cardiac type, which we now call high-altitude pulmonary edema. Again, the descriptions are accurate and vivid. Here is part of the description of the first case.

He seemed in good health on arrival, and said that he felt quite well, but nevertheless he kept quiet, ate sparingly, and went to bed early. He woke next morning feeling ill, with symptoms of the normal type of puna.

As the day drew on he began to feel very ill indeed. In the afternoon his pulse-rate was 144, respirations 40. Later in the evening he became very cyanosed, had acute dyspnoea, and evident air hunger, all the extraordinary muscles of respiration being called into play. The heart sounds were very faint, the pulse irregular and of small tension. He seemed to present a typical picture of a failing heart. This condition persisted during the night; he coughed up with difficulty. He vomited at intervals. This condition persisted during the night; he had several inhalations of oxygen; strychnine and digitalis also were given. Towards morning he recovered slightly, and as there was luckily a train going down to Antofagasta in the early morning, he was sent straight down.

I heard that when he got down to 12,000 ft. he was considerably better, and at 7000 ft. he was nearly well. It seemed to me that he would have died had he stayed in the altitudes for another day.

A surprising omission is auscultation of the chest, which presumably would have been rewarded with rales. Indeed, these were reported in the third case. The rapid improvement on descent is typical of high-altitude pulmonary edema.

Ravenhill did not personally see the second case, a patient who died while being carried down from a neighboring mine. The third case is interesting because it was a young Turk, aged 23, who had lived in the district for some months before but who had been at sea level for some weeks and developed puna the day after his return to high altitude. Ravenhill was asked to see him on the fifth day, when he was "profoundly dyspnoeic, respirations being 60, pulse 144, and hardly perceptible." He died that night. The case is interesting, because the illness occurred in someone who had been living at high altitude but went to sea level and then returned to altitude. This sequence of events has been described many times since.

These are the first convincing descriptions of high-altitude pulmonary edema. Angelo Mosso, in his book "Fisiologia dell'Uomo Sulle Alpi", 1897 [31] [English translation: "Life of Man on the High Alps", 1898 [32]] described two cases of what he called "inflammation of the lungs," and while these may have been high-altitude pulmonary edema, the descriptions are not typical. The first was Dr. Jacottet who died on Mont Blanc in 1891, and the description included "violent shivering fits" (fort frissons). The diagnosis at postmortem examination was "capillary bronchitis and lobular pneumonitis," with the immediate cause of death being "suffocative catarrh accompanied by acute edema of the lung".[4]

Mosso's other case was a fit young soldier named Pietro Ramella who developed "inflammation of the lungs" at the Capanna Margherita, altitude 4559 m. The man survived without being taken down, and no pathology is available. The clinical description could fit high-altitude pulmonary edema, although the rectal temperature rose to 39.9 °C, and the sputum was thought to be typical of a lung infection.[5] Thus

---

[4] The postmortem report by Dr. Wizard includes "*poumon* couleur violet, gonflé, foncé, congestion bilatérale, oedème considérable, muqueuse bronchique injectée fortement. Le liquide de la coupe est écumeux. Congestion égale partout." Although the description of "considerable edema" is consistent with high-altitude pulmonary edema, it is also consistent with pneumonia.

[5] The relevant section is "La tosse quasi mancante, le qualità fisiche dello sputo che aveva l'aspetto tipico, rugginoso, sanguigno, consistente e vischioso–la mancanza di altri sintomi caratteristici dei catarri bronchiali–ci fanno ammettere che si trattasse veramente di una infezione per il

Ravenhill may not have seen the first case of high-altitude pulmonary edema but he did give the first typical clinical description.

It is not surprising that Ravenhill attributed high-altitude pulmonary edema to cardiac failure, although we now know that this was an error. The pulmonary arterial wedge pressure has repeatedly been shown to be normal [53], and the left atrial pressure is also normal [15]. In fact, recent work shows that the normal left ventricle tolerates the hypoxemia of extreme altitude extraordinarily well, maintaining normal contractility up to altitudes of 8000 m [42, 47]. Interestingly, early climbers on Mt. Everest who became fatigued were sometimes diagnosed as having "dilatation" of the heart. Indeed, as late as 1934, Leonard Hill stated that "degeneration of the heart and other organs due to low oxygen pressure in the tissues, is the chief danger which the Everest climbers have to face" [19].

The second divergent type of puna described by Ravenhill was puna of a nervous type. He noted that this was a "rare divergence from the normal." However, he added that "the nervous symptoms may develop to such a degree as to become alarming." Ravenhill goes on

> The most marked case I had was a young Chileno, aged 19. He arrived at the neighboring mine in the usual way; three days later I was called to see him. He was then unable to speak, there were violent spasmodic movements of the limbs, and he resisted examination. The face was blanched, the lips almost white, the pupils slightly dilated. Temperature and respiration were normal; the pulse 140. He was unable to stand or walk. I was told that he had been in this condition almost since his arrival, and that he had been delirious, talking all sorts of nonsense. I could find nothing organically wrong on physical examination. He was sent down the same day; three days later, *i.e.,* by the time he had reached the coast, he had quite recovered.

This seems to be a typical description of high-altitude cerebral edema, except for the spasmodic movements of the limbs. The rapid recovery on going to lower altitude is certainly characteristic. Ravenhill described two other cases. One was a young man whom he did not see personally. The patient had marked convulsions but recovered on being sent down and had no recollection of being at high altitude at all. In another case, vertigo was the most prominent symptom.

Again, Ravenhill must be given credit for first recognizing the condition that we now call high-altitude cerebral edema. Indeed, this was not described again until many years later [14], notably when a large series of Indian troops were airlifted to altitudes of ~5500 m in the 1960s [46].

The remainder of Ravenhill's paper deals with conditions that influence puna, its treatment, and the effects of acclimatization. He was of the opinion that puna was most common in the first 3 months of the year, when the weather was often stormy. Imbibing alcohol increased the severity as did physical exertion. Aspirin was very useful for puna of a normal type, but he was not impressed by herbal remedies such as "Chacha Como" and "Flor de Puna," which were used by local Indians. Giving oxy-

---

pneumococco del Fraenkel." This is translated in the English edition of 1898 as "The exceedingly slight cough, the physical qualities of the sputum which had the typical appearance, being rusty-coloured, mixed with blood, firm and glutinous, the lack of other symptoms characteristic of bronchial catarrh encourage the statement that this was a case of infection with the Fraenkel pneumococcus." The clinical description of Ramella's illness is detailed and occupies six pages. A more extensive account of the same case is in Abelli (1896).

gen was not particularly efficacious, although one reason may have been that he had to prepare his own by heating a mixture of potassium chlorate and manganese dioxide! By far the best treatment was descent to lower altitude, and this is current practice today. He recognized the value of acclimatization, although he made the point that everyone at these altitudes experienced dyspnea on exertion, whether acclimatized or not. Visitors to the mine who took the fast train from the coast (which took only 42 h) were more likely to develop puna than those who ascended over a period of a week.

One of the most remarkable features of Ravenhill's paper was how it disappeared from sight to be discovered some 50 years later. This is all the more remarkable because high-altitude pulmonary edema began to be recognized in the 1930s and 1940s in the Peruvian Andes [6–9, 24, 25, 28], not so very far from where Ravenhill worked. By contrast, the condition was not described in the North American and European literature until the 1960s.

Herbert Hultgren, a cardiologist at Stanford University School of Medicine, visited Peru in early 1959 and deserves credit for alerting the English-speaking world to the work that was being done there on high-altitude pulmonary edema. He visited the Chulec General Hospital at La Oroya at an altitude of 3700 m in the Andes not far from Lima. This was the central medical facility for the United States-owned corporation, the Cerro de Pasco Corporation. There he had the opportunity of reviewing the clinical records of a series of patients who had developed pulmonary edema after arriving at high altitude [21].

Hultgren pointed out that Hurtado had possibly recognized the condition in 1937 [24]. In that publication, he described a resident of Casapalca (altitude 4200 m), who developed acute pulmonary edema on return from a trip to Lima. However, signs of cardiac failure persisted, and it is likely that the patient had underlying cardiac disease. Hurtado also referred to the same patient in a later publication [25]. In 1952, Lundberg, who was Chief of Medicine at Chulec General Hospital, reported six cases to the Asociacion Medica de Yauli, but this was not published. Shortly afterward, Lizarraga carried out a study of 14 patients at the Chulec General Hospital for his Bachelor's Thesis, which was subsequently published [28]. Other Bachelor Theses by Lopez and Marticorena evaluated many cases in adults and children but were not otherwise published. Bardález published additional cases [6, 8, 9], and in 1956 a brief English summary appeared as a "foreign letter" in the Journal of the American Medical Association [7]. This last reference has caused some confusion, because the full name of the author was Arturo Bardález Vega and the report is sometimes cited by his mother's maiden name Vega rather than the correct Bardález. With the exception of this brief report, which was generally overlooked, all the papers were in Spanish, and the condition was essentially unknown in the English literature.

In 1960, Charles S. Houston, an internist in Aspen, Colorado, reported a case of "acute pulmonary edema of high altitude" in the New England Journal of Medicine and thus brought the condition into prominence in the English-speaking literature [20]. At the time, he was apparently unaware of the South American studies because these were not cited. However, in the following year, Hultgren, Spickard, Hellriegel, and Houston published an extensive study giving full credit to the Peruvian work [22]. Many other reports including hemodynamic measurements in the condition soon followed [5, 15, 23, 35, 44, 45].

All this time, Ravenhill's beautiful description lay unnoticed, waiting for its rediscovery. This occurred in June 1964 when William H. Hall of the US Army Research Institute of Environmental Medicine carried out an extensive literature search in Boston prior to writing a paper on acute mountain sickness. Together with colleagues, he had studied Indian troops deployed in Ladakh during the India-China border war. The publication [17] was the first to cite Ravenhill's article according to the Science Citation Index. At about the same time, Drummond Rennie apparently independently came across Ravenhill's article while he was doing a literature search for his M.D. thesis, but he did not cite it. Ravenhill's paper was subsequently cited on many occasions, and it was included in an anthology of high-altitude papers in 1981 [54].

It is not clear what prompted Ravenhill to go to the mines in north Chile in 1909. A search for advertisements in medical journals of the time has not been successful. However, it was not uncommon for young men to go abroad at that time; as indicated earlier, Ravenhill's younger brother spent a period of several years in Canada on a farm. A living relative has suggested that life in general practice in Bordesley must have been rather grim, with living-in assistants being trained and four sisters at home. Perhaps it is not surprising that Ravenhill wanted to get away. Nevertheless, there was presumably some event or some person who prompted this. A detailed study of his period at King Edward's School by one of the librarians there has pointed out that the most popular school club was the Natural History Society, and while Ravenhill was at school the Society made seven visits to local mines, which, according to accounts in the school chronicle, greatly impressed those who participated. Another event that possibly influenced Ravenhill was that in 1901 Dr. Walter Myers, who was an Old Edwardian, died while leading a yellow fever expedition in Brazil for the Liverpool School of Tropical Medicine. He was only 28. A commemorative plaque was unveiled at Birmingham University in the same year, and Ravenhill was there at the same time.

Although north Chile might seem very remote from Birmingham in the early part of this century, there were many links between Britain and the mines of north Chile in the late nineteenth and early twentieth centuries. The railway from Antofagasta to La Paz, which Ravenhill describes in his article, was built by a British company. The nitrate mines in north Chile had strong British links, and there was an English enclave in Iquique on the coast. An English entrepreneur, John Thomas North, had invested heavily in the nitrate mines in the late 1880s and became known as the Nitrate King (although he was actually a colonel). At the turn of the century, about 20 million £ were invested from Britain in the nitrate industry. This declined when German chemists developed processes for fixing nitrogen from the air during the 1914–1918 war.

For three quarters of a century, up to the outbreak of the 1914–1918 war, Great Britain held first place in Chilean foreign trade, in both imports and exports. The main product of the Collahuasi and Poderosa mines was copper, and the export of this mineral from Chile rose from 42 to 106 million metric tons from 1913 to 1918 [4]. Ravenhill went to the Collahuasi-Poderosa area at a time when the mines were growing rapidly. The 40 mile-long spur railway line joining the mines to the main Antofagasta-La Paz line had been completed in 1908 [13], and a contemporary

Fig. 18.2 Collahuasi mine at the time of Ravenhill. (From Packard [34])

report stated that Collahuasi had a visible supply of 500,000 tons of copper [33]. The Compania Minera de Collahuasi had been formed in 1899 [55]. Thus the mines were burgeoning when Ravenhill was there (Figs. 18.2 and 18.3).

Ravenhill apparently left the mines in August 1911, and his whereabouts for the next 2 years are unclear. According to a Birmingham newspaper (*Birmingham*

Fig. 18.3 Commemorative stamp from the Collahuasi mine purportedly issued in the same year that Ravenhill arrived. Translation: "Commemoration of the two million years since the arrival of the first geologists in Collahuasi." (From Ulloa [48] with permission)

Gazette, October 16, 1935), he spent some time in Mexico, the United States, and in Canada where his brother resided. His article on mountain sickness was written in January 1912. His niece believes he was in general practice in Leicester when the 1914–1918 war broke out.

18.3   War Experiences

Ravenhill volunteered in September 1914 in the first week of World War I and became a lieutenant in the Royal Army Medical Corps. Initially he was a medical officer in the 63rd Brigade, but from February to June 1915 he was attached to the British Military Mission in Serbia where he served in Nish, Kragugevatz, and Belgrade. There he was awarded the prestigious Cross of St. Sava of Serbia, a splendid enamel decoration. June 1915 found him at the Tigue General Hospital in Malta, and the following month he was sent to Gallipoli, attached to the 2nd Casualty Clearing Station. He contracted sandfly fever and was sent back to England, the sole doctor on a ship full of sick and wounded, himself delirious most of the time and helped only by a nurse, according to one of his friends [51]. He spent several months at the Military Hospital, Tidworth, and then in March 1916 was sent to France with the 96th Field Ambulance of the 30th Division [18].

There his record shows that he was involved with some of the grimmest battles of the war.[6] He was at the battle of the Somme in 1916, then Arras, Messines, Ypres, and Passchendaele Ridge. He was promoted to captain in 1915 and awarded the Military Cross for his exploits in June 1917. The citation reads: "For conspicuous gallantry and devotion to duty in collecting and evacuating wounded under heavy shell fire, notably during a gas attack, when his coolness and judgment saved very many casualties."[7]

A war record like this would have shattered many men, and its effect on Ravenhill is not altogether clear. The eminent Birmingham surgeon, Seymour Barling, who was a friend of Ravenhill's, wrote in his obituary in the *British Medical Journal* that "the fires of war were too much for his gentle character, and.... left him broken in spirit." Barling was a colonel in the Army Medical Services, Consultant Surgeon,

---

[6] A detailed summary of Ravenhill's war record was published in Service Record of King Edward's School, Birmingham, 1914–1919, edited by C. H. Heath, Birmingham, Cornish Brothers, 1920.

[7] Poison gas was first used by the Germans on April 22, 1915, when 6000 cylinders were opened to release 150 tons of chlorine gas along a 7000 m front within 10 min. The yellowish-green cloud drifted across to the French. Algerian lines causing many deaths and general panic. The Allies soon retaliated in kind. By July of 1917 (the month after Ravenhill was awarded the M.C.), more sophisticated types of poison gas such DS mustard gas (dichlordiethyl sulfide) were in use and these were delivered by shells. On July 12–13, 1917, the Germans fired 50,000 mustard gas shells into an area just east of Ypres, and on July 20–21, there were 6400 civilian casualties caused by gas [16]. Among the doctors who studied the effects of mustard gas in the field was C. G. DougJas, who communicated his results to J. S. Haldane and others in England.

British Expeditionary Force, and joint author of the book *Manual of War Surgery* [11]. These army links may explain the emphasis of the obituary on Ravenhill's war experiences and the complete omission of his high-altitude and archeological interests. Certainly, Ravenhill gave up medicine within a few years. However, he then had a productive second career as a painter, with a parallel interest in archeology, and it is arguable whether his spirit was broken.

On the outbreak of the World War II in 1939, he again volunteered and was made Officer Commanding Typhus in the Midlands. However, no cases of the disease occurred, and THR spent some time helping a friend in Cannock, near Birmingham, with his general practice. His friend gradually became disabled by alcoholism, and Ravenhill took over most of the practice.

18.4   Post World War I and Archeology

After the war, Ravenhill returned to general practice at Handsworth, a poor suburb of Birmingham. His patients were in "the lower income groups, poorly housed, with scant reserves, in a time of industrial depression," as Cranston Walker,[8] one of his friends, described. "They were not easy patients, but Ravenhill had deep insight, almost amounting to genius, which discerned and honoured the courage and elementary virtues beneath unornamental exteriors" [51]. Ravenhill worked very hard, and his health, no doubt influenced by the war, was affected. He bought a small house, White Cottage, in Little Rollright, a tiny village on the border of Warwickshire and Oxfordshire, where he could spend time recuperating. Nearby were the Rollright Stones (Fig. 18.4), an ancient monument rather like a miniature Stonehenge, and Ravenhill became interested in the archeology of the Stones, which probably date back to early Neolithic times, ~2,000 B.C.

In the early 1920s, he decided to give up medicine and devote himself to painting, with an additional interest in archeology. A small legacy from his mother who died in 1923 made this practicable. In 1925, he published a book, *The Rollright Stones and the Men Who Erected Them: An Outline of Recent Evidence* [37]. It is a scholarly work that shows a considerable background in archeology. The Rollright Stones comprise a circle of stones illustrated in Fig. 18.4 and two other groups, the large monolith called the King Stone and the cluster called the Whispering Knights. The Stones are thought to have an old religious significance, and the group of Whispering Knights, at least, is associated with a burial chamber originally covered by a long barrow. Ravenhill carried out some excavations there and found a human skull fragment within the chamber. The Stones are situated alongside a very old road running east-west along the crest of a hill, and the track probably dates from Celtic times. The two editions of Ravenhill's book were dedicated to M. H. R., probably

---

[8] Cranston Walker had a serious interest in physiology and wrote an impressive paper on the tension in the wall of a hollow organ distended by internal pressure [50].

Ravenhill's sister, Mary Holmes Ravenhill, who died of cancer in 1925. The book incorporates a paper read before the Birmingham Archaeological Society.

Ravenhill moved his residence to White Cottage in 1925. However, he must have spent time at Little Rollright before then, because his book was published in that year. His brother lived in the same house from 1932 to 1937, and when he offered his services to the military during the World War II and was assigned to the Observer Corps, his post was the Rollright Stones.

An expanded second edition of Ravenhill's book appeared in 1932 [40], and the book was highly praised by George Lambrick, who wrote more modern accounts of the Rollright Stones in 1983 and 1988 [26, 27]. Ravenhill published at least two papers or notes in the transactions of the Birmingham Archaeological Society [38, 41]. There is also an extensive paper in the Reports of the Oxfordshire Archaeological Society in 1926 [39].

Fig. 18.4 Main ring of the Rollright Stones. Photograph was taken from a kite controlled by the man in the foreground. (From Lambrick [27])

18.5   Painting

In 1927 Ravenhill moved to London where he devoted the last third of his life to painting. His mother was a talented amateur watercolorist, and Ravenhill had begun sketching and painting during his war years. He studied at the Chelsea School of Art and went on to be a productive and reasonably successful painter. In 1933 he held a one-man exhibition at the Cooling and Sons Gallery in London, where 73 of his paintings were on display. Two years later, there was a one-man exhibition at the Graves Gallery in Birmingham. One of his paintings, "The Thames at Chelsea," was selected for the prestigious Royal Academy exhibition in 1940.

Ravenhill was particularly fond of watercolors and was said to be influenced by the French impressionists, and by Constable, particularly for skies and trees. He did many paintings of the Thames and its shipping, of the Cotswolds around Little Rollright, of Cambridge, Bristol, Somerset, the west coast of Ireland, and especially of north Wales near Harlech, where the Ravenhills spent their holidays away from the Birmingham slum where they lived. Ravenhill's grandfather (Thomas Holmes Ravenhill) was the rector of Arlingham on the Severn for many decades, Ravenhill also produced lovely oils of Skye.

He also painted nudes using watercolors, which was rather unusual at the time. He painted a number of portraits, and Fig. 18.5 shows part of a self-portrait that was exhibited at Birmingham in 1935. A landscape near Cannock is shown in Fig. 18.6.

Ravenhill had a circle of artistic friends in London, including Hester Dowden and Evelyn Clutton-Brock, who met for music making and painting in a house in Cheyne Gardens in Chelsea. His first address in London in 1927 was 28 Danvers

Fig. 18.5 Self-portrait of Ravenhill that was exhibited in Graves Gallery, Birmingham, in 1935. (From the *Birmingham Evening Despatch,* November 18, 1935)

Fig. 18.6 Watercolor by Ravenhill; near Cannock, north of Birmingham

Street, Chelsea,[9] and he later moved to 15 Cheyne Gardens and then to Albert Studios off Albert Bridge Road. This road is a busy north-south artery that runs along the side of Battersea Park, but Albert Studios consists of eight small attached houses in a peaceful garden setting, protected from the noise of the main road by a large block of Edwardian flats. The studios would provide an idyllic setting for a painter, with the special charm that can be found in London. Ravenhill's paintings never received wide recognition, but his needs were small and, according to his relatives, he lived in great simplicity in the Albert Studios. He was a member of the Chelsea Arts Club and bequeathed his "studio equipment such as easels, brushes and similar articles" to Stanley Grayson at the Club. Cranston Walker notes that Ravenhill advocated painting for occupational therapy.

Ravenhill was also musical and played the cello with skill. His father had apparently hoped that the six children would form a musical group, but this did not transpire. However, Ravenhill derived much pleasure from playing music with his Chelsea circle.

18.6 Conclusion

Anyone who spends time pondering Ravenhill's life cannot but be impressed by two aspects of his character. On the one hand, he was a top class rugby player who had the enterprise to spend 2 years at very high altitude in a remote part of Chile

---

[9] Sir Alexander Fleming, discoverer of penicillin, lived only three houses away on Danvers Street at the same time. He painted in oils, was also a member of the Chelsea Arts Club and, according to a biographer, "entertained several artists who lived in the neighbourhood" [29]. It is possible that one of these was Ravenhill.

and subsequently survived 4 years of World War I in some of the grimmest battles, for which he was awarded a high military decoration. On the other hand, he was a sensitive, very perceptive person who spent one-third of his life as a productive artist and who made useful contributions to archeology. It remains unclear to what extent his war experiences brought about his remarkable career change. Certainly 2 years on the western front in France was more than enough to shatter many young people's lives. As indicated earlier, one of his biographers, Seymour Barling, was of this opinion and stated that after the war "he never took up the broken thread of his own damaged life." He goes on to say that "painting was his solace in later years…"

However, another interpretation is possible. As stated earlier, his mother was a watercolorist, and Ravenhill apparently started sketching and painting when he could during the war. Perhaps he was never particularly fond of medicine, and we can imagine that there was some parental pressure to pursue this, his father being a surgeon. It seems likely that in the 1920s, with the help of a small legacy, he was able to take up what he really loved best, which was painting.

Ravenhill never married, and we do not know much about his private life, although he certainly had many friends, particularly in London. One of his friends was Dr. Cranston Walker, who obtained his medical degree in Birmingham 4 years after Ravenhill and apparently knew him well. In his obituary of Raven hill in the *Old Edwardian's Association Gazette,* Walker writes that Ravenhill was "one of the finest men the School ever turned out" [51]. He goes on "as a rugby player he is even yet remembered in the Midlands, perhaps with special affection because of his unshakable good humor and good will." After referring to his war service, he continues "his more intimate work in medical practice was known only to his medical friends and his patients, by whom it is warmly remembered. His still more intimate life was probably known to few."

Ravenhill died in London at 66 Onslow Gardens, South Kensington, on March 24, 1952. A funeral service was held in the lovely Christ Church, Flood Street. His will, which was prepared in 1947, is an informative document because it lists many of his large circle of friends. Since Ravenhill had no children, all bequests were made to his brother, his two surviving sisters, his niece and nephew, and his artistic friends. Individual bequests included his books of illuminated manuscripts, an edition of Horace Walpole's letters, various pieces of silver, and, of course, his paintings.

The only easily accessible obituary of Raven hill is the one paragraph in the *British Medical Journal* by Seymour Barling [10]. Another, by Cranston Walker, is in the *Birmingham Medical Review* [52], and there are short statements in the *Old Edwardian's Association Gazette of* July and December, 1952. A substantial number of his paintings are still extant, chiefly with relatives. A niece is still living, as well as the widow of his nephew and two of her children and several grandchildren.

Acknowledgements I am indebted to Margaret Rosenthal; Marjorie Rosenthal; S. R. Jenkins, Barnes Librarian, University of Birmingham; Dr. Tilli Tansey, Wellcome Institute for the History of Medicine; Kerry York, Resources Centre Librarian, The Schools of King Edward the Sixth in Birmingham; Margaret Ramsden; and George Lambrick, Oxford Archeological Unit, for the photograph used as Fig. 18.4.

References

1. Abelli V. Développment et guérison d'une pulmonite sur Ie sommet du Mont Rosa (4650 mètres). Arch Ital Rial. 1896;26:1–10.
2. Acosta I. de Historia Natural y Moral de las Indias. Seville: Juan de Leon; 1590.
3. Acosta I. de The Naturall, and Morall Historie of the East and West Indies. London: ward Blount and William Aspley; 1604. pp. 144–9.
4. Anonymous. Economic development of Chile. Chilean Rev Lond. 1921;1:3–10.
5. Alzamora-Castro V, Garrido-Lecca G, Battilana G. Pulmonary edema of high altitude. Am J Cardiol. 1961;7:769.
6. Bardález A. Algunos cases de edema pulmonar agudo por soroche grave. Anal Fac Med Lima. 1955;38:232–43.
7. Bardález VA. Edema of the lung in mountain sickness. J Am Med Assoc. 1956;160:698.
8. Bardález VA. Edema pulmonar agudo por soroche grave. Rev Peruana Cardiol. 1957a;6:115–39.
9. Bardález VA. Edema pulmonar agudo por soroche grave. Rev Asoc Med Provincia Yauli. 1957b;2:279–305.
10. Barling SG. Obituary of Dr. Thomas Holmes Ravenhill. Br Med.J. 1952;1:1032.
11. Barling S, Morrison JT. A manual of war surgery. London: Hodder & Stoughton; 1919.
12. Bert P. La Pression Barométrique. Paris: Masson; 1878. (English translation by M. A. Hitchcock and F. A. Hitchcock. College Book Co., 1943.)
13. Blakemore H. From the Pacific to La Paz. London: Lester Crook Academic Publishing; 1990. p. 44.
14. Fitch RF. Mountain sickness: a cerebral form. Ann Int Med. 1964;60:871–6.
15. Fred HL, Schmidt AM, Bates T, Hecht HH. Acute pulmonary edema of altitude. Clin Physiol Obs Circ. 1962;25:929–37.
16. Haber LF. The poisonous cloud. Oxford: Clarendon; 1986.
17. Hall WH, Barila TG, Metzger EC, Gupta KK. A clinical study of acute mountain sickness. Arch Environ Health. 1965;10:747–53.
18. Heath CH, editor. Service record of King Edward's school. Birmingham: Cornish Brothers; 1920.
19. Hill L. Foreword. In: Campbell A, Poulton EP, editor. Oxygen and carbon dioxide therapy. London: Oxford University Press; 1934.
20. Houston CS. Acute pulmonary edema of high altitude. N Engl J Med. 1960;263:478–80.
21. Hultgren H, Spickard W. Medical experiences in Peru. Stanford M Bull. 1960;18:76–95.
22. Hultgren H, Spickard W, Hellriegel K, Houston C. High altitude pulmonary edema. Medicine. 1961;40:289–313.
23. Hultgren HN, Lopez CE, Lundberg E, Miller H. Physiologic studies of pulmonary edema at high altitude. Circulation 1964;29:393–408.
24. Hurtado A. Aspectos Fisiólogicos y Patológicos de la Vida en la Altura. Lima: Empresa it. Rimac; 1937.
25. Hurtado A. Pathological aspects of life at high altitudes. Mil Med. 1955;117:272–84.
26. Lambrick G. The Rollright Stones: the archaeology, and folklore of the stones and their surroundings, a survey and review. Oxford: Oxford Archaeological Unit; 1983.
27. Lambrick G. The Rollright Stones: megaliths, monuments, and settlement in the prehistoric landscape. London: Historic Buildings and Monuments Commission for England; 1988.
28. Lizárraga L. Soroche agudo: edema agudo del pulmón. An Fac Med Lima. 1955;38:244–74.
29. Maurois A. The life of Sir Alexander Fleming, discoverer of penicillin. London: Cape; 1959.
30. Meyer-Ahrens C. Die Bergkankheit oder der Einfluss des Ersteigens grosser Höhen auf den thierischen Organismus. Leipzig: Brockhaus; 1854.
31. Mosso A. Fisiologia dell'Uomo Sulle Alpi; Studii Fatti sul Monte Rosa. Milan: Treves; 1897.
32. Mosso A. Life of man on the high Alps. London: Fisher Unwin; 1898. pp. 42–7.
33. Ortuzar A. Chile of today. New York: Tribune Assoc;1907.

34. Packard RO. The copper-silver mines of the Collahuasi district. Chilean Rev Lond. 1924;382–5 (October 1924).
35. Peñaloza D, Sime F. Circulatory dynamics during high altitude pulmonary edema. Am J Cardiol. 1969;23:369–78.
36. Ravenhill TH. Some experiences of mountain sickness in the Andes. J Trop Med Hyg. 1913;16:313–20.
37. Ravenhill TH. The Rollright Stones, and the men who erected them: an outline of recent evidence. Birmingham: Birmingham Printers;1926a.
38. Ravenhill TH. Notes on the Rollright Stones. Birmingham Archaeol Soc Trans. 1926b;51:43–4.
39. Ravenhill TH. The Rollright Stones: some facts and problems. Rep Oxfordshire Archaeol Soc. 1926c;121–43.
40. Ravenhill TH. The Rollright Stones, and the men who erected them: an outline of recent evidence. 2nd ed. Birmingham: Birmingham Printers;1932.
41. Ravenhill TH, Chatwin PB. The Rollright Stones. Birmingham Archaeol Soc Trans Proc. 1927;52:305.
42. Reeves JT, Groves BM, Sutton JR, Wagner PD, Cymerman A, Malconian MK, Rock PB, Young PM, Houston CS. Operation Everest II: preservation of cardiac function at extreme altitude. J Appl Physiol. 1987;63:531–9.
43. Rotta A, Canepa A, Hurtado A, Velasquez T, Chavez R. Pulmonary circulation at sea level and at high altitudes. J Appl Physiol. 1956;9:328–36.
44. Roy SB, Guleria JS, Khanna PK, Manchanda Sc, Pande JN, Subba PS. Haemodynamic studies in high altitude pulmonary oedema. Br Heart J. 1969;31:52–8.
45. Singh I, Kapila CC, Khanna PK, Nanda RB, Rao BDP. High-altitude pulmonary oedema. Lancet. 1965;1:229–34.
46. Singh I, Khanna PK, Srivastava MC, Lal M, Roy SB, Subramanyam CSV. Acute mountain sickness. N Engl J Med. 1969;280:175–84.
47. Suarez J, Alexander JK, Houston CS. Enhanced left ventricular systolic performance at high altitude during Operation Everest II. Am J Cardiol. 1987;60:137–42.
48. Ulloa P. Rincon filatelico. Revista La Candela. 1994;5:14.
49. Von Euler US, Liljestrand G. Observations on the pulmonary arterial blood pressure in the cat. Acta Physiol Scand. 1946;12:301–20.
50. Walker C. The relation of the curvature of vessels and of hollow viscera to their internal pressure. Br Med J. 1922;1:260–2.
51. Walker C. Obituary of T. H. Ravenhill. Old Edwardian's Assoc. Gazette December, 1952a.
52. Walker C. Obituary of Thomas Holmes RavenhilL. Birmingham Med Rev. 1952b;17:231–32.
53. Ward MP, Milledge JS, West JB. High-altitude medicine and physiology. 2nd ed. London: Chapman and Hall; 1995. pp. 393–4.
54. West JB. High altitude physiology. Stroudsburg: Hutchinson Ross Publishing;1981.
55. Yunge G. Estadistica Minera de Chile en 1906 i 1907. Santiago de Chilo: Imp. Litografia i Encuadernacion Barcelona; 1909.

Chapter 19
George I. Finch and His Pioneering Use of Oxygen for Climbing at Extreme Altitudes

Abstract George Ingle Finch (1888–1970) was the first person to prove the great value of supplementary oxygen for climbing at extreme altitudes. He did this during the 1922 Everest expedition when he and his companion, Geoffrey Bruce, reached an altitude of 8320 m, higher than any human had climbed before. Finch was well qualified to develop the oxygen equipment because he was an eminent physical chemist. Many of the features of the 1922 design are still used in modern oxygen equipment. Finch also demonstrated an extraordinary tolerance to severe acute hypoxia in a low-pressure chamber experiment. Remarkably, despite Finch's desire to participate in the first three Everest expeditions in 1921–1924, he was only allowed to be a member of one. His rejection from the 1921 expedition was based on medical reports that were apparently politically biased. Then, following his record ascent in 1922, he was refused participation in the 1924 expedition for complex reasons related to his Australian origin, his forthright and unconventional views, and the fact that some people in the climbing establishment in Britain saw Finch as an undesirable outsider.

At the end of 1920, a telegram arrived in the India Office in London to say that the Tibetan government had given the long-awaited permission to explore the Everest region from Tibet, and the Alpine Club and Royal Geographical Society (RGS) in London lost no time in planning a reconnaissance expedition. In fact, there were three expeditions in the space of 4 years, the reconnaissance in 1921 and the attempts to reach the summit in 1922 and 1924. In thinking about the best people to make a summit bid, J. P. Farrar, the influential President of the Alpine Club, consulted widely and came to the conclusion that George I. Finch and his brother Maxwell "were two of the best mountaineers we have ever seen" and that they would be his first choice for the summit party [2].

At about the same time, the issue of climbing oxygen was discussed at length, and George Finch took a leading role. This was natural because he was an eminent physical chemist with a special interest in properties of gases. Furthermore, in experiments carried out in a high-altitude chamber in Oxford, Finch demonstrated an extraordinary tolerance to acute severe hypoxia during exercise. In fact, his performance in the low-pressure chamber in 1921 has apparently not yet been surpassed.

From all this, it might be expected that Finch would play a key role in the first three Everest expeditions. However, this was not to be, although during the 1922 expedition he and his companion Geoffrey Bruce reached the altitude record of 8320 m and in the process clearly demonstrated the great value of supplementary oxygen at extreme altitude. By an extraordinary series of events, Finch was first invited to join the 1921 expedition and then the invitation was withdrawn after he was declared unfit by two physicians whose reports were apparently politically in-fluenced. Even more remarkable was that, after his great success on the 1922 expe-dition and after being asked to modify the oxygen sets for 1924, he was not invited to be a member of that expedition.

The story is a fascinating one of personality conflicts and how Finch, an Aus-tralian who was educated in Europe, did not fit well with the English climbing establishment in the early 1920s to the great disadvantage of everyone concerned.

19.1   Use of Oxygen at High Altitude Before 1921

The beneficial effects of inhaling oxygen at high altitude were predictable when Paul Bert proved that low partial pressure of oxygen was responsible for physiolog-ical impairment [3]. Indeed, Bert showed that inhaling oxygen dramatically reduced the symptoms of hypoxia when he exposed himself to a low barometric pressure in a pressure chamber. For example, in one experiment, he reduced the barometric pressure to 418 mmHg, equivalent to that on the summit of Mont Blanc, and showed that breathing oxygen immediately relieved his nausea and reduced his pulse rate. In another remarkable experiment, the chamber pressure was reduced to as low as 248 mm Hg, equivalent to that on the Everest summit, and Bert demonstrated that he could remain conscious by breathing supplementary oxygen. Bert also recom-mended that oxygen be taken on the famous but ill-fated flight of the Zénith balloon in 1875, but unfortunately too little oxygen was carried and the equipment made it difficult to inhale, resulting in two of the three balloonists dying from hypoxia.

Bert did not carry out any experiments on mountains; in fact, one of the first proposals to test the effects of inhaling oxygen at high altitude in the field preceded Bert's experiments by ~50 years. A Russian, Joseph Hamel (1788–1862), was prob-ably the first person to suggest the use of supplementary oxygen when he made an attempt on Mont Blanc in 1820 with the intention of using compressed oxygen, although in the event suitable equipment was not available [27]. His expedition was also ill-fated because of an avalanche that killed three guides.

The first use of supplementary oxygen in the Himalayas was apparently in 1907 when A. L. Mumm, Thomas Longstaff, and Charles Bruce went to the Garhwal and made the first ascent of Trisul (7127 m), which remained the highest summit to be climbed for 21 years. Small oxygen generators were taken along on this expedition, which was partly done to celebrate the Golden Jubilee of the Alpine Club [25]. However, no serious assessment of the value of oxygen was possible.

By far the most important forerunner in the use of oxygen at extreme altitudes was Alexander Mitchell Kellas (1868–1921). He carried out the first rigorous tests of the value of supplementary oxygen during climbing at high altitude on Kamet in the autumn of 1920 [19, 20]. He arranged for 74 tanks of oxygen to be sent to him, and each of these weighed nearly 9 kg with its regulator. Kellas carried out a series of preliminary experiments at an altitude of ~6,400 m on Kamet and more systematic studies at ~5,500 m on the Bagini glacier. His conclusion was that the cylinders were "too heavy for use above 18,000 ft [5486 m], and below that altitude they are not required. They would be quite useless during an attempt on Mt. Everest" [20].

Fortunately, Kellas also took along rubber bags and Oxylithe (sodium peroxide), an oxygen generator suggested by Prof. Leonard Hill of University College in London. In one set of experiments, he breathed oxygen from a bag and timed an ascent of ~10 min while breathing air and then compared this with another done without previous oxygen breathing. He stated "the times were practically identical" and that "the excess amount [of oxygen] in the lungs at starting was of negligible value in promoting ascent." Kellas then pointed out that this conclusion could have been predicted based on the very small oxygen stores of the body. Nevertheless, it is interesting that this technique has been repeatedly used, for example by the Swiss in their attempt on Mt. Everest in 1952 [6].

Kellas then carried out a third set of experiments using oxygen from the Oxylithe bag while climbing. He carried the bag under the arm, which he stated was inconvenient. However, the results were dramatic, and Kellas stated that "the gain while using oxygen was quite decisive, the advantage being up to 25 %. This again was to be expected, and clearly indicates that the light oxygen cylinders suggested above might be of considerable value as regards increase of rate of ascent at high altitudes" [20].

The first expedition to Mt. Everest took place only a few months later in the spring of 1921. Kellas was a member of this expedition, but although oxygen tanks were taken in, they were never used [30]. The main reason for this was that Kellas died during the approach march, and there was insufficient interest on the part of the other members of the expedition. An additional reason was that the tanks were heavy because there was not enough time to prepare new equipment, mainly because Kellas did not return to England between his tests on Kamet in 1920 and his joining the Everest expedition in the spring of 1921.

As discussed below, the first extensive use of supplementary oxygen for climbing at high altitude took place on the second Everest expedition in 1922. Curiously, the pioneering experiments of Kellas were virtually ignored, partly perhaps because Kellas was a very private person and usually climbed on his own, except for a few native porters accompanying him. An exception to this was his expedition to Kamet in 1920 when he was accompanied by Major H. T. Morshead. The fact that his results were published in the *Geographical Journal* in February 1921 [20] should have made them available; remarkably, however, in the preparations for the use of oxygen in 1922 his name was hardly mentioned.

19.2 Early History of G. I. Finch

George Ingle Finch (1888–1970) (Fig. 19.1) was born in Orange, New South Wales, Australia, then a small country town ~200 km west of Sydney. He was the eldest son of Charles Edward Finch (1843–1933) who owned a sheep and cattle property.

Fig. 19.1 George I. Finch during the 1922 expedition. He is wearing a jacket of his own design filled with eiderdown. This was ridiculed by A. R. Hinks as a "patent climbing outfit" but in fact was the forerunner of the down clothing that is now standard for high-altitude climbing. (From Ref. [9].)

Scott Russell, son-in-law of G. I. Finch, wrote a memoir about him [14] and made the point that Charles Finch had been born in Australia only 6 years after Queen Victoria had ascended to the British throne and that he regarded himself as entirely English. The emergence of Australian nationalism lay far in the future. Charles Finch was very interested in local affairs, and his legal training made him a suitable chairman of the district Land Board Court. George had a brother Maxwell, about 1 year younger, and also a sister, Dorothy.

In his beautifully written and influential book *The Making of a Mountaineer*, George Finch described how at the age of 13 he found himself on the summit of a hill outside Orange entranced by the view and that "after this, my first mountain ascent, I had made up my mind to see the world; to see it from above, from the tops of mountains" [9]. The family moved to Europe in 1902 when George was only 14 and spent some time in England and France. George's mother Laura was more cosmopolitan than her husband and, it is said, was bored by the country life in England that her husband so much enjoyed. In Paris, Laura had a large and interesting circle of friends, including Charles Richet, a Nobel laureate for his work on anaphylaxis, and Sir Oliver Lodge, an eminent physicist. Something of a crisis occurred when George's education was being planned. It would have been natural for him to go to an English public school, but there was a feeling that the discipline and restrictions were too repressive. Instead, Laura arranged for the boys to be privately tutored in Paris, and Charles Finch returned to Australia without her. It is said that Charles Finch never saw his wife again, but his daughter Anne disputes this.

George and his brother wasted no time in testing their climbing abilities when they reached Europe. The following two extracts are from *The Making of a Mountaineer*.

My brother Maxwell and I, now proud possessors of Edward Whymper's *Scrambles in the Alps* emulated our hero's early exploits by scaling Beachy Head [a steep cliff in southern England] by a particularly dangerous route, much to the consternation of the lighthouse crew and subsequent disappointment of the coast guards who arrived up aloft with ropes and rescue tackle just in time to see us draw ourselves, muddy and begrimed, over the brink of the cliff into safety.... A few weeks later, an ascent of Notre-Dame [the cathedral in Paris] by an unorthodox route might well have led to trouble, had it not been for the fact that the two gendarmes and the kindly priest who were the most interested spectators of these doings did not lack a sense of humour and human understanding.

George spent a short period at the École des Médecine in Paris and became fluent in French. However, he felt that he would be more comfortable in a more exact science and soon switched to the physical sciences. At the suggestion of Laura's friend, Sir Oliver Lodge, George moved to the Eidgenossische Technische Hochschule in Zurich and soon became fluent not only in proper German but also in the Swiss dialect. He was at the Eidgenossische Technische Hochschule from 1906 to 1911 and was awarded the Gold Medal at the end of his course for the diploma in technical chemistry. The weekends and summer vacations were spent climbing extensively in the Alps, and George became an outstanding mountaineer and president of the prestigious Zurich Academischer Alpen Club.

In 1912, he returned to England and in the following year became associated with the Imperial College of Science and Technology in London, which remained his scientific base for the next 40 years. During the First World War, he served with the Royal Field Artillery in France and was later attached to the Ordinance Corps and worked on explosives in Salonica. He was awarded the Military M.B.E., was mentioned in dispatches, and was demobilized with the rank of captain. He then returned to Imperial College. He was married briefly in 1916 and there was a son, Peter Finch, who became a successful actor. The paternity of Peter Finch has been the subject of much discussion [7, 8]. A second marriage in 1921 was very happy, and there were three daughters, one of whom, Anne, married Scott Russell.

19.3 Preparations for the Expedition of 1921

Although, as indicated above, Kellas' experiments with oxygen equipment on Kamet in 1920 were subsequently ignored, another of his observations was responsible for a renewed interest in oxygen equipment in 1921 and 1922. In discussing the reasons for their failure to reach the summit of Kamet, Kellas and Morshead referred to the difficulties with stoves at high altitude. Morshead wrote, "I was unaware until too late that the large Primus stove, on which I had been relying, would not work in the rarified atmosphere of 20,000 ft, beyond which point methylated spirit is the only possible fuel; while Dr. Kellas had only one small spirit stove which took an hour to thaw sufficient snow to fill a teapot" [20].

With the aim of improving the Primus stoves for the 1921 expedition, Finch went to Oxford in March 1921 to see G. M. B. Dobson, a lecturer in meteorology, who was carrying out tests in a low-pressure chamber in the laboratory of Georges Dreyer, F.R.S., the professor of pathology. Finch was accompanied by P. J. H. Unna, another member of the Oxygen Subcommittee set up by the main Everest Committee and who wrote an informative account of the visit [29]. Dreyer (1873–1934) had been educated in Copenhagen and obtained his medical degree there in 1900. He subsequently worked as a bacteriologist and virologist and coauthored a report with the physiologist Mabel FitzGerald [15]. Dreyer had been a consultant to the Royal Flying Corps (to become the Royal Air Force) during World War I and had probably carried out more research with oxygen for aviators than anyone else in the United Kingdom. His design of oxygen equipment for aircraft was very successful and was used, for example, by the Air Service, U.S. Army [1]. Dreyer's low-pressure chamber was a steel cylinder 2.1 m in diameter with glass windows and was evacuated by means of an electric pump.

The stoves were duly modified for high-altitude use, and then Finch and Unna switched their attention from the effects of oxygen on the stoves to the possible value of oxygen for climbers. Dreyer had strong views on this, stating "I do not think you will get up [Everest] without it [supplementary oxygen], but if you do succeed you may not get down again" [29]. Finch, Unna, and J. P. Farrar, former President of the Alpine Club, went to Oxford on March 25, 1921, to see the tests on the stoves, and Dreyer convinced Farrar that the question of oxygen for climbers should receive serious consideration. Dreyer's contributions to the use of oxygen on the 1922 expedition were therefore considerable. Finch stayed in Oxford overnight, and Dreyer carried out experiments on him in the low-pressure chamber the next day. The simulated altitude of the chamber was set at 21,000 ft (6400 m), and Finch stepped up on a chair, first with one foot and then with the other, 20 times in succession while carrying a load of 35 lb. (16 kg) slung over his shoulder. The stepping rate was chosen to correspond to a fairly rapid climbing pace.

The results showed that while Finch was standing still and holding the load at the simulated altitude of 6,400 m his pulse rate was 104 beats/min. Immediately after the exercise during which he stepped up onto the chair 20 times in 2.5 min, the pulse rate was 140 beats/min. However, when Finch breathed supplementary oxygen, the pulse rate while standing was only 77 beats/min, only slightly more than his normal pulse rate of 68 beats/min while standing at sea level. Immediately after the exercise, which this time was accomplished in only 2 min, his pulse rate was not above 100 beats/min. As Dreyer stated, "The striking effect of taking oxygen is obvious from these figures." Dreyer went on to add, "Apart from the effect upon pulse, there was a marked change in his whole condition and appearance shortly after he began to breathe oxygen. His expression and colour became normal, and his elasticity of movement returned, as shown by the fact that although he attempted to maintain the same rate for his exercises in both cases, he unconsciously shortened the time from 2 1/2 to 2 min" [29].

Dreyer reported his medical examination of Finch and the results of the low-pressure chamber studies in a four-page letter to Farrar dated March 28, 1921, which is in the archives of the RGS. Part of the summary reads as follows [26]

1. Captain Finch is slightly under weight at present, otherwise his physique is excellent.
2. He has an unusually large vital capacity. This indicates a high degree of physical fitness, and he should therefore be able to stand great exertion at high altitudes better than most persons.
3. Furthermore, the tests in the low pressure chamber proved that Captain Finch possesses quite unusual powers of resistance to the effects of high altitudes. Among the large number of picked, healthy, athletic young men which we have examined, more than 1,000 in all, we have not come across a single case where the subject possessed the resisting power to the same degree.
4. The administration of oxygen to Captain Finch at 21,000 ft at once restored him to normal, as measured by colour, expression, pulse and elasticity of movement.

The improvement shown by Finch while breathing oxygen was impressive, but even more remarkable was his ability to do this large amount of work at this very considerable simulated altitude of 6,400 m while breathing air. An analysis of the exercise test is given in Appendix 1. This extraordinary demonstration of fitness by Finch contrasted greatly with subsequent events in 1921.

All members of the 1921 expedition were required to have a routine medical examination, and on March 17 and 18, just 1 week before his exemplary performance in the low-pressure chamber in Oxford, Finch was examined by two London physicians. Amazingly, they concluded that he was unfit, and soon after this his invitation to join the expedition was rescinded by the selection committee less than 1 month before he was scheduled to sail! The medical reports are in the archives of the RGS and make astonishing reading today. They are reproduced in Fig. 19.2; a transcription is given in Appendix 2 (Table 19.1) because the handwriting is difficult to read. Additional information about the authors of the two reports and the expedition doctor, A. F. R. Wollaston, is in Appendix 3. The role of Wollaston in the decision to reject Finch is not spelled out, although Mallory stated in a letter to Winthrop Young that "Wollaston told me that there could be no question of taking Finch after the doctors' report" [26]. This statement suggests that Wollaston was biased toward excluding Finch in that he was willing to ignore Dreyer's report.

By any reasonable assessment, the two medical reports are totally inadequate and cannot possibly justify rejecting Finch at this stage. The surgeon's report is absurdly short and replete with vague terms, such as sallow, nutrition poor, spare, and flabby. The medical report (that is, by an internist) is a little longer but contains no convincing data for rejecting Finch. The only positive findings were a mild degree of anemia (now known to be a frequent finding in athletes), some missing teeth, and apparently some glucosuria as indicated by the positive Fehling test. This last finding is curious and suggests that the test should have been repeated immediately by Larkins. When Dreyer performed the same test 1 week later, it was normal; when it was repeated in November (later that year) by Larkins it was also normal. The con-

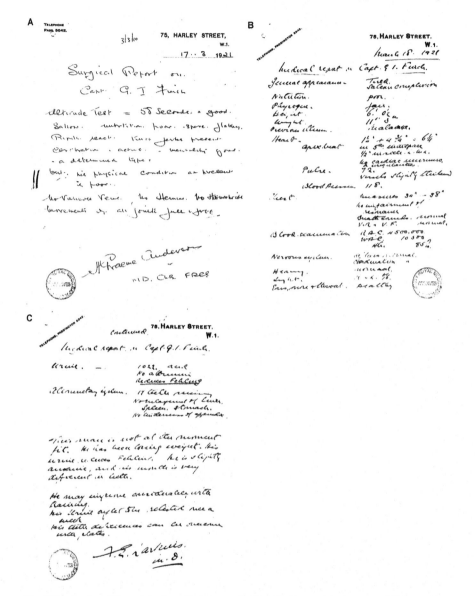

Fig. 19.2 Copies of the original medical reports by Drs. H. Graeme Anderson and F. E. Larkins. The handwriting is transcribed in Table 19.1. The originals are in the archives of the Royal Geographical Society (reference RGS/EE3/14) and are reproduced with permission

trast between these two inadequate medical reports and the remarkable performance recorded by Dreyer in the low-pressure chamber 1 week later is dramatic indeed.

How is it possible to explain Finch's rejection on such flimsy grounds? The Everest Committee, as it was known, consisted of representatives from the RGS and the

Alpine Club. Finch had strong supporters, such as John Percy Farrar (1857–1929), former President of the Alpine Club, but powerful enemies, including Arthur Robert Hinks (1873–1945), Secretary of the RGS [28]. There had been several instances in the past where Hinks' animosity toward Finch had been demonstrated, and it seems likely that he was looking for an excuse to reject Finch. In fact, it is natural to wonder whether the opinions of the two physicians who provided the medical reports were influenced by Hinks or some other member of the Everest Committee.

There may have been other reasons for the antipathy of some members of the Everest Committee. A photograph of Finch had been published in an illustrated paper with some information about the forthcoming expedition, and this annoyed Hinks who had a great dislike of publicity in the press. And there may have been other factors, too. The 1921 Everest expedition, and indeed the subsequent ones for the next 30 years, were dominated by men who had been to English public schools and were educated in the Oxford and Cambridge universities or who had a military background. The elitist character of the Alpine Club was amusingly referred to by Scott Russell when he described his first visit to the Club shortly after he had been elected a member. He met Sydney Spencer, one of the vice presidents, who remarked, "I hope your proposers told you that in addition to being the oldest mountaineering club in the world, the Alpine Club is a unique one—a club for gentlemen who also climb." Then, pointing out of a window at a street sweeper in Savile Row where the club was then situated he added, "I mean that we would never elect that fellow even if he were the finest climber in the world" [14]. Naturally, such an attitude would not set well with the Australian Finch who had been educated in Europe, was not part of the English establishment, and was a great believer in each person's right to determine his own destiny.

The old-fashioned views of some members of the Alpine Club were highlighted in an anecdote related by C. J. Morris, who was also a member of the 1922 expedition. Col. E. L. Strutt was the deputy leader of the expedition and a power in the Alpine Club. As stated by Morris, Strutt was aware that Finch [24]

> had been educated in Switzerland and had acquired a considerable reputation for the enterprise and skill of his numerous guideless ascents [many members of the Alpine Club were against ascents without guides]. Besides, he was by profession a research chemist and therefore doubly suspect, since in Strutt's old-fashioned view the sciences were not a respectable occupation for anyone who regarded himself as a gentleman. One of the photographs which particularly irritated him depicted Finch repairing his own boots. It confirmed Strutt's belief that a scientist was a sort of mechanic. I can still see his rigid expression as he looked at the picture. "I always knew the fellow was a shit," he said, and the sneer remained on his face while the rest of us sat in frozen silence.

In spite of this, Finch and Strutt later became firm friends. Morris also remarked on the rigorous scientific approach of Finch, which contributed to other logistical aspects of the expedition. Finch's empirical attitude contrasted with the more romantic approach of other members of the expedition, for example, that of Mallory.

The allegation that the two medical reports were biased is of course a very serious one, and it is pertinent here to summarize the events in the decision to rescind Finch's invitation to be a member of the 1921 expedition. The responsibility for the membership of the expedition was vested in the Everest Committee, which was made up of four members of the RGS and four from the Alpine Club. Hinks was the Secretary of the RGS and became the main secretary of the Everest Committee. The other three members from the RGS were Sir Francis Younghusband, Col. E. M. Jack, and Edward Somers-Cocks. From the Alpine Club, there were J. E. C. Eaton, Norman Collie, Capt. J. P. Farrar, and C. F. Meade. The Everest Committee quite reasonably required that candidates for the expedition should undergo a medical examination. The two doctors who examined Finch were presumably chosen by Wollaston, the official expedition doctor. Details of these three people are in Appendix 3. The choice of Wollaston as expedition doctor might have raised eyebrows because his choice to study medicine was made reluctantly and, as pointed out in Appendix 3, he disliked its practice. On the other hand, he had been on several expeditions to remote areas and had distinguished himself as a surgeon in the Royal Navy during World War I when he was awarded a D.S.C.

His choice of Dr. Graeme Anderson as one of the physicians seems reasonable because Anderson had written the first book on the medical and surgical aspects of aviation and was interested in the selection of aviators. The reasons for the choice of Dr. F. E. Larkins are less clear because he was a pediatrician chiefly interested in the health of school children. It is possible that Wollaston asked Anderson's advice and he recommended Larkins. Larkins and Anderson had the same address and telephone number (Appendix 2) and presumably were partners in a practice. This raises the question of whether the two reports were independent.

The medical reports themselves (see Table 19.1 in Appendix 2) definitely seem to be biased in that the conclusions do not fit with the facts of the examination. Anderson's report states that Finch's "physical condition at present is poor," and Larkins' report concludes, "This man is not at the moment fit." However, nothing else in the reports supports these conclusions with the possible exception of the positive Fehling test, which should have been repeated. The allegation of bias is based on the fact that the conclusions of the two medical reports are not consistent with the details of the reports.

These internal inconsistencies are further highlighted by the enormous discrepancy between the medical tests and Finch's extraordinary demonstration of fitness 1 week later in the low-pressure chamber (Appendix 1). Of course, Finch's performance during extreme hypoxia was far more relevant to his potential on Everest than the routine medical examinations and is an example of the value of "specificity of testing" now often applied to elite athletes. The results of these tests were sent by Dreyer to Farrar on March 28 only 6 days after the medical reports were received by Hinks. It is inconceivable that Farrar, who was a strong supporter of Finch, did not communicate this information to the other members of the Everest Committee. Yet, as indicated earlier, Wollaston apparently stated that there could be no question of taking Finch after the doctors' reports. The most likely scenario, unpleasant as it is, is that the wishes of the Committee to rescind Finch's invitation were made known to the two examining physicians and, furthermore, that the evidence from Dreyer's low-pressure chamber experiments was deliberately suppressed.

19.4 Expedition of 1922

In November 1921, Finch was reexamined by Dr. Larkins and declared fit (presumably to no one's surprise). The report read, "I have also reexamined G. I. Finch today. He is now absolutely fit and has lost his glucosuria. In my first report on him I stated that I thought all he needed was to get into training" [14]. Presumably, Finch still lacked the 17 teeth. Do we detect a slightly defensive air on the part of the good Dr. Larkins?

In any event, Finch was selected for the 1922 expedition and on January 13, 1922, found himself in Oxford again with Farrar, Unna, and this time the surgeon T. H. Somervell. Another test was carried out in the low-pressure chamber, this time at the higher altitude of 23,000 ft (7010 m). Finch with a 30 lb. (14 kg) load stepped on and off a chair 20 times without apparent difficulty at the rate of 8 s per step. Somervell, similarly loaded, started at 5 s a step, but was stopped after the fifth step, and oxygen was forcibly administered to prevent him from fainting [29]. When the group returned to London, a detailed report was made to the Everest Committee who agreed that oxygen should be obtained for the expedition.

It is interesting that, although Finch visited Oxford on several occasions in 1921 and 1922, there is no mention of the eminent physiologist J. S. Haldane who worked in Oxford at the time. Haldane is famous as the leader of the 1911 Anglo-American Pikes Peak Expedition that laid the groundwork for much of the physiology of high-altitude acclimatization. Haldane was certainly active in 1921–1922; in fact, the first edition of his influential book *Respiration* was published in 1922 [16]. The explanation may be that, after Haldane was passed over in preference to C. S. Sherrington for the Chair of Physiology at Oxford in 1913, he left the department and built his own laboratory in the garden of his house in north Oxford. There is also evidence from correspondence of some tension between Haldane and Dreyer. Haldane did take part in a discussion of the physiological problems of extreme altitude after the return of the 1924 expedition [17] when he made some curious disparaging remarks about the "somewhat disappointing effects of oxygen on well-acclimatized persons."

Dreyer naturally had links with the Air Ministry who were the experts in supplying compressed oxygen in light cylinders. A series of steel tanks 0.53 m long and 7.6 cm in diameter were made that could safely contain oxygen at a pressure of 150 atm. Tests showed that dropping the charged bottles from a height of 9 m on to a concrete floor resulted in nothing more serious than leakage at the valve; however, to allow for a further margin of safety, it was decided that the pressure would be reduced from 150 to 120 atm. Because the water content of the tank was 2 l, each held 240 l of oxygen at normal temperature and pressure. The total weight including the valve was 2.6 kg, of which 0.36 kg was the oxygen. A drawing and photograph of the tanks are shown in Figs. 19.3 and 19.5.

Dreyer was insistent that the supply of oxygen, once started, should not be interrupted, and this meant that the line going to the mask was always connected to two bottles so that it was possible to switch from one bottle to another without stopping

Fig. 19.3 Oxygen equipment designed for the 1922 expedition. This was the first successful oxygen apparatus for high-altitude climbing. Much modern equipment uses essentially the same principles. (Modified from Ref. [10].)

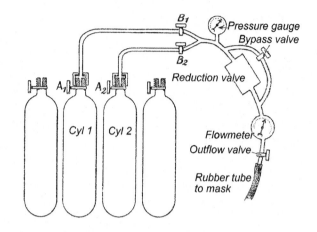

the supply. Subsequent experience has shown that this was a needless concern. The reducing valve, made by Siebe Gorman, was based on one used in high-altitude flying during the war. Two kinds of masks were designed, the "standard" and the "economizer." The latter can be seen in Fig. 19.4 and includes valves and a corrugated tube in which oxygen was stored during expiration. The tube was bent in the form of a U so that "the oxygen, being heavier than air, cannot escape through the open end" [29]. One wonders how effective this was because the difference in density between air and 100 % oxygen is only 11 %. Pipe mouthpieces were also made available in case anyone using a mask should find this suffocating. The climber put the rubber tube in his mouth and clamped it with his teeth during expiration. The whole apparatus complete with four full bottles of oxygen weighed 14.5 kg, and 10 sets were made available to the expedition.

Fig. 19.4 Drawing showing the mask, inspiratory and expiratory valves, and the economizer that was designed for the 1922 expedition. The economizer stores oxygen during the climber's expiration so that it can be subsequently inspired. The right-hand end of the economizer is open to the air. (Modified from Ref. [9])

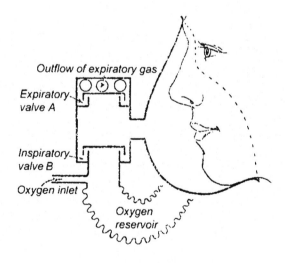

Fig. 19.5 Finch wearing the oxygen equipment of the 1922 expedition. The face mask is of the "standard" type without the economizer. (From Ref. [10].)

The oxygen flow rates advocated by Dreyer varied from 2 l/min at 7,000-m altitude to 2.4 L min at the Everest summit. He recommended that oxygen not be used below 7,000 m but always inspired above that altitude. He also suggested that sleeping oxygen be available at a flow rate of 1 l/min. Subsequent experience showed that sleeping oxygen at only 1 l/min was indeed appropriate; however, at altitudes of 8,000 m and above, a flow rate as high as 4 l/min was valuable. At the flow rates recommended by Dreyer, four bottles would provide for a climb lasting for 7 h. One of the oxygen sets is in the Science Museum (London, UK), although probably not on display.

Because Finch was a trained physical scientist with considerable technical skills, he took charge of the oxygen arrangements on the actual expedition. He instituted periodic oxygen drills, but these were unpopular and indeed there was a good deal of antagonism toward the use of oxygen. Finch stated that, "By the time we reached Base Camp, I found myself almost alone in my faith in oxygen" [14]. The anti-oxygen group, which included George Mallory, saw the use of oxygen as unnecessary, and they tended to ridicule the equipment. There were also objections on moral grounds, and Mallory referred to its use as "damnable heresy," which mystified Finch who pointed out that the mountaineers adopted "other scientific measures which render mountaineering less exacting to the human frame" [9]. Finch's rigorous professional attitude derived from his scientific training was evident in other logistical aspects of the expedition as well [24].

Finch and Geoffrey Bruce carried out the first trials of the oxygen equipment at high altitude toward the end of May 1922 and immediately obtained convincing results. It was clear that climbers who used oxygen could outpace those who were breathing air. For example, when Finch and Bruce climbed from Camp 3 (6400 m) to the North Col (7010 m), this took only 3 h with oxygen; this was much less than the usual time without oxygen. Subsequently, they used supplementary oxygen to put in a camp at 7770 m, which they expected would allow them to reach the summit the next day. However, a fierce storm developed, and they spent much of the night and the following day trying to prevent the tent from being blown away with themselves inside it. This unexpected delay reduced morale to a low level, and then Finch thought of inhaling oxygen at a low flow rate while they were sitting in the tent, whereupon their spirits improved. They used oxygen during the night and slept well, and Finch stated that he had no doubt that oxygen had saved their lives [11].

On the following morning, Finch and Bruce set off hoping to reach the summit, but the weather worsened. At an altitude of 8,320 m, they had some difficulties with the equipment and it soon became clear that the long period when they were delayed by the storm had weakened them too much. They reluctantly decided to go down, although they were only a little over 500 m from the summit. However, Finch and Bruce had reached a higher altitude than any human being had previously attained.

19.5   1924 and After

Finch had clearly demonstrated the value of supplementary oxygen for climbing at extreme altitude and indeed had, with Bruce, attained the altitude record. Subsequent calculations showed that the climbing rate of Finch and Bruce up to 7,770 m exceeded that of Mallory, Norton, and Somervell without oxygen by up to 50 % [21] or as much as 300 % [5]. Furthermore, as a professional scientist with strong engineering skills and someone who had been intimately involved with the design of the equipment for the 1922 expedition, Finch knew more about the technical aspects than any of the other climbers. In fact, on June 16, 1923, when the plans for the 1924 expedition were discussed, Finch was asked to advise the Everest Committee on the use of oxygen. It would be natural to think that, in light of these qualifications, Finch would be an obvious choice for the next Everest expedition, to take place in 1924. However, he was not selected by the Everest Committee and indeed never again had an opportunity to test the value of oxygen at extreme altitude.

The reasons for Finch's rejection have been discussed in detail by Scott Russell [14] and will only be summarized here. First, there was a dispute with Hinks regarding lectures that Finch was invited to give in Switzerland about the 1922 expedition. There was a limitation on the amount of lecturing that expedition members could do, and, according to Finch, Hinks claimed that this agreement would be enforced until Everest was climbed! Because this would not be for another 30 years, it does seem unreasonable. But there were other tensions as well along the lines of those referred to earlier. Finch had the reputation for being an outspoken, unconventional

Australian in a setting where these characteristics created antagonism. He was very much a square peg in a round hole in the setting of the Alpine Club in the early 1920s.

It is interesting that subsequent events proved the key importance of developing good oxygen equipment for the first ascent of Mt. Everest. One major reason for the failure of the Swiss on Everest in 1952 was their inadequate oxygen equipment [31]. John Hunt, leader of the successful British expedition of 1953, wrote after the expedition [18]

> Among the numerous items in our inventory, I would single out oxygen for special mention. Many of our material aids were of great importance; only this, in my opinion, was vital to success…. But for oxygen, without the much-improved equipment which we were given, we should certainly not have got to the top.

It is noteworthy that the oxygen equipment of 1922 (Figs. 19.3–19.5) was similar in many ways to that used on the successful 1953 expedition. Technical advances meant that the later tanks were lighter and could be filled to a higher pressure, but the reducing valve and economizer used the same principles. Finch saw the equipment in 1952 and referred to it as "splendid" [13]. It is natural to wonder whether E. F. Norton, who reached an altitude of 8570 m during the 1924 Everest expedition without supplementary oxygen, would have been able to accomplish the last 300 m if Finch's expertise had been available. Mallory and Irvine used oxygen equipment of a new design based on Finch's ideas, but whether they reached the summit will presumably never be known.

Finch went on to become an eminent scientist, particularly on the surface chemistry of metals [4]. His early work was on the mechanism of combustion in gases as a result of electrical discharges. He subsequently used electron diffraction to examine surface structures and developed the Finch camera for this purpose, which was a major advance. He also worked on the physical chemistry of lubricants, the electrodeposition of metals on surfaces, and electron diffraction in small crystals. He was elected a Fellow of the Royal Society in 1938.

In 1952, Finch retired from Imperial College to become Director of the National Chemical Laboratory in Poona, India. There, he stimulated research in two new fields related to solid state physics and to surfaces and thin films. He returned to England in 1957. Happily, he was eventually elected President of the Alpine Club and served in that position from 1959 to 1961 immediately following the term of John Hunt. It is not clear to what extent the square peg and round hole had changed their shapes, but probably a little of both occurred.

Finch wrote a book on his Everest experiences in German [10] and also a shorter account in English [12]. However, his most enduring monument is his book *The Making of a Mountaineer* [9]. Farrar's review in the *Alpine Journal* ended with the sentence

> All in all, the book is worthy to be set alongside Whymper's immortal "Scrambles [Amongst the Alps]" and Mummery's "My Climbs [in the Alps and Caucasus]" as indicating, as they did in their day, the high-water mark of mountaineering of the period.

The book has inspired generations of mountaineers. Scott Russell edited a reprint in 1988 [14] with a fine 116-page memoir of Finch at the beginning. John Hunt wrote the foreword, which begins "My copy of George Finch's *The Making of a Mountaineer* was a present from my paternal grandmother at Christmas 1924. I was 14 at the time, and it is that book which sowed the seed of my desire to climb." He went on, "The chapter on the Everest expedition in 1922 added a dimension of realism to the heroic and romantic character of the attempt on the summit which followed 2 years later. I felt that George Finch, who had done so much to show how the physiological problems might be solved, might well have played an important—perhaps a decisive—part in the ill-fated 1924 expedition."

Thus Finch, who was acknowledged as one of the strongest mountaineers of his generation, who had demonstrated extraordinary tolerance to hypoxia in a low-pressure chamber, and who had the scientific and technical expertise to develop oxygen equipment for the first three Everest expeditions, was denied the opportunity to participate in two of them because of the complexities of human nature. The story is one more in the never-ending romance of the ascent of Everest.

Acknowledgements I am indebted to the staff of the archives of the Royal Geographical Society and the Alpine Club and of the Biomedical Library at the University of California (San Diego, CA). Anne Russell kindly provided an interview.

The work was supported by National Heart, Lung, and Blood Institute Grant RO1 HL-60698.

Appendix 1

Analysis of the Exercise Tests of Finch in the Low-Pressure Chamber

Finch performed two exercise tests in Dreyer's low-pressure chamber in Oxford [26, 29]. The first was on March 26, 1921, and the second was on about January 14, 1922.

For the first test, the simulated altitude of the low-pressure chamber was 21,000 ft (6400 m), and Finch stepped up on to a chair, first with one foot and then with the other 20 times in succession in 2.5 min. This description allows us to make a reasonable estimate of his work rate. His body weight was 11 stone 3 lb. (see Appendix 2), and he carried a 35-lb. load, giving a total mass of 87.3 kg. If we assume a normal chair height of 46 cm, the work rate is $(87.3 \times 0.46 \times 20)/2.5$ or 321 kg/m = 52 W. The barometric pressure was 341 mmHg.

For the second test, the simulated altitude of the chamber was 23,000 ft (7,010 m). Finch carried a load of 30 lb. and stepped on and off the chair 20 times at 8 s per step. If we assume the same body weight, the work rate is $(85 \times 0.46 \times 20)/2.67$ or 293 kg/m = 48 W. The barometric pressure was 323 mmHg.

Almost no similar experiments have been reported with such a severe degree of acute hypoxia. However, it is clear that Finch's performance was exceptional. The only comparable study in the literature is that of Margaria [22, 23], in which three students aged 22 years and Margaria himself aged 27 years exercised on a bicycle ergometer in a low-pressure chamber at several simulated altitudes, including 6,500 m, barometric pressure 330 mmHg. Under these conditions, two of the students were unable to sit on the bicycle and ended up lying on the floor of the chamber where their skin became pale, their lips turned blue, and they lost consciousness! The third student was not studied at this altitude, and only Margaria was able to exercise under these conditions. In fact, Margaria was able to exercise up to an altitude of 7,000 m, barometric pressure 316 mmHg, although he developed spastic contractions of his hand and facial muscles. Margaria concluded that, on the average, work capacity falls to zero at just over 7,000-m altitude. These results are consistent with the statement of Dreyer that, of the 1,000 selected athletic young men exposed to simulated high altitude that he had examined, Finch showed the highest resistance to acute hypoxia. Incidentally, it is extremely unlikely that such an experiment would be allowed under present guidelines for human investigations because of the dangers of the extremely severe hypoxia. Also, it is probable that most of the 1,000 men referred to by Dreyer were exposed to hypoxia by using a rebreathing circuit, not a low-pressure chamber.

Appendix 2

These are the two medical reports that resulted in the withdrawal of the invitation to Finch to join the 1921 expedition (see Fig. 19.2). These medical reports are transcribed in Table 19.1.

Table 19.1 Transcribed version of the two medical reports shown in Fig. 19.2

Report by Dr. H. Graeme Anderson
Telephone
PADD. 5042.

75, Harley Street,
W.1.
17 .. 3 1921

Surgical Report on Capt. G. I. Finch

Altitude Test = 58 Seconds. = good [this was a breath-holding test]. Sallow. Nutrition poor. Spare. Flabby. Pupils react. Knee jerks present. Cerebration active. Mentality good - a determined type. But his physical condition at present is poor. No varicose veins. No hernias. No haemorrhoids. Movements of all joints full and free.

[Signed] H. Graeme Anderson, M.D., Ch.B., F.R.C.S.

Report by Dr. F. E. Larkins
Telephone, Paddington 5042

75, Harley Street,
W.1.
March 18, 1921

Medical Report on Capt. G. I. Finch

General appearance –	Tired. Sallow complexion
Nutrition.	Poor.
Physique.	Fair.
Height.	6' 02 in
Weight	11st 3 [157 lbs]
Previous illness.	Malaria
Heart – apex beat	$1\frac{1}{4}$" + $4\frac{3}{4}$" = $6\frac{1}{4}$" in 5th interspace
	$\frac{1}{2}$" inside? mcl
	no cardiac murmurs or irregularities.
Pulse –	72. Vessels slightly thickened
Blood Pressure.	118.
Chest	measures 34"–38"
	no impairment of resonance
	breaths sounds normal
	V.R. & V.F. normal
Blood examination	R.B.C. 4500,000
	W.B.C. 10,500
	Hb. 85%
Nervous system.	Reflexes normal. Coordination normal.
Hearing.	Normal.
Sight.	R & L 6/6.
Ears, nose & throat.	Healthy
Urine.–	1022. acid
	No albumin
	Reduces Fehling
Alimentary system.	17 teeth missing
	No enlargement of liver, spleen, stomach.
	No tenderness of appendix

This man is not at the moment fit. He has been losing weight. His urine reduces Fehling. He is slightly anaemic, and his mouth is very deficient in teeth.

He may improve considerably with training. His urine ought to be retested once a week. His teeth deficiences can be overcome with plates.

[Signed] F. E. Larkins, M.D.

Appendix 3

The medical reports by Drs. H. Graeme Anderson and F. E. Larkins were pivotal in Finch's exclusion from the 1921 expedition. More information on these two people and A. F. R. Wollaston, the expedition doctor, is reported below.

Major Henry Graeme Anderson (1882–1925), M.B.E., M.D., Ch.B., F.R.C.S. (England), was born in Scotland and was educated at Glasgow University, Kings College London, and the London Hospital. He was house surgeon at St. Mark's Hospital (London, UK) from 1907 to 1908 and surgical registrar at several hospitals, including the Royal National Orthopaedic Hospital (1909–1912). He joined the Royal Navy in 1914 and was attached as surgeon to the original Royal Navy Air Service Expeditionary Force. He gained his aviator's certificate at the Royal Aero Club in 1916 and was later transferred from the Royal Navy to the Royal Air Force as Major. In 1919, he authored a 255-page book (*The Medical and Surgical Aspects of Aviation*), which was apparently the earliest textbook on this subject. This book includes a section on the influence of acute exposure to high altitude on the body and so he was well-informed on this topic.

Francis Edmond Larkins (1880-?1962), M.B., Ch.B., M.D. (Edinburgh), D.P.H., R.C.P.S. (Edinburgh and Glasgow), was born in Kingston-on-Thames (Surrey, UK) and took his medical degree at the University of Edinburgh. He was a house physician at the Edinburgh Royal Infirmary and had other appointments at the Infectious Disease Hospital (Leith, UK) and the Children's Hospital at Newcastle-on-Tyne. He published several articles on children's diseases and was particularly interested in caries in elementary school children and children's nutrition. Incidentally, his interest in caries may explain his detailed comments on Finch's lack of teeth.

Alexander Frederick Richmond Wollaston (1875–1930), M.A., B.Ch., M.R.C.S., L.R.C.P., F.R.G.S., was born in Clifton near Bristol and was educated at Clifton College, Kings College (Cambridge, UK) and the London Hospital Medical School. He was a gifted naturalist who early made up his mind to become an explorer and only studied medicine to facilitate his chances of travel. In fact, he once wrote to his father, "medical practice as a means of a livelihood does not attract me—in fact I dislike it extremely." In a letter to a young man who was considering going into medicine, Wollaston wrote "I made such a horrible mistake when I went in for a profession which I loathe, and I am so constantly regretting it." He took part in several expeditions, including the British expedition to Ruwenzori, Central Africa, 1905–1907, and two expeditions to Dutch New Guinea, 1910–1913. He was a surgeon in the Royal Navy during World War I and medical officer and naturalist on the 1921 Everest expedition. In later life he was elected a fellow of Kings College, where tragically he was shot dead by a demented undergraduate. It is interesting that his entry in *Who Was Who* lists

F.R.G.S. (Fellow of the Royal Geographical Society) as his primary degree.

References

1. Air Service Medical. War Department (Air Service), Division of Military Aeronautics. Washington, DC: Government Printing Office; 1919.
2. Alpine Club Archives. B13 Farrar Letters. 1919.
3. Bert P. La Pression Barométrique. Paris: Masson; 1878.
4. Blackman M. George Ingle Finch. Biogr Mems Fellows R Soc. 1972;18:223–239.
5. Bruce CG. The assault on mount Everest. London: Arnold; 1923.
6. Dittert R, Chevalley G, Lambert R. Forerunners to Everest. London: George Allen and Unwin; 1954.
7. Dundy E. Finch, bloody Finch: a life of Peter Finch. New York: Holt, Rinehart, and Winston; 1980.
8. Faulkner T. Peter Finch, a biography. New York: Taplinger; 1979.
9. Finch GI. The making of a mountaineer. London: Arrowsmith; 1924.
10. Finch GI. Der Kampf um den Everest. Leipzig: F. A. Brockhaus; 1925.
11. Finch GI. Climbing Mt. Everest. London: Philip; 1930.
12. Finch GI. Man at high altitudes. Proc Roy Inst. 1952;35:349–57.
13. Finch GI. The Making of a mountaineer. 2nd edn. Bristol: Arrowsmith; 1988.
14. FitzGerald MP, Dreyer G. The unreliability of the neutral red method, as generally employed, for the differentiation of *B. typhosus* and *B. coli*. In: Contributions From the University Laboratory for M edical Bacteriology to celebrate the inauguration of the State Serum Institute. Copenhagen: Univ. Laboratory for Mical Bacteriology; 1902.
15. Haldane JS. Respiration. New Haven: Yale University Press; 1922.
16. Hingston RWG. Physiological difficulties in the ascent of Mount Everest. Geogr J. 1925;65:4–23.
17. Hunt J. The ascent of Everest. London: Hodder & Stoughton; 1953.
18. Kellas AM, Morshead HT. Dr. Kellas' expedition to Kamet. Alpine J. 1920/1921;33:312–319.
19. Kellas AM, Morshead HT. Expedition to Kamet. Geogr J. 1921;57:124–130, 213–219.
20. Longstaff TG. Some aspects of the Everest problem. Alpine J. 1923;35:57–68.
21. Margaria R. Capacita' di lavoro dell'uomo nell'aria rarefatta. Boll Soc Ital Biol Sper. 1929;4:691–3.
22. Margaria R. Die Arbeitsfahgkeit des Menschen bei vermindertem Luftdruck. Arbeitsphysiologie. 1930;2:261–72.
23. Morris J. Hired to kill. London: Hart-Davis; 1960.
24. Mumm AL. Five months in the Himalaya; a record of mountain travel in Garhwal and Kashmir. London: Arnold; 1909.
25. Royal Geographical Society. Everest Archives. Box 3. 1 A.D.
26. Simons E, Oelz O. Mont Blanc with oxygen: the first rotters. High Alt Med Biol. 2001;2:545–9.
27. Spencer Jones H, Fleure HJ. Arthur Robert Hinks. Obituary Notices Fellows Roy Soc. 1947;5:717–32.
28. Unna PJH. The oxygen equipment of the 1922 Everest expedition. Alpine J. 1923;34:235–50.
29. West JB. Alexander M. Kellas and the physiological challenge of Mt. Everest. J Appl Physiol. 1987;63:3–11.
30. West JB. High life: a history of high altitude physiology and medicine. New York: Oxford University Press; 1998.
31. West JB. Failure on Everest: the oxygen equipment of the spring 1952 Swiss expedition. High Alt Med Biol. 2003;4:39–43.

Chapter 20
Joseph Barcroft's Studies of High Altitude Physiology

Abstract Joseph Barcroft (1872–1947) was an eminent British physiologist who made contributions to many areas. Some of his studies at high altitude and related topics are reviewed here. In a remarkable experiment he spent 6 days in a small sealed room while the oxygen concentration of the air gradually fell simulating an ascent to an altitude of nearly 5500 m. The study was prompted by earlier reports by J.S. Haldane that the lung secreted oxygen at high altitude. Barcroft tested this by having blood removed from an exposed radial artery during both rest and exercise. No evidence for oxygen secretion was found and the combination of 6 days incarceration and the loss of an artery was heroic. In order to obtain more data, Barcroft organized an expedition to Cerro de Pasco, Peru, altitude 4300 m, that included investigators from both Cambridge UK and Harvard. Again oxygen secretion was ruled out. The protocol included neuropsychometric measurements and Barcroft famously concluded that all dwellers at high altitude are persons of impaired physical and mental powers, an assertion that has been hotly debated. Another colorful experiment in a low-pressure chamber involved reducing the pressure below that at the summit of Mt. Everest but giving the subjects 100 % oxygen to breathe while exercising as a climber would on Everest. The conclusion was that it would be possible to reach the summit while breathing 100 % oxygen. Barcroft was exceptional for his self-experimentation under hazardous conditions.

20.1   Introduction

Joseph Barcroft (1872–1947) (Fig. 20.1) was an eminent respiratory physiologist and information is available about him in an extensive obituary [21] and a biography [13]. This article deals with some of his studies of high altitude physiology that are not well known.

Barcroft was born in Northern Ireland and spent almost the whole of his working life in the University of Cambridge, UK. He carried out extensive studies in five different areas including metabolism of the submaxillary gland, physiology of hemoglobin including factors affecting the oxygen affinity, high altitude physiology particularly the determinants of the arterial PO_2, physiology of the spleen, and neonatal physiology. He made important contributions in all of these areas and re-

© American Physiological Society 2015
J. B. West, *Essays on the History of Respiratory Physiology,*
Perspectives in Physiology, DOI 10.1007/978-1-4939-2362-5_20

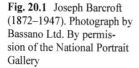

Fig. 20.1 Joseph Barcroft
(1872–1947). Photograph by
Bassano Ltd. By permis-
sion of the National Portrait
Gallery

ceived many honors during his lifetime including Fellowship of the Royal Society
and recipient of its Copley Medal, honorary doctorates from several universities,
and a knighthood.

The present article concentrates on four projects and relates some aspects that
have not previously received much attention.

20.2   Glass Chamber Experiment of 1920

This is one of the most colorful episodes in the history of high altitude physiology.
Barcroft was the subject himself, and he lived continuously in a small sealed room
for 6 days while subjecting himself to increasingly severe hypoxia. The conditions
were heroic. In the process he surrendered about 2.5 cm of his left radial artery,
nearly fainted during the blood-drawing procedure needing brandy to revive himself, and all for a result with an *n* of 1.

20.2.1   Rationale for the Experiment

This is clearly set out at the beginning of the article [6]. The study was a direct
response to the claim by John Scott Haldane (1860–1936) that the lung actively secreted oxygen across the blood-gas barrier. Much of the evidence for this came from
the Pikes Peak expedition of 1911 that included Haldane and three other physiolo-

gists [12], and the results were published in 1913, 7 years before Barcroft's experiment. The report from that expedition stated that on the first day of the period of a month on the summit, altitude 4300 m, the oxygen partial pressure in the blood exceeded that in the alveolar gas by 7 mm Hg. Furthermore as Barcroft noted, after 2 or 3 more days, and during exercise that doubled the oxygen consumption, the arterial PO_2 exceeded that in alveolar gas by an average of 32 mm Hg.

Haldane had originally become interested in the possibility of oxygen secretion by the lung following a visit to Christian Bohr (1855–1911) in Copenhagen. Haldane and Lorrain Smith carried out a series of experiments on animals that seemed to support the secretion hypothesis. Interestingly, Haldane retained his belief in oxygen secretion in spite of mounting evidence against, it throughout his life [22].

Barcroft was not the only person who was skeptical of oxygen secretion but he was determined to test the secretion theory. He pointed out that Haldane and colleagues had calculated the arterial PO_2 from the oxygen and carbon monoxide concentrations in the blood after the subject had inhaled a small amount of carbon monoxide, and he argued that this indirect measurement of PO_2 might be liable to error. For this reason Barcroft made direct measurements of the oxygen concentration of arterial blood by exposing the blood to a partial vacuum [4]. He then calculated the oxygen saturation, and derived the PO_2 from his own oxygen dissociation curve that he had measured in 1910 during an expedition to Tenerife [1].

It should be noted that in spite of this direct challenge by Barcroft to Haldane's belief, the two men were friends and in fact had previously worked together on methods of measuring the PO_2 in small samples of blood [4]. Although Haldane had his base in Oxford and Barcroft was in Cambridge, and Haldane held to the theory of oxygen secretion by the lung for the remainder of his life, these two eminent scientists remained cordial and spoke highly of each other's work on several occasions.

20.2.2   Background of the Glass Chamber

Barcroft's procedure of placing humans in a small closed glass chamber where they could inspire an altered oxygen mixture can be traced back to his work on gas poisoning during World War I [3]. Barcroft's family had been members of the Society of Friends (Quakers) for generations and were therefore non-combatants. Barcroft resolved this dilemma during the war by acting as a civilian attached to the army. One of his studies was to test the effects of poisoned gases in a closed chamber. Chlorine gas was initially used by the German army starting in April 1915 although apparently this new form of warfare did not have the support of the German High Command but came about because of a directive from the Kaiser himself [13]. The initial physiological studies using the chambers were made on rabbits exposed to chlorine or phosgene, but later soldiers who had been gassed in the trenches were treated with about 40–50 % oxygen in the chambers. Interestingly the idea of giving these patients an increased oxygen mixture to breathe was not immediately accept-

Fig. 20.2 Glass chamber in which Barcroft exposed himself to increasingly low concentrations of oxygen over 6 days. A bed, cycle ergometer and other equipment can be seen. The man is Sergeant-Major Secker who was Barcroft's longtime assistant. From [13]

ed although today we would automatically use oxygen in the treatment of patients with acute inflammation of the lung caused by a noxious gas.

The glass chambers were relatively small having capacities of 10 or 12 m³ and they were constructed of plate glass supported by iron frames. Three of the chambers could be connected together. Carbon dioxide that accumulated in the chamber was removed by soda lime and excess water vapor was taken out by calcium chloride. Typically the patients remained in the chamber overnight but left it for 8 h during the day. The size of the chamber used in the experiment described here was not given but the photograph (Fig. 20.2) shows that it was large enough to contain a bed and other equipment such as a cycle ergometer [19].

In his account of the experiment in 1920, Barcroft contrasts his use of a glass chamber in his laboratory with existing high altitude facilities. For example he describes the Margherita Hut on the Monte Rosa, altitude 4559 m, as "but an improvisation as compared with the modern physiological laboratory". He goes on to say that "at such places there is usually a certain amount of indigestion, sometimes constipation, sometimes diarrhoea, usually extreme cold. It is therefore open to dispute how far such symptoms as headache are due to disorders not directly connected with anoxemic conditions". By contrast he claimed that the glass chamber in his laboratory meant that "every sort of convenience was accessible". However the limited hygiene facilities for a continuous stay of 6 days and nights in his glass chamber suggests that while laboratory equipment was no doubt accessible, the conditions in the chamber must have been spartan.

The long period of 6 days was chosen because Haldane and his colleagues had reported that the oxygen secretion ability of the lung increased over time at high altitude. For the same reason, Barcroft exercised on a stationary bicycle on the last day because this was also reported to increase oxygen secretion. Throughout the 6 days and nights Barcroft's activities were closely monitored by eleven university students whose names are duly acknowledged at the end of the article. An interesting sidelight was that although the students reported that while Barcroft appeared to sleep well at night, his own belief was different. He stated "My own view of the matter was quite otherwise. I thought I had been awake half the night and was unrefreshed in the morning ... the slumber was very light and fitful with incessant dreams" [2].

Nowhere does Barcroft refer to the difficulties of living continuously in the chamber for 6 days although one wonders whether the British stiff upper lip attitude was a factor. He claimed that he "had been thoroughly well fed—a rather light breakfast, tea, eggs, bread and butter cooked by the attendant and lunch and dinner sent from the college kitchens". An exception was when he awoke on the morning of the 6th day with what he refers to as "the typical symptoms of mountain sickness, vomiting, intense headache and difficulty of vision". He very reasonably attributed these symptoms to the hypoxia because by then he was at an equivalent altitude of 18,000 ft (5486 m).

20.2.3 Experimental Procedure

Barcroft entered the chamber on February 1, 1920 and the operative procedures to sample arterial blood were carried out during the evening of February 7. When he entered the chamber, the PO_2 of the air was given as 163 mm Hg and this gradually fell as nitrogen was pumped in and oxygen was consumed by the subject. Barcroft analyzed the air in the chamber himself throughout the 6 days using a Haldane gas analyzer [14]. The lowest inspired PO_2 was 84 mm Hg on the morning of February 7. As stated above this was equivalent to an altitude of 18,000 ft (5486 m) which was far higher than any existing high altitude laboratory at the time. The removal of CO_2 from the chamber by soda lime was more efficient at the end than the beginning of the experiment, and during the last 48 h the PCO_2 varied from 3 to 5 mm Hg.

The actual surgical procedure is described in detail. The operation on the radial artery began at 7:28 p.m. but by 7:30 to 7:40 it was recorded that the subject was "inclined to be faint" and as a result he was given tea and brandy. As a result he was "sufficiently himself to breathe through valves without the samples being invalid through looseness of grip with the lips". The left radial artery was exposed for an inch and a half, a ligature was applied to the distal end, and a clip was placed on the proximal end. Then an incision was made in the artery to receive the cannula from which blood was withdrawn. This was done both during rest and exercise on the cycle ergometer, the total volume of blood removed being 83 ml. The exercise protocol was impressive, the total time being 37 min with the work rate varying between 350 and 386 kg m min^{-1} (57–63 watts). During the blood draws, alveolar gas samples were collected thus allowing the difference between the PO_2 in alveolar gas and arterial blood to be measured.

Some details of the operative procedure are puzzling. This must have been carried out in the chamber while Barcroft was exposed to the hypoxia. but it seems unlikely that the person could do the operative procedure when acutely exposed to an altitude of 5400 m. Perhaps he used an oxygen mask, but in this case it would be difficult to avoid contaminating the chamber with oxygen.

20.2.4   Experimental Results

These are described in Sect. 4 of the manuscript with the details given in Table XII. To be frank, the description of the results is not as clear as one would like. First there is a qualitative statement in the first paragraph that might raise eyebrows today. Barcroft stated that with the measured PO_2 in the alveolar gas of 57–68 mm Hg, the oxygen saturation of blood with the same PO_2 would be expected to be between 80 and 90%. He then went on to say that when the first sample of arterial blood was drawn, "the blood looked dark" and from this he argued that this was evidence against oxygen secretion since if the PO_2 of the blood were a few mm Hg higher than that of the alveolar gas as predicted by Haldane, the blood would be expected to have its usual red color. This is hardly convincing.

Table XII then shows the actual numbers. For the resting subject, the alveolar PO_2 is given as 68.4 mm Hg. This comes from a procedure that Barcroft referred to as the Krogh method, and consisted of collecting the last 2 ml of gas from each of a series of expirations in a glass tonometer. This value was then compared with the calculated arterial PO_2 of 60 mm Hg. This PO_2 was derived from the measured arterial oxygen saturation of the blood sample and then using a previously determined oxygen dissociation curve from Barcroft's own blood [1]. In this way for the resting subject the arterial PO_2 was calculated as 8.4 mm Hg less than the alveolar value.

For the exercising subject, the alveolar PO_2 was given as 56.5 mm Hg using the Krogh technique but 54 mm Hg using a slight modification of this called the Haldane method. The PO_2 in arterial blood was again calculated from the measured oxygen saturation and oxygen dissociation curve and the value was 48 mm Hg. Therefore the arterial PO_2 was some 6–8 mm Hg lower than that in alveolar gas. Barcroft went on to say that the disparity between the PO_2 in alveolar gas and arterial blood was greater during work than during rest although the actual numbers do not seem to support this.

Barcroft then calculated what he called the "diffusion constant" during work. This was done by dividing the measured oxygen consumption of 750 ml per minute by the alveolar-arterial PO_2 difference. The value was given as 107 ml (actually

107 ml min^{-1} mmHg^{-1}). However the basis of this calculation is questionable because the PO$_2$ of the blood in the pulmonary capillary during the loading of oxygen is not the arterial value but the gradual transition between the PO$_2$ of mixed venous blood and the arterial blood.

It was concluded that the results of these experiments disproved Haldane's secretion theory because both at rest and during exercise the arterial PO$_2$ was less than the alveolar value. But in the final paragraph of the manuscript Barcroft gallantly remarked "We yield to none in appreciation of the accuracy of the work on this subject of oxygen secretion published by Haldane and his colleagues, and we cannot but think that the difference between us and them has to be sought in the assumptions which Haldane makes, rather than the accuracy of his observations".

The reader of today comes away from this paper with somewhat conflicting emotions. Clearly this was a heroic experiment entailing what must have been 6 grueling days in a small enclosed space with the unpleasantness of the worsening hypoxia, not to mention the loss of one radial artery. On the other hand the method of measuring the arterial PO$_2$ was questionable with the value being derived from an oxygen dissociation curve measured several years previously. Perhaps the biggest limitation of all was that only two measurements were made, one at rest and another at exercise, and in only one subject. It is easy to imagine that Barcroft was looking for an opportunity to obtain additional data, and this was provided by the expedition to Cerro de Pasco that is described below. However before moving to that here is a brief description of another remarkable experiment carried out by Barcroft in one of his glass chambers.

20.3 Toxicity of Hydrocyanide Gas

As indicated earlier, Barcroft worked on gas poisoning during World War I. One of the gases being considered was hydrogen cyanide (HCN) and Barcroft volunteered to test the effects of this gas on himself and a dog. Although the experiment was carried out in 1917, the full report was published until 1931 [3].

Barcroft and a dog weighing about 12 kg were simultaneously exposed to an atmosphere containing a concentration of what was estimated to be 1 part in 2000 of hydrogen cyanide gas in air. The results of the experiment were set out in Table V which is reproduced here as Fig. 20.3. As the table shows, the dog became unsteady by 50 s and was unconscious by 1 min and 15 s. By 1 and a half minutes it developed tetanic convulsions and was thought to be nearly dead. A few seconds later Barcroft came out of the chamber, put on a respirator, and removed the dog. Five minutes after the initial exposure Barcroft felt a brief period of nausea and a little later noticed difficulty in maintaining attention during conversation. As the note at the bottom of the table indicates, the apparently dead dog was set aside for burial but in fact was found walking around next morning and fully recovered.

The reason for the striking difference in response of the dog and human to hydrogen cyanide is apparently still not understood. However the paper states that

Table V.

Time from zero	Dog	Man
50 sec.	Became unsteady	—
1 min. 15 sec.	On floor unconscious	—
1 min. 30 sec.	Crying sounds and tetanic convulsions sufficiently established to render it probable that animal was in extremis	—
1 min. 31 sec.	—	Came out of chamber and put on respirator having felt no symptoms. No apparent dyspnoea
1 min. 33 sec.	Respiration apparently* ceased, animal believed to be dead, was pulled out by lead	Re-entered chamber in respirator for purpose of pulling out dog. Having done this he remained outside
5 min.	—	Momentary feeling of nausea
10 min.	—	Attention difficult to concentrate in close conversation

* Although the corpse was set aside for burial about 6.30 p.m., the dog did in point of fact recover, and was found walking about next morning. It showed no further symptoms.

Fig. 20.3 Table summarizing the events when Barcroft and a dog were exposed to hydrocyanide gas in a glass chamber during experiments on poisonous gases prompted by World War I. From [3]

there are species difference in the toxicity of HCN and, for example, the guinea pig is unusually tolerant. The toxicity of HCN is principally due to its inhibition of the enzyme cytochrome oxidase c in mitochondria and this is highly conserved so the species difference is puzzling. Although this remarkable experiment is not strictly related to Barcroft's high altitude studies, it vividly demonstrates his willingness for self-experimentation and his courage.

20.4 International High Altitude Expedition to Cerro De Pasco, Peru

Barcroft was the principal organizer of this expedition although it was bi-national with the group from Cambridge collaborating with a number of physiologists from Harvard. In all there were eight members of the expedition and the initial planning took place in the early summer of 1921, only a year and a half after the glass chamber experiment described above. The expedition itself was quite short. Barcroft and colleagues from Cambridge arrived in Lima, Peru on December 18, 1921, and they left Peru exactly a month later. The group from Harvard arrived in Peru about three weeks before the Cambridge team.

Cerro de Pasco is a mining town in the Andes about 200 km to the northeast of Lima. The town is on a slope but the altitude is often given as 4300 m which is the same as the summit of Pikes Peak. An advantage of the venue was a railway line from Callao, the port of Lima, all the way to Cerro which greatly facilitated access

Fig. 20.4 Laboratory used during the expedition to Cerro de Pasco. This was set up in a baggage van of the railway from Lima to Cerro de Pasco. From [7]

to this high altitude. This arrangement is similar to that of the Pikes Peak expedition where the cogwheel railway to the summit made that an attractive venue. The railway in Peru was particularly convenient because a laboratory was set up in one of the baggage vans, and accommodation for expedition members was also available in other passenger cars. Figure 20.4 shows the laboratory.

The account of the expedition was published in the *Philosophical Transactions of the Royal Society of London* [7], and by today's standards was extremely long with a considerable amount of narrative that is interesting but largely irrelevant from a scientific viewpoint. The table of contents on the first page essentially sets out all the scientific objectives which numbered sixteen. However it is clear that a major thrust was the issue of oxygen secretion, that is the relationship between the alveolar and arterial PO_2 at high altitude. Barcroft subsequently wrote a book about the expedition findings and added some new material [2].

20.4.1   Oxygen Secretion

In the Cerro de Pasco study the arterial PO_2 was determined by the "bubble" tonometric method described by Barcroft and Nagahashi [5]. This was a modification of the technique originally introduced by Krogh [16]. Table I in the report shows the arterial and alveolar PO_2 values that were only measured on five people at high altitude. In the table Category I refers to the expedition members but only the results

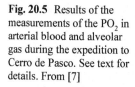

Fig. 20.5 Results of the measurements of the PO$_2$ in arterial blood and alveolar gas during the expedition to Cerro de Pasco. See text for details. From [7]

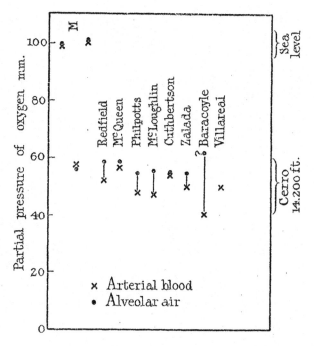

for Meakins are shown. Category II refers to mining engineers who were originally from low altitude but had been at Cerro for months or years. Category III refers to the high altitude natives who had been at high altitude for generations. In only one of these was the alveolar PO$_2$ given. As can be seen in the table the alveolar PO$_2$ exceeded the arterial value by 2, 7, 9, −1 and 1 mm Hg.

The results were also shown in Fig. 11 reproduced here as Fig. 20.5. On the extreme left under the M are the results for Meakins, at sea level, Cerro, and then on returning to sea level. The alveolar-arterial differences were small and in fact the measurement at high altitude showed an arterial value slightly above the alveolar value. To the right of Meakins' values are those for five engineers. To the right of these are the results of three people who were permanent residents of high altitude. Note that for seven subjects, the arterial value was less than the alveolar value. In the case of Villareal the alveolar value was not available.

The paper goes on to state that two main facts stand out. (1). There is no suggestion of oxygen partial pressures in the blood of 90–120 mm Hg such as have previously been alleged to exist in dwellers at high altitudes. (This was a reference to the results of the Pikes Peak expedition.) (2). The pressures of oxygen in arterial blood are such as would naturally be brought about by diffusion.

Measurements of the arterial oxygen saturation were also made and were shown in Table II in the report. In five members of the expedition where measurements were made at Cerro during rest, the saturation varied between 81.5 and 91 %. In two members measurements were also made during work the values being 79 and 90.5%. In four mining engineers the saturations were between 86 and 91%. The three people who were permanent residents of the mountains had saturations between 82.3 and 86%. These last low values surprised the investigators because, as they stated in the article, they presumably indicated the low arterial oxygen saturations that these people lived with all their lives.

20.4.2 Other Measurements of Respiration and Circulation

The increased pulmonary ventilation at high altitude was studied by measuring the alveolar PCO_2 on many occasions and the mean value was about 28 mm Hg as opposed to 40 mm Hg at sea level. The discussion of possible causes for the increased ventilation is rather tortured and is reminiscent of a similar discussion by Haldane and his colleagues [12] on the results obtained on Pikes Peak. We now know that the mechanism was the stimulation of the peripheral chemoreceptors by the low PO_2 in the arterial blood but the chemoreceptors were not discovered by the Heymans father and son team until 4 years later [15].

There is a long section on the position of the oxygen dissociation curve at high altitude, and the various factors influencing the oxygen affinity of hemoglobin. This was one of Barcroft's primary interests. The results were somewhat confusing and part of the reason is that the investigators were not aware of the role of 2,3-diphosphoglycerate. Indeed this was not discovered until 1967 [9, 11]. However the consensus was that the dissociation curve was shifted to the left, probably because of the reduced arterial PCO_2.

The shape and size of the chest in high altitude permanent residents was studied in some detail including X-rays of the upper body. It was concluded that many of the highlanders had larger chests than would be expected from their height which was generally less than that of the investigators.

In a novel series of studies, measurements of the diffusing capacity of the lung for carbon monoxide were made using the technique developed by Marie Krogh [17]. The equipment for this had been made available from Krogh's laboratory. It was concluded by the investigators that there was no change in the diffusing capacity compared with sea level. However it was argued that the tendency of the arterial oxygen saturation to fall with exercise could be explained by diffusion limitation across the blood-gas barrier. Barcroft carried out these calculations using the Bohr integration technique [10] and this was the first clear demonstration of the diffusion limitation of oxygen uptake at high altitude.

Extensive studies were made of the increase in red cell concentration at high altitude. These essentially confirmed previous measurements made on the Pikes Peak expedition and elsewhere.

20.4.3   Neuropsychometric Measurements

These studies were some of the most original of this expedition. A large series of mental tests were employed including recognition of duplicate letters in a sequence, tests of handwriting, memorizing and multiplication, and a general assessment of mental fatigue. The conclusion was that it was possible to carry out complicated cognitive tests accurately but this took more time than at sea level. Barcroft himself felt that concentration was more difficult and time was wasted in "bungling", that is being unable to organize tasks as effectively as at sea level. To quote the study "Judged by the ordinary standards of efficiency in laboratory work, we were in an obviously lower category intellectually at Cerro than at sea level. By a curious paradox this was most apparent when it was being least tested. For perhaps what we suffered from chiefly was the difficulty of maintaining concentration. When we knew we were undergoing a test, our concentration could by an effort be maintained over the length of time taken for the test, but under ordinary circumstances it would lapse. It is, perhaps, characteristic that, while each individual mental test was done as rapidly at Cerro as at the sea level, the performance of the series took nearly twice as long for its accomplishment".

It was interesting that the engineers at the mine who were originally from sea level, but had spent several years at high altitude in some cases, were adamant that they were definitely incapable of doing their own sea-level standard of work at high altitude, whether mental or physical.

20.4.4   Acclimatization and Its Effectiveness

In the final section on "Summary and Conclusions" the three main factors that had a positive influence in acclimatization were listed as

1. An increase in total ventilation
2. The leftward shift of the oxygen dissociation curve such that at any given PO_2 the oxygen concentration is increased
3. The increase in concentration of the red blood cells

Factors that were not helpful in acclimatization were listed as oxygen secretion, alterations in the lung volumes, diffusion properties of the lung, and the cardiac output although measurements of the last were made by very indirect methods and conflicting results were obtained.

The final topic that was dealt with is of particular interest because the results were novel and sparked considerable controversy. The issue was whether the acclimatization process allowed the physiological state of people at high altitude to equal that of sea-level residents. As we have seen, Barcroft was able to make measurements on three groups of people at high altitude: the expedition members, the engineers, and the permanent residents.

Barcroft made the famous statement "All dwellers at high altitude are persons of impaired physical and mental powers". There seemed to be good experimental evidence for this in the expedition members. Although formal measurements of work capacity were not carried out, it was clear to the members when they tried to walk uphill that their exercise ability was substantially reduced. In addition, as we have seen, there was impairment of mental ability.

No formal mental testing was done on the mining engineers. However, as mentioned above, the engineers themselves were adamant that their standard of both mental and physical work at Cerro was clearly less than at sea level.

When we come to the permanent residents at high altitude, the situation is not nearly so clear. Barcroft and his colleagues were very impressed by the ability of young people to carry heavy loads, and the fact that they played football at Cerro. However no comparisons could be made of either physical or mental performance at high altitude and sea level. Therefore it was not reasonable for the expedition to extrapolate their own physical and mental limitations to the permanent residents. In fact the eminent Peruvian physician, Carlos Monge Medrano took great exception to Barcroft's remarks on the permanent residents and wrote "For our part, as early as 1928 … we proved … that Professor Barcroft was himself suffering from a subacute case of mountain sickness without realizing it. His substantial error is easy to explain as resulting from an improper generalization on his part of what he himself felt and applying his reactions to Andean man in general" [20].

It should be added that Barcroft took part in two brief earlier expeditions to high altitude but these will not be described in detail here. The first was in1910 when he was a member of an expedition to Tenerife in the Canary Islands led by Nathan Zuntz (1847–1920). Barcroft chiefly worked on factors affecting the position of the oxygen dissociation curve [1]. However he also made an interesting observation relevant to the contention of Mosso that the deleterious effects of high altitude were caused by the low PCO_2. During this expedition Barcroft who had an almost normal PCO_2 and therefore a low PO_2 at high altitude developed obvious Acute Mountain Sickness. By contrast Douglas who remained well had a much lower PCO_2 and higher PO_2. Barcroft pointed out that this anecdotal observation supported Paul Bert's contention that hypoxia was the most important factor rather than Mosso's suggestion of a low PCO_2.

Barcroft also spent some time at the Capanna Margherita in 1911 where again he worked mainly on the oxygen affinity of hemoglobin and the position of the oxygen dissociation curve.

20.5   Exercise at Extreme Altitude While Breathing 100% Oxygen

In 1931 Barcroft participated in a remarkable chamber experiment that has received little attention. Unlike the two projects already discussed, the glass chamber experiment and the Cerro de Pasco expedition, Barcroft was not the primary instigator of this study. It was the brainchild of Rodolfo Margaria (1901–1983) who was Professor of Physiology at the University of Milan and who made many important contributions to the physiology of exercise, including exercise at altitude. The study was carried out in a low-pressure chamber in Oxford by Margaria, Barcroft, Douglas and Kendall [8]. Douglas was a longtime associate of J.S. Haldane and was an important member of the Pikes Peak expedition referred to earlier. L.P. Kendal was a colleague of Douglas's at Oxford.

The low-pressure chamber in which the studies were carried out had an interesting history. It was originally constructed in Lancaster, Pennsylvania, to test the effects of low barometric pressures on US aviators during World War I and train them to deal with these. The chamber was 2.7 m high and 2.1 m in diameter and was shipped to a U.S. Air Service base in Issoudun in central France where it arrived in September 1918. However the Armistice was signed on November 11 of that year and so little use was made of the chamber. With the end of the war the question arose of what to do with it.

Georges Dreyer (1873–1934) was Professor of Pathology at Oxford University and had worked extensively on oxygen equipment for aviators including periods with the U.S. Air Service. The upshot was that the chamber was donated to Oxford in 1919 where it was set up in the Department of Pathology. Interestingly, a series of experiments were carried out on George Ingle Finch (1888–1970) in 1921 because he planned to take part in the Everest Reconnaissance Expedition of that year [23]. However the chamber was not extensively used for physiological measurements and indeed the work of Dreyer on the use of oxygen at high altitude has been largely forgotten [24].

Margaria's interest in exercise at extreme altitude was prompted in part by the early British expeditions to Everest. In 1924 E.F. Norton reached an altitude only 300 m short of the summit (8848 m) without supplementary oxygen. During the same expedition, Mallory and Irvine set out for the summit using supplementary oxygen but never returned. In 1930 Margaria had studied the effects of reducing barometric pressure on maximum exercise in a low-pressure chamber and concluded that at a barometric pressure of 300 mm Hg, the work rate of humans fell to zero [18]. He therefore argued that it would be impossible for climbers to reach the summit of Mt. Everest (barometric pressure about 250 mm Hg) without supplementary oxygen, but he wondered to what extent the maximal work rate would be increased during oxygen breathing.

The actual experiment was carried out on February 20, 1931 [8]. The exercise protocol consisted of stepping on and off a box 33 cm high every 4 s for an hour. This procedure was chosen because it was thought to be a reasonable estimate of the work rate required for a climber to reach the summit. Barcroft reported the results of the experiment at a meeting of the Physiological Society on March 14, 1931, and Franklin [13] reproduced the blackboard drawing made by Barcroft at the time (Fig. 20.6). Note that Margaria is stepping on and off the box while breathing 100% oxygen, and Barcroft, also breathing oxygen, is keeping watch within the chamber. Douglas is outside the chamber smoking his pipe.

Fig. 20.6 Experimental arrangements for the study of exercise at extreme altitude while breathing 100% oxygen. This was a blackboard drawing that Barcroft made on March 14, 1931 during a meeting of the Physiological Society in University College, London when he described the experiment. From [13]

Two protocols were studied. In the first, the barometric pressure was 240 mm Hg which was equivalent to an altitude above that of the Everest summit where the pressure is about 250 mm Hg. Margaria exercised for 1 h and at the end of that time stated that he could continue indefinitely. However he complained of acute pain in the extensor muscles of both legs and both knees although he had only been exercising with one leg. Kendall repeated the protocol and also reported the same leg pain.

For the second protocol the pressure was reduced to 170 mm Hg and both men were able to complete the 1 h exercise although they felt that they could not have done this at a lower pressure. Near the end of the exercise, Margaria's pulse rate was only 88 min^{-1} whereas Kendal's was 164.

The pain reported by both men is fascinating. In retrospect this was almost certainly due to decompression sickness since they went into the chamber with a full load of nitrogen in their tissues and the combination of the low barometric pressure and oxygen breathing would have greatly accelerated nitrogen elimination. This may have been the first reported instance of decompression sickness caused by a low barometric pressure.

Barcroft went on to calculate the amount of oxygen that would be necessary to reach the Everest summit and return and came up with a figure of 700 L. He stated that the weight of this would only be a kilogram, a trifling amount, but he recognized that the weight of the cylinders would be substantial. He was not to know that Everest would not be climbed using oxygen for another 22 years, and that it would be another 25 after that before the first ascent was made without supplementary oxygen.

References

1. Barcroft J. The effect of altitude on the dissociation curve of blood. J Physiol. 1911;42:44–63.
2. Barcroft J. The respiratory function of the blood, part I: lessons from high altitudes. Cambridge: Cambridge University Press; 1925.
3. Barcroft J. The toxicity of atmospheres containing hydrocyanic acid gas. J Hyg. 1931;31:1–34.
4. Barcroft J, Haldane JS. A method of estimating the oxygen and carbonic acid in small quantities of blood. J Physiol. 1902;28:232–40.
5. Barcroft J, Nagahashi M. The direct measurement of the partial pressure of oxygen in human blood. J Physiol. 1921;55:339–45.
6. Barcroft J, Cooke A, Hartridge H, Parsons TR, Parsons W. The flow of oxygen through the pulmonary epithelium. J Physiol. 1920;53:450–72.
7. Barcroft J, Binger CA, Bock AV, Doggart JH, Forbes HS, Harrop G, Meakins JC, Redfield AC, Davies HW, Duncan Scott JM, Fetter WJ, Murray CD, Keith A. Observations upon the effect of high altitude on the physiological processes of the human body, carried out in the Peruvian Andes, chiefly at Cerro de Pasco. Philos Trans R Soc Lond Ser B. 1923;211:351–480.
8. Barcroft J, Douglas CG, Kendal LP, Margaria R. Muscular exercise at low barometric pressures. Arch Sci Biol (Napoli). 1931;16:609–15.
9. Benesch R, Benesch RE. The effect of organic phosphates from the human erythrocyte on the allosteric properties of hemoglobin. Biochem Biophys Res Commun. 1967;26:162–7.
10. Bohr C. Über die spezifische Tätigkeit der Lungen bei der respiratorischen Gasaufnahme und ihr Verhalten zu der durch die Alveolarwand stattfindenden Gasdiffusion. Skand Arch Physiol. 1909;22:221–80. English translation in West JB editor. Translations in respiration physiology. Stroudsburg: Dowden, Hutchinson and Ross; 1975.
11. Chanutin A, Curnish RR. Effect of organic and inorganic phosphates on the oxygen equilibrium of human erythrocytes. Arch Biochem Biophys. 1967;121:96–102.
12. Douglas CG, Haldane JS, Henderson Y, Schneider EC. Physiological observations made on Pike's Peak, Colorado, with special reference to adaptation to low barometric pressures. Philos Trans Roy Soc Lon B. 1913;203:185–318.
13. Franklin KJ. Joseph Barcroft, 1872–1947. Oxford: Blackwell; 1953.
14. Haldane JS. Methods of air analysis. 3rd edition. London: C. Griffin & Company, Limited; 1920.
15. Heymans J-F, Heymans C. Sur les modifications directes et sur la regulation réflexe de l'activité du centre respiratoire de la tête isolée du chien. Arch Int Pharmacodyn Ther. 1927;33:273–372.
16. Krogh A. On micro-analysis of gases. Skand Arch Physiol. 1908;20:279.
17. Krogh M. The diffusion of gases through the lungs of man. J Physiol. 1915;49:271–96.
18. Margaria R. Die Arbeitsfähigkeit des Menschen bei vermindertem Luftdruck. Arbeitsphysiologie. 1930;141:233–61.
19. Martin CJ. A simple and convenient form of bicycle ergometer. J Physiol. 1914;48:xv.
20. Monge MC. Acclimatization in the Andes: historical confirmations of "climatic aggression" in the development of Andean Man. Baltimore: Johns Hopkins University Press; 1948.
21. Roughton FJW. Joseph Barcroft. Obituary Not R Soc. 1949;6:315–245.
22. West JB. Centenary of the Anglo-American high-altitude expedition to Pikes Peak. Exp Physiol. 2012;97:1–9.
23. West JB, George I. Finch and his pioneering use of oxygen for climbing at extreme altitudes. J Appl Physiol. 2003;94:1702–13.
24. West JB, Sidebottom E. Georges Dreyer (1873–1934) and a forgotten episode of respiratory physiology at Oxford. J Med Biogr. 2006;14:140–9.

Chapter 21
The Physiological Legacy of the Fenn, Rahn and Otis School

Abstract Extraordinary advances in respiratory physiology occurred between 1941 and 1956 in the Department of Physiology, University of Rochester. These were principally the result of a collaboration between Wallace Fenn, Hermann Rahn and Arthur Otis. Remarkably all three scientists had worked in very dissimilar areas of physiology before and, by their own admission, were largely ignorant of respiratory physiology. However because of the exigencies of war they were brought together to study the physiology of pressure breathing. The result was that they laid much of the foundations of pulmonary gas exchange and pulmonary mechanics and some of their work is still cited today. In pulmonary gas exchange they exploited the new oxygen-carbon dioxide diagram, clarified the effects of changes of altitude, hyperventilation and pressure breathing, and pioneered the analysis of ventilation-perfusion relationships. In respiratory mechanics, they carried out ground-breaking work on the pressure-volume behavior of the lung and chest wall, and went on to analyze aspects of gas flow and work of breathing. This explosion of ideas from what initially appeared to be a poorly prepared group has lessons for us today.

21.1 Introduction

One of the most productive periods of modern respiratory physiology occurred during and shortly after World War II. The trio of Wallace Fenn, Hermann Rahn and Arthur Otis (Fig. 21.1) at the University of Rochester NY beginning in about 1941 was responsible for some of the most important concepts of respiratory physiology in the last century and some of their work is still cited today. The story is particularly interesting because these three scientists, as they relate, initially knew very little about respiratory physiology and were brought together by the exigencies of World War II. However in spite of their initial ignorance, they were responsible for laying much of the foundations of modern respiratory gas exchange and respiratory mechanics. Few members of the large group who were trained by this school are still active. This article is based mainly on archival material [3, 13, 14, 17, 22] but includes some personal experience. Some of the material, for example that in reference [14] is difficult to find and this short account should be helpful to students interested in the history of their discipline.

© American Physiological Society 2015
J. B. West, *Essays on the History of Respiratory Physiology,*
Perspectives in Physiology, DOI 10.1007/978-1-4939-2362-5_21

Fig. 21.1 From *left* to *right*: Arthur Otis, Hermann Rahn, and Wallace Fenn at the 1963 Fall meeting of the American Physiological Society in Coral Gables, Florida. (From [31])

21.2 Unlikely Beginnings

Wallace Fenn (1893–1971) was the leader of the group and he was some 20 years older than his two colleagues. His father was a Unitarian minister who became Professor of Theology at Harvard, and Wallace entered Harvard planning to prepare for the ministry [22]. However his interests turned to physiology and he obtained his PhD at Harvard in plant physiology. He then spent a period in the laboratory of A.V. Hill in London where a major interest was muscle energetics. There he described the "Fenn effect" which refers to the heat produced by muscle during contraction [2]. He is also well known for his work on potassium efflux from contracting muscle [4]. In addition his group showed for the first time that both oxygen poisoning and the biological effects of X-irradiation exert their damaging effects by the action of free radicals [9]. Fenn was a towering figure in American and international physiology being president of both the American Physiological Society and the International Union of Physiological Sciences, and a member of the National Academy of Sciences.

Hermann Rahn (1912–1990) was born in Lansing, Michigan, and his early life was remarkably disruptive [17]. His father, Otto, was on the faculty of Michigan State University having emigrated from Germany to the US in 1907. In 1914 Otto and his wife took the 2-year-old Hermann to Germany to visit his grandparents but World War I began while they were there. Otto was not allowed to return to the US in spite of the fact that he was in the process of becoming a US citizen but was drafted into the German army. After the war he was not invited to return to Michigan because of the anti-German sentiment at that time and the family suffered severe deprivation. It was not until 1926 that the 14-year-old Hermann found himself in Ithaca, NY where his father had obtained a faculty appointment at Cornell University. Hermann then had the task of learning English. Not surprisingly he had strong links with Germany and at one stage it was uncertain whether he would return or decide to stay in the US. Fortunately he remained, and after graduating from Cornell he obtained his PhD at the University of Rochester where his work was on the placenta and corpus luteum of snakes [18] and the development of the pituitary gland in birds [10]. He then went to the University of Wyoming at Laramie to study the reproductive behavior of rattlesnakes [19]. In 1941 he stopped briefly in Rochester en route from Woods Hole to Laramie and called on Fenn whom he admired. At the end of their conversation Rahn accepted a job as instructor of physiology in Rochester at the princely salary of $ 60 per month.

Arthur Otis (1913–2008) received his PhD from Brown University where his dissertation was on the study of the effects of drugs and ions on the oyster heart. Subsequently he spent a post-doctoral year at Iowa working on the development of grasshopper eggs [1]. He then moved to Rochester in 1942 to work in Fenn's department. Later he spent a period at Johns Hopkins Medical School and finally he became chairman of the department of physiology at the new medical school at the University of Florida in Gainesville [12].

So here we have a trio of scientists who are working on very different areas of cellular physiology in a variety of animals. Fenn has made important contributions to muscle contraction, Rahn is working on reproduction in rattlesnakes, and Otis is studying the activation of the enzyme tyrosinase in grasshopper eggs. None of them has had any serious exposure to respiratory physiology. On the face of it this seems an unlikely group to lay the foundations of both respiratory gas exchange and mechanics as they exist today. However this is what happened as a result of the demands of the war.

21.3   The Initial Research Topic

Rahn relates that he started work in Fenn's department on September 1, 1941 on hormone assays, but on December 7 when the attack on Pearl Harbor occurred, Fenn decided that the research should be switched to problems related to the war effort [14]. Apparently Fenn had previously been apprised of the likely need of research on aviation physiology and after the war began he was contracted by A.N. Richards of the Office of Scientific Research and Development (OSRD) to work on the physiologic effects of pressure breathing. Fenn later wrote that he "timidly obtained" a grant of $ 500 (sic) from the OSRD and that this was the first research grant he had ever received in 17 years as chairman of the department of physiology [3] Fenn was always ambivalent about federal grant support and after mentioning the $ 500 stated "It must be admitted, however, that there are dangers in this easy money regime which must be recognized and guarded against".

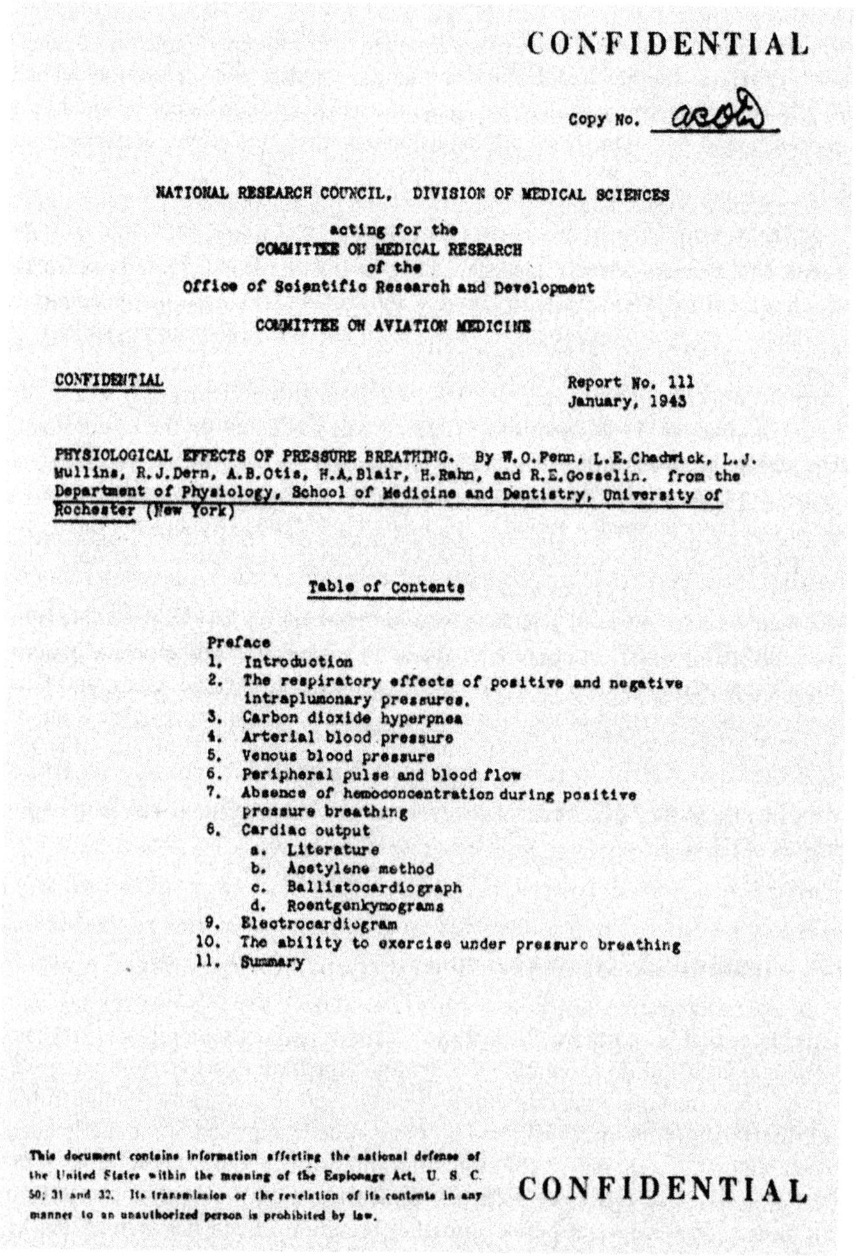

CONFIDENTIAL

Copy No. _aao&i_

NATIONAL RESEARCH COUNCIL, DIVISION OF MEDICAL SCIENCES

acting for the
COMMITTEE ON MEDICAL RESEARCH
of the
Office of Scientific Research and Development

COMMITTEE ON AVIATION MEDICINE

CONFIDENTIAL   Report No. 111
                                               January, 1943

PHYSIOLOGICAL EFFECTS OF PRESSURE BREATHING. By W.O.Fenn, L.E.Chadwick, L.J.
Mullins, R.J.Dern, A.B.Otis, H.A.Blair, H.Rahn, and R.E.Gosselin. from the
Department of Physiology, School of Medicine and Dentistry, University of
Rochester (New York)

Table of Contents

Preface
1. Introduction
2. The respiratory effects of positive and negative
    intrapulmonary pressures.
3. Carbon dioxide hyperpnea
4. Arterial blood pressure
5. Venous blood pressure
6. Peripheral pulse and blood flow
7. Absence of hemoconcentration during positive
    pressure breathing
8. Cardiac output
    a. Literature
    b. Acetylene method
    c. Ballistocardiograph
    d. Roentgenkymograms
9. Electrocardiogram
10. The ability to exercise under pressure breathing
11. Summary

This document contains information affecting the national defense of
the United States within the meaning of the Espionage Act, U. S. C.
50; 31 and 32. Its transmission or the revelation of its contents in any
manner to an unauthorized person is prohibited by law.

CONFIDENTIAL

Fig. 21.2   Title page of the first report made after 6 months of work principally by Fenn, Rahn and Otis. See text for details. (From [5])

The pace of work must have been exceptionally fast because on January 7, 1943 the first report appeared [5] and Fenn stated that it covered the first 6 months of the work. The cover page of the report is shown in Fig. 21.2 and as the table of contents indicates, a number of projects were reported on. The principal authors were Fenn, Rahn and Otis but several other investigators were included.

In retrospect it seems odd that pressure breathing should have been given such emphasis when there were so many other important physiological issues in aviation physiology. The reason was the belief at the time that if fighter aircraft could fly as high as 40,000 ft they would have a commanding advantage. Interestingly this belief was also shared by the air forces of the US, UK and Canada. In fact most of the sorties in the war were carried out at altitudes of 20,000–25,000 ft although reconnaissance flights were made higher. Pressurization of aircraft cabins had begun in about 1939 with the first commercial example being the Boeing 307 Stratoliner but essentially no fighter aircraft were pressurized during World War II.

The development of equipment for the new research project makes entertaining reading. Otis and Rahn described [14] how Fenn bought a steel tank designed for the transport of beer, borrowed a tree sprayer pump from the university grounds department, reversed its valves, and thus provided a high-altitude chamber that could decompress at the rate of 5000 ft/min (Fig. 21.3). Subjects entered the chamber by

Fig. 21.3 The high altitude chamber used for the work by Fenn, Rahn and Otis. The original tank was for transporting beer. Arthur Otis is the observer. (From [14])

lowering themselves through a small circular hatch at the top. A Fleisch pneumo-tachograph was made from a cluster of soda straws inside a brass tube with the pressure difference between the two ends being measured with a sensitive membrane manometer. A later version replaced the soda straws with glass wool enclosed in a ladies' hair net. Fenn was endowed with Yankee ingenuity as well as parsimony and loved to improvise. A key piece of equipment for the studies of gas exchange was an end-tidal gas sampler using a condom [27].

The report itself [14]) makes fascinating reading even today. One's first impression is that it is extraordinary how much was accomplished in 6 months. The report begins by stating that it has been shown that positive-pressure breathing is a very effective way of gaining altitude in that for each 5 cm H_2O increase in pressure, the tolerable altitude was increased by about 1000 ft. The authors noted that there were a number of measurements of very brief periods of pressure breathing such as the Valsalva maneuver in the literature as well as studies on animals, but few if any on humans. In the report, studies were made at positive pressures as high as 40 cm H_2O, and negative pressure breathing was also investigated down to -30 cm H_2O. The periods of pressure breathing varied from 15 s to 2 h. The general conclusion was that pressure breathing of 20–25 mm Hg (27–33 cm H_2O) was well-tolerated by normal men.

One of the early figures in the report shows the variation in maximal inspiratory and expiratory pressure with changes in lung volume (Fig. 21.4). This figure remains part of the classical teaching today and emphasizes how the research into

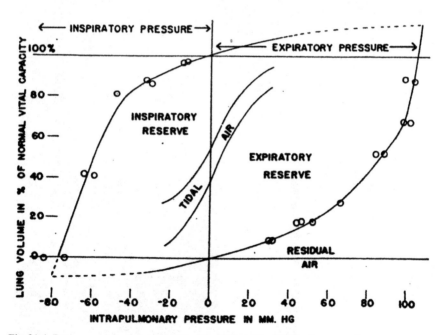

Fig. 21.4 Pressure-volume behavior of the lung. This is Fig. 21.4 in the report shown in Fig. 21.2. (From [5])

a very practical problem such as pressure breathing throws light on general physiological principles. The report then goes on to calculate the work of breathing and they discuss both the elastic resistance of the lung and chest wall and the viscous losses. The oxygen cost of pressure breathing was also measured. Incidentally it is sometimes stated that this group were unaware of previous studies of respiratory mechanics especially those reported in the German literature. To some extent this is true (see later) but the list of references on page 11 includes four in German, a language that Rahn was fluent in.

There is a short section on the effects of raising the inspired Pco_2 during pressure breathing which shows the expected increase in ventilation. One finding was that during positive-pressure breathing the alveolar Pco_2 tends to increase which the authors attribute to an increase in dead space. The effects of positive-pressure breathing on arterial blood pressure were studied and both systolic and diastolic pressure were shown to increase. It was also found that peripheral venous pressure rose with positive-pressure breathing presumably because of obstruction to venous return. Because of the possibility that the increased venous pressure could cause movement of fluid out of the circulation as a result of disturbing the Starling equilibrium, the hemoglobin concentration was measured but shown not to be affected.

The effect of positive-pressure breathing on cardiac output was examined. The authors pointed out that there are numerous studies showing that in animals cardiac output is decreased with increased intrathoracic pressure presumably because of restricted venous return. However cardiac output in humans proved to be difficult to study. Measurements were made using the uptake of acetylene during rebreathing and these appeared to show a reduction in cardiac output although the scatter of the results was high. Some measurements were also made with a homemade ballistocardiograph and these seem to show that pressure breathing caused a decrease in stroke volume but that this was compensated by an increase in heart rate. A few measurements of stroke volume were also made using images of the heart obtained by chest x-ray. These showed a small reduction in stroke volume.

There is a remarkable sentence on page 35 of the report that states "The large decrease [of stroke volume] found for W.O.F. seems very improbable although this subject has collapsed three times under pressure breathing and might have been near this when the film was taken". That rather ominous sentence fits with a short section in the account by Otis and Rahn [14]. "In the early days of our altitude chamber operations, Wallace Fenn was always the first to volunteer when a new procedure was to be tried out, but when the dean of our school, Dr. George Whipple, one day came by at an opportune time to see Wallace passing out from a mask leak at high altitude, he immediately gave strict orders to us that Fenn should henceforth not be allowed to enter the chamber".

This initial report dealing with the first 6 months of research emphasizes the flying start of the research group. Following this there was a very fertile period of some 15 years devoted mainly to pulmonary gas exchange and pulmonary mechanics although some other topics were investigated as well. Initially all the work was classified and could not appear in the open literature. The result was a series of classified technical reports, several of them titled "Studies in respiratory physiology" that were put out by the US Air Force, Dayton, Ohio [7, 21, 23]. It was through contracts with this organization that most of the work was done. The reports continued to appear after Rahn moved to Buffalo and eventually there were a total of seven [31]. They make good reading even today.

In the summer of 1945 shortly after the end of World War II, the material was declassified, and the authors began to prepare articles for the open literature. The first public oral presentations were made at the FASEB Meeting in Atlantic City in the spring of 1946, and in that year the first publications occurred in the American Journal of Physiology and the Journal of Aviation Medicine [7, 15, 26]. At about this time it was felt that a new journal might be more appropriate than the American Journal of Physiology for these studies in applied physiology. This was one of the reasons why the American Physiological Society started the Journal of Applied Physiology in 1949. The Foreword to volume I of the Journal includes the sentence "In connection with the Journal, the term 'applied' will broadly connote human physiology, with particular emphasis on man in relation to his environment". The early volumes of this journal contain much of the work of this group. However not all of the 100 or so reports in the "Studies in respiratory physiology" series ended up in the open literature.

21.4   Pulmonary Gas Exchange

One of the most productive areas of the research was pulmonary gas exchange. It was not surprising that the investigators tackled this topic because changes in altitude exposure, pressure breathing, and hyperventilation clearly affected the composition of alveolar gas. This work was accelerated by the development of the end-tidal gas sampler referred to earlier [27] and later by the introduction of a rapid infrared analyzer for carbon dioxide by Fowler [8].

According to Rahn it was Fenn who introduced the oxygen-carbon dioxide diagram with P_{O_2} on the horizontal axis and P_{CO_2} on the vertical axis [6]. This diagram was a major breakthrough both for understanding the changes in alveolar gas that occur with altitude and ventilation, but also later for analysis of the effects of ventilation-perfusion ratio inequality.

Figure 21.5 shows one of the early uses of the O_2-CO_2 diagram with the lines for different respiratory exchange ratios at various altitudes, alveolar ventilation on the vertical axis, and what was called "average alveolar air". This last was measured using the end-tidal sampler in subjects who were acutely exposed to high altitude, and later Rahn and Otis [25] expanded this to show the alveolar gas composition in acclimatized subjects at high altitude (Fig. 21.6). This classical diagram is still used today.

The O_2-CO_2 diagram led to one of the most productive lines of research of this group, that is the difficult problem of analyzing ventilation-perfusion inequality. Depicting alveolar gas concentrations on the diagram under various conditions such as changes of altitude and ventilation including the effects of changing the respiratory exchange ratio was relatively simple. Once the exchange between gas

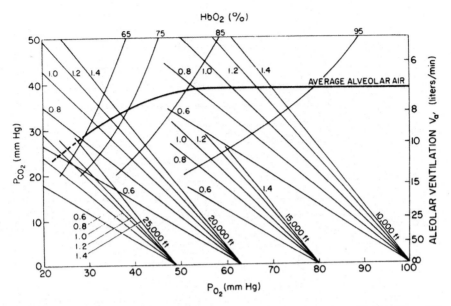

Fig. 21.5 O_2-CO_2 diagrams with respiratory exchange ratio lines for various altitudes. (From [7])

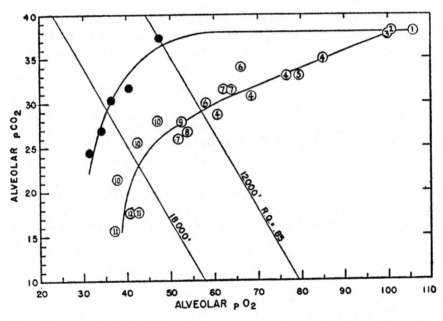

Fig. 21.6 Oxygen-carbon dioxide diagram showing the alveolar gas composition in subjects acutely exposed to high altitude (*upper line*) and after acclimatization (*lower line*). (From [25])

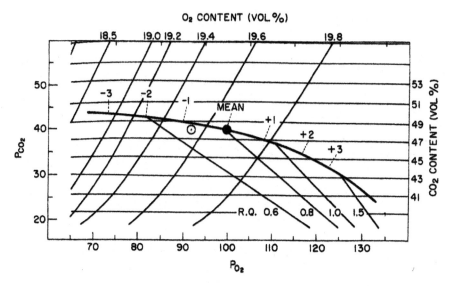

Fig. 21.7 Oxygen-carbon dioxide diagram with the *curved line* showing the composition of alveolar gas and blood as the ventilation-perfusion ratio is altered. The distribution of ventilation-perfusion ratios is assumed to have a log standard deviation of 1 to 3. The numbers +1, +2 etc. show the composition for the appropriate standard deviations from the mean. (From [20])

and blood was added, the problem became much more complicated because of the nonlinear and interdependent O_2 and CO_2 dissociation curves. During the 15 or so years of the work of the Fenn, Rahn and Otis group, it was appreciated that algebraic solutions to the problem were impossible and that the only way of tackling the problems was by graphical analysis.

One elegant example of this is shown in Fig. 21.7 where a ventilation-perfusion ratio line is added to the O_2-CO_2 diagram [20]. In this diagram the curved line shows calculated arterial Po_2 and Pco_2 values for a distribution of ventilation-perfusion ratios with a log standard deviation of 1 to 3, and the points +1, +2, etc. show the results for the stated standard deviations from the mean value. Analysis of ventilation-perfusion inequality has progressed a long way since then, but Fig. 21.6 represents a sea change in thinking. Much of the work by Fenn and Rahn was brought together in a small monograph titled "A graphical analysis of the respiratory gas exchange" [24] and this was the bible for many of us in the late 1950s. In fact it persuaded me to apply to spend a year in Rahn's laboratory in 1961–1962.

21.5 Pulmonary Mechanics

The other area of respiratory physiology where the Fenn, Rahn and Otis group made fundamental advances was pulmonary mechanics, particularly the pressure-volume behavior of the lung and chest wall. The major publication in this area was titled

"The pressure-volume diagram of the thorax and lung" by Rahn, Otis, Chadwick and Fenn [26].

This paper was devoted to three main areas. The first was the maximum expiratory and inspiratory pressures at different lung volumes, and this expanded on the work already mentioned in Fig. 21.4 of their first report and shown here in Fig. 21.4. The 1946 publication covers the same area but includes additional data. As indicated previously, Fig. 21.4 is part of the classical literature today.

The next major advance was a description of the relaxation pressure-volume curve for the total respiratory system, that is the lung together with the chest wall, and the individual relaxation curves for the chest wall and the lung separately. This is shown in Fig. 21.8. It remains a classical diagram which in slightly different forms is part of the teaching to medical and graduate students in their course on respiratory physiology even today. Among other things, the diagram emphasizes that the relaxation pressure of the total system is zero at functional residual capacity (FRC) shown as Vr in Fig. 21.8. As the volume of the system is increased the relaxation pressure increases and the opposite occurs as the volume is decreased. In addition the relaxation pressure-volume curve of the lung shows a positive pressure of about 4 mmHg at FRC, and this is balanced by an equal and opposite outward expanding pressure developed by the chest wall. 4 mmHg corresponds to about 5 cm H_2O and this is the value that we use today. Note that the relaxation curve for

Fig. 21.8 Relaxation pressure-volume curves for the lung plus chest wall (*solid line*) and the lung and chest wall separately (*broken lines*). (From [26])

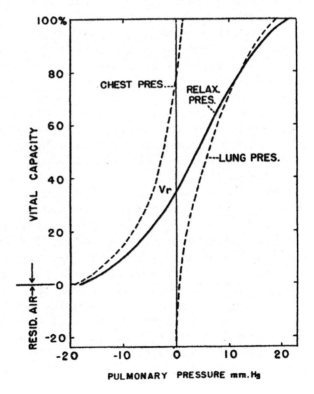

the chest wall alone crosses the zero pressure line at about 70% of vital capacity. In other words below this volume the chest wall continues to develop a negative pressure, that is tends to expand, whereas above that volume a positive pressure is required to expand the chest wall. The authors noted that some of the values in this diagram were estimated from other published studies. Also they acknowledged that although they developed this diagram independently, they subsequently recognized that Rohrer had described a similar situation in one subject in 1916 [28]. Neergaard and Wirz [29, 30] had also worked along similar lines.

Incidentally few people are able to relax their respiratory muscles sufficiently to generate the data shown in Fig. 21.8. However Rahn became an expert in doing this. Rahn also acknowledges that Fenn was responsible for the idea of the relaxation pressure-volume curve [22].

The third part of this important paper dealt with the various lung volume compartments such as vital capacity, FRC and residual volume (RV) at different total lung volumes and postures of the subject. It was found that both total lung capacity and RV (here called residual air) changed little with posture but FRC was substantially higher in the sitting rather than the supine posture. The difference is caused by the weight of the abdominal contents pulling the diaphragm down in the sitting position and pushing the diaphragm head-ward in the supine position. The small difference in the supine volumes of the corrected curve is because additional measurements showed a small change in RV. Again somewhat similar measurements had been made by Rohrer [28] although the authors were not aware of this. Rohrer's data were obtained in only one subject whereas the Rochester group made measurements in 14 subjects for part of the study and 10 for other parts.

The paper described above concentrated on the pressure-volume behavior of the lung and chest wall. Today we call this statics. Another important paper was published in 1950 authored by Otis, Fenn and Rahn [16] and this was devoted to dynamics, that is the pressures and flows during breathing. The data included flow rates measured by a pneumotachograph, and pressures measured by a sensitive manometer. The necessary instrumentation had been largely developed in the Rochester laboratory. An important advance was the ability to measure the sudden change in pressure in the airway when flow was briefly interrupted with a shutter. Many of the measurements were made in relaxed subjects that were ventilated by reducing the pressure around the upper part of the body using a tank respirator. This had been devised by Drinker and was called a Drinker respirator.

Figure 21.9 shows the relationship between flow rate as measured at the mouth and the pressure difference between the alveoli and the mouth. Alveolar pressure was inferred from the airway pressure following a sudden short interruption of flow. The authors went on to consider many aspects of the pressure-flow relationships although much of this discussion was theoretical and could not be based on actual data. For example they calculated the amount of pressure that was required for what they called air viscance and air turbulence, where viscance referred to the pressure drop proportional to the flow rate and turbulence was the pressure drop proportional to the square of the flow rate. These are oversimplified concepts and would be later revised in the area of fluid dynamics. The authors also calculated the forces required

Fig. 21.9 Flow rate at
the mouth plotted against
alveolar pressure in a subject
ventilated by a Drinker respirator. Alveolar pressure is
inferred from the interrupted
technique. (From [16])

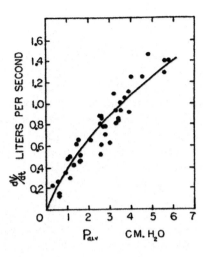

for tissue deformation although again a number of simplifications were necessary.
They then went on to calculate the work of breathing including the elastic forces
required to expand the lung and chest wall, the viscous forces (those proportional
to flow rate), and the turbulent forces. The result was a calculated work rate against
breathing frequency shown in Fig. 21.10. The actual numbers here are not reliable
but the concept of breaking down the work of breathing into elastic and flow related

Fig. 21.10 Calculated
work of breathing with the
contributions made by elastic
forces and what were called
viscous and turbulent forces.
See text for details. (From
[16])

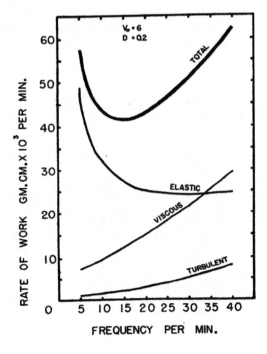

elements was important. Otis went on to study the work of breathing in greater detail [11].

Rahn moved to Buffalo in 1956 to become chairman of the Department of Physiology there. The center of gravity of the work moved with him and many people including myself benefited from the hive of activity that surrounded him. However Fenn remained in Rochester although he frequently visited Rahn, and Otis moved first to Johns Hopkins University and eventually become the first Professor of Physiology in the new Medical School of the University of Florida in Gainesville.

This summary concentrates on work done in Rochester between 1941 and 1956. However it should be emphasized that all three of the trio went on to make important contributions in other areas. For example Fenn became fascinated by two new frontiers that were beginning to unfold, man's exploration of space and the ocean depths. He worked extensively on oxygen toxicity and also the effects of high inert gas pressures on the metabolism of unicellular organisms. Rahn also branched out into several new areas. One of these was diving and he became particularly interested in the physiology of the Ama, the diving women of Korea and Japan. He also became fascinated by gas exchange in fishes and pointed out that they necessarily have a very low arterial Pco_2. The reason is that the solubility of oxygen in water is so low that a very large amount of water flows over the gills and this removes most of the carbon dioxide from the blood. Finally he made seminal contributions to the topic of gas exchange in avian eggs. Otis became interested in carbon monoxide poisoning, and also made important contributions to the history of respiratory physiology, particularly mechanics of breathing.

The remarkable blossoming of ideas that occurred in Rochester between 1941 and 1956 constitute a striking example of how three very intelligent scientists who were faced with a problem they had never seen before were able to make very special contributions to our knowledge within it. Many of us have benefited from this remarkable trio's contributions.

References

1. Allen TH, Otis AB, Bodine JH. The pH stability of protyrosinase and tyrosinase. J Gen Physiol. 1942;26:151–5.
2. Fenn WO. A quantitative comparison between the energy liberated and the work performed by the isolated sartorius muscle of the frog. J Physiol (Lond). 1924;58:175–203.
3. Fenn WO. Born fifty years too soon. Annu Rev Physiol. 1962;24:1–10.
4. Fenn WO, Cobb DM. The potassium equilibrium in the muscle. J Gen Physiol. 1934;17:629–56.
5. Fenn WO, Chadwick LE, Mullins LJ, Dern RJ, Otis AB, Blair HA, Rahn H, Gosselin RE. Physiological effects of pressure breathing. National Research Council, Division of Medical Sciences, Office of Scientific Research and Development, Report No. 111. January, 1943.
6. Fenn WO, Rahn H, Otis AB. A theoretical study of the composition of the alveolar air at altitude. Am J Physiol. 1946;146:637–53.
7. Fenn WO, Otis AB, Rahn H. Studies in Respiratory Physiology. Tech. Rep. No. 6528. U.S. Air Force, Dayton, Ohio, 1951.
8. Fowler RC. A rapid infra-red gas analyzer. Rev Sci Instrum. 1949;20:175–8.
9. Gerschman R, Gilbert DL, Nye SW, Dwyer P, Fenn WO. Oxygen poisoning and x-irradiation: a mechanism in common. Science. 1954;119:623–6.

10. Kleinholz LH, Rahn H. The distribution of intermedin in the pars anterior of the chicken pituitary. PNAS. 1939;25:145–7.
11. Otis AB. The work of breathing. Physiol Rev. 1954;34:449–58.
12. Otis AB. Monologue. Samuel Proctor Oral History Project. University of Florida Health Center, 1977. http://ufdc.ufl.edu/UF00006261/00001.
13. Otis AB. My initiation into respiratory physiology. Am J Respir Crit Care Med. 2000;161:345–6.
14. Otis AB, Rahn H. Development of concepts in Rochester, New York in the 1940s. In: West JB, editor. Pulmonary gas exchange. Ventilation, blood flow and diffusion. Vol. I. New York: Academic Press; 1980.
15. Otis AB, Rahn H, Epstein MA, Fenn WO. Performance as related to composition of alveolar air. Am J Physiol. 1946;146:207–21.
16. Otis AB, Fenn WO, Rahn H. Mechanics of breathing in man. J Appl Physiol. 1950;11:592–607.
17. Pappenheimer J. Hermann Rahn biographical memoirs. Vol. 69. Washington D.C.: National Academy of Sciences; 1996. pp. 242–67.
18. Rahn H. Structure and function of placenta and corpus luteum in viviparous snakes. Proc Soc Exp Biol Med. 1939;40:381–2.
19. Rahn, H. The reproductive cycle of the prairie rattler. Copeia. 1942;4:233–40.
20. Rahn H. A concept of mean alveolar air and the ventilation-bloodflow relationships during pulmonary gas exchange. Am J Physiol. 1949;158:21–30.
21. Rahn H. Studies in Respiratory Physiology, 3rd Ser., WADC Tech. Rep. 56-466, ASTIA Doc. No. AD 110487. U.S. Air Force, Dayton, Ohio, 1956.
22. Rahn H. Wallace Osgood Fenn biographical memoirs. Vol. 50. Washington D.C.: National Academy of Sciences; 1979. pp. 140–173.
23. Rahn H, Fenn WO. Studies in Respiratory Physiology. 2nd Ser., WADC, Tech. Rep. 55-357. U.S. Air Force, Dayton, Ohio, 1955a.
24. Rahn H, Fenn WO. A graphical analysis of the respiratory gas exchange. Washington D.C.: Am Physiol Soc. 1955b.
25. Rahn H, Otis AB. Man's respiratory response during and after acclimatization to high altitude. Am J Physiol. 1949;157:445–62.
26. Rahn H, Otis AB, Chadwick LE, Fenn WO. The pressure-volume diagram of the thorax and lung. Am J Physiol. 1946a;146:161–78.
27. Rahn H, Mohney J, Otis AB, Fenn WO. A method for the continuous analysis of alveolar air. J Aviat Med. 1946b;17:173–8.
28. Rohrer F. Der Strömungswiderstand in den menschlichen Atemwegen und der Eifluss der unregelmässigen Verzweinung des Brochialsystems auf den Atmungsverlauf in verschiedenen Lungenberzirken. Plüger's Archive für die gesamte Physiologie des Menschen und der Tiere. 1915;162:225–99. On a method of measuring lung elasticity in living human subjects, especially in emphysema. [English trans. in Translations in Respiratory Physiology, edited by West J.B. Stroudsburg: Dowden, Hutchinson and Ross, 1975.]
29. Von Neergaard K, Wirz K. Uber eine Methode zur Messung der Lungenelastizität am lebenden Menschen, insbesondere beim Emphysem. Zeitschrift für klinische Medizin. 1927a;105:35–50. Flow resistance in the human air passages and the effect of irregular branching of the bronchial system on the respiratory process in various regions of the lung. [English trans. in Translations in Respiratory Physiology, edited by West J.B. Stroudsburg: Dowden, Hutchinson and Ross, 1975.]
30. Von Neergaard K, Wirz K. Die Messung der Strömungswiderstände in den Atemwegen des Menschen, insbesondere bei Asthma und Emphysem. Zeitschrift für klinische Medizin. 1927b;105:51–82. Measurement of flow resistance in human air passages, particularly in asthma and emphysema. [English trans. in Translations in Respiratory Physiology, edited by West J.B. Stroudsburg: Dowden, Hutchinson and Ross, 1975.]
31. West JB. High life: a history of high-altitude physiology and medicine. New York: Oxford University Press; 1998.

Chapter 22
The Physiological Challenges of the 1952 Copenhagen Poliomyelitis Epidemic and a Renaissance in Clinical Respiratory Physiology

Abstract The 1952 Copenhagen poliomyelitis epidemic provided extraordinary challenges in applied physiology. Over 300 patients developed respiratory paralysis within a few weeks, and the ventilator facilities at the infectious disease hospital were completely overwhelmed. The heroic solution was to call upon 200 medical students to provide round-the-clock manual ventilation using a rubber bag attached to a tracheostomy tube. Some patients were ventilated in this way for several weeks. A second challenge was to understand the gas exchange and acid-base status of these patients. At the onset of the epidemic, the only measurement routinely available in the hospital was the carbon dioxide concentration in the blood, and the high values were initially misinterpreted as a mysterious "alkalosis." However, pH measurements were quickly instituted, the PCO_2 was shown to be high, and modern clinical respiratory acid-base physiology was born. Taking a broader view, the problems highlighted by the epidemic underscored the gap between recent advances made by physiologists and their application to the clinical environment. However, the 1950s ushered in a renaissance in clinical respiratory physiology. In 1950 the coverage of respiratory physiology in textbooks was often woefully inadequate, but the decade saw major advances in topics such as mechanics and gas exchange. An important development was the translation of the new knowledge from departments of physiology to the clinical setting. In many respects, this period was therefore the beginning of modern clinical respiratory physiology.

In 1952, Copenhagen was struck by a severe epidemic of poliomyelitis that included a large number of cases of bulbar polio resulting in respiratory paralysis. During the period from August to December, about 3000 patients with polio were admitted, mainly to one infectious disease hospital, the Blegdam Hospital, and of these, about 1250 had some type of paralysis (Fig. 22.1). Some 345 patients had bulbar polio affecting the respiratory and swallowing muscles [20, 22]. For several weeks, 30–50 patients with bulbar symptoms were admitted daily and 6–12 of these were desperately ill. During the first 3 week of the epidemic 27 of 31 patients with bulbar polio died, 19 of them within 3 days of admission. Clearly a catastrophe was in the making. Indeed Henry Cai Alexander Lassen (1900–1974), chief physician at the hospital stated "Although we thought we knew something about the management

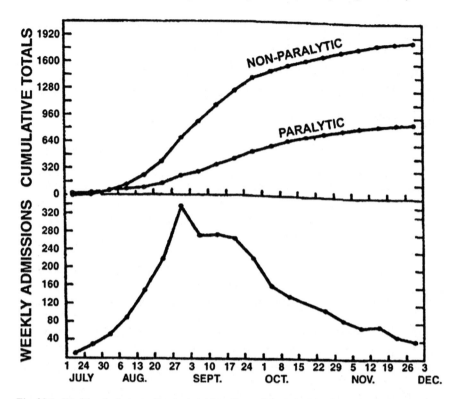

Fig. 22.1 Weekly admissions of patients with poliomyelitis to the Blegdam Hospital in Copenhagen in July to December 1952. Note the very rapid increase in the first part of August. The figure also shows that about half of the patients had some paralysis. (From Lassen [20])

of bulbar and respiratory poliomyelitis it soon became clear that only very little of what we did know at the beginning of the epidemic was really worth knowing" [22].

The epidemic resulted in two enormous challenges in applied physiology. The hospital lacked ventilators. The stunningly innovative solution was to use manual positive pressure administered by a roster of 200 medical students who repeatedly squeezed a rubber bag attached to a tracheostomy tube around the clock. The second challenge was understanding the life-threatening abnormalities of pulmonary gas exchange and acid-base status. At the start of the epidemic, the only laboratory test available was the total carbon dioxide concentration of the blood, and the high values were interpreted as a mysterious "alkalosis." Aspects of the epidemic are discussed elsewhere [3, 21, 22, 29, 31, 34, 35].

The problems posed by the epidemic are interesting in their own right. However, the thesis here is that the situation was symptomatic of the parlous state of clinical physiology in the early 1950s. Great advances in respiratory physiology had been made in the 1940s, partly in response to the demands of World War II, but many had not been translated into the clinical setting. The coverage of respiratory physiology in medical student textbooks around 1950 was generally abysmal.

But in the early part of the decade a renaissance occurred that can be exemplified by the publication in 1955 of *The Lung* by Julius Comroe and his coauthors [9]. Indeed the decade ushered in a revolution in applied respiratory physiology that lasted for much of the remainder of the century.

22.1 The Poliomyelitis Epidemic

Mechanical ventilation. In the initial stages of the epidemic there was some confusion about the need to ventilate patients with bulbar polio and respiratory insufficiency. Some patients were said to have polioencephalitis or "cerebralia" with a constellation of symptoms and signs, including haziness of consciousness, increased secretions in the airways, long periods of apnea punctuated by occasional inspirations, and obtunding of consciousness from which they could be aroused by verbal stimulation. The relative importance of hypoxia, carbon dioxide retention, fever, and uremia as a cause of this condition was debated [22]. There was an impression that some patients had an overwhelming viral bulbar infection for which little could be done, and this led to therapeutic nihilism [21]. This group had a very high mortality, and, in retrospect, many of these patients should probably have been ventilated much earlier than they were, if indeed they were ventilated at all.

When it was recognized that many of these patients could not survive without mechanical ventilation, the lack of machines became a serious crisis. At the outbreak of the epidemic the Blegdam Hospital had only one Emerson tank respirator and six cuirass respirators. The latter consisted of jackets that fitted around the chest and assisted ventilation by changing the pressure outside the thorax. Although these were useful for patients with mild respiratory impairment, they were utterly inadequate for cases with respiratory paralysis.

The bold solution was to manually ventilate the patients by squeezing a rubber bag attached to a tracheostomy tube inserted through an incision just below the larynx. The bag was connected to a tank of 50 % oxygen in nitrogen together with a soda lime absorber to remove carbon dioxide (Figs. 22.2 and 22.3). The logistical problem was solved by having a roster of 200 medical students who operated in relays. At the height of the epidemic, 70 patients had to be manually ventilated around the clock. The medical students worked 6–8 h shifts so that three or four shifts were needed in the 24 h. It is daunting to think of the responsibility of these students who were essentially ventilating blind with only the patient's appearance to guide them, at least in the initial stages. One account refers to a patient rolling her eyes up to signal that she needed more ventilation. Nevertheless the mortality rate is said to have dropped from ~90 to ~25 % as a result of this heroic intervention.

The demands were so great that the supply of medical students dwindled, and a number of lectures were given at the university to encourage more to volunteer [24]. Students from the dental school were also recruited. One report states that 1500 students in all took part in this activity with a total of 165,000 h [17].

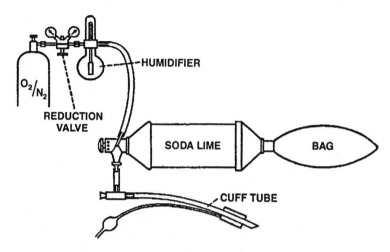

Fig. 22.2 Apparatus for manual mechanical ventilation. The tank was 50 % O_2– 50 % N_2. At the bottom is the cuffed endotracheal tube that was inserted through a tracheotomy. (From Lassen [20])

The introduction of manual bag ventilation early in the epidemic was due to the anesthesiologist Bjørn Ibsen (born: 1915) (Fig. 22.4). He was a Dane who had spent a period in the Department of Anesthesia at the Massachusetts General Hospital under Henry Knowles Beecher (1904–1976). Earlier in 1952, Ibsen had been involved in the treatment of a child with tetanus who was curarized and ventilated manually

Fig. 22.3 Photograph of a patient being manually ventilated with the apparatus shown in Fig. 22.2. (From Lassen [22])

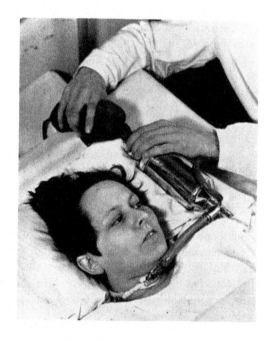

Fig. 22.4 Bjørn Ibsen (1915-), Danish anesthesiologist who suggested positive pressure ventilation for the patients with respiratory paralysis. (From Zorab [37])

through a tracheostomy [35]. There was a dramatic meeting at the Blegdam Hospital on August 25, 1952, when, as stated earlier, 31 patients with bulbar polio had been treated with the tank and cuirass respirators in the preceding 3 week but 27 had died. On that day alone, four patients were autopsied, one of them a 12-year-old boy who died with a "bicarbonate level in the serum far above the normal level." The attendees included Lassen, Poul Astrup (1915–2000) (Fig. 22.5), chief of the hos-

Fig. 22.5 Poul Astrup (1915–2000) who ushered in the modern period of clinical acid-base physiology. (From Severinghaus and Astrup [28])

pital laboratory, Mogens Bjørneboe, a member of the hospital's medical staff, and Ibsen, who had apparently somewhat reluctantly been invited to attend by Lassen at the urging of Bjørneboe. Ibsen soon recognized that the high blood bicarbonate levels were not an alkalosis of unknown origin but were caused by severe carbon dioxide retention ([3], p. 258). He recommended immediate manual ventilation via a tracheostomy. It is interesting that no successful treatment by positive pressure ventilation given continuously over a long period had been reported by 1950 [12].

Ibsen related how the first patient was a 12-year-old girl who had paralysis of all four extremities, had atelectasis of the left lung, and who was gasping for air and drowning in her own secretions [1, 18]. She was pyrexic, cyanotic, and sweating. A tracheotomy was done under local anesthesia, a cuffed endotracheal tube was placed, and she was eventually ventilated satisfactorily. This event was a turning point in critical care medicine, partly because it was one of the first occasions when an anesthesiologist moved out of the operating room into another environment. Positive pressure ventilation had previously been used for short periods in a polio epidemic in Los Angeles in 1948–1949 [7, 8], but this work had been published in an obscure journal and was not well known.

The ventilation circuit shown in Figs. 22.2 and 22.3 was a semi-closed system similar to that used in many anesthetic machines with the advantage that it required less fresh gas from the tank than an open circuit. The physicians also felt that the buffering effect of the large volume of the circuit made it easier for inexperienced medical students to maintain the appropriate ventilation. The flow meter was set to 5–10 L min. However, the CO_2 absorber incorporated in the circuit caused some problems because of potential aspiration of soda lime particles into the lung. In later versions, the absorber was removed from the circuit and there was simply an inspiratory/expiratory valve at the tracheostomy tube so that expired gas was vented directly into the outside air. This required a higher flow rate from the oxygen/nitrogen tank.

It is interesting to look back at a contemporary discussion of the principles of mechanical ventilation in the clinical setting, for example, Rattenborg [27]. There was confusion about the mode of action of positive pressure vs. negative pressure ventilation, and the role of airway resistance on the one hand and lung and chest wall compliance on the other in limiting inflation of the lung. The analysis included a curious statement that negative pressure inflation was unsatisfactory because it did not ensure a constant rate of inspiratory airflow. The reader today is aware of the great contrast between this discussion and the work done in departments of physiology a few years before when the relevant principles of the mechanics of respiration had been clearly enunciated by such groups as Fenn, Otis, and Rahn in the University of Rochester, New York [13], and Mead and Whittenberger and their colleagues at the Harvard School of Public Health [23]. This is an example of the prevailing dissociation in the early 1950s between the advances made in departments of physiology on the one hand and their application to clinical situations on the other.

Pulmonary gas exchange and acid-base status. As indicated earlier, a major difficulty in the successful ventilation of these patients was the almost complete lack of laboratory data about pulmonary gas exchange and acid-base status. The clinical symptoms and signs of respiratory insufficiency were vague or simply caused by intense anxiety. Patients felt suffocated and had difficulty in coping with their secretions because they could not swallow, there was cyanosis if they were not being given oxygen, and a clammy skin and hypertension were sometimes seen, presumably the results of increased blood catecholamine levels. The only routine laboratory investigation available was the total carbon dioxide concentration in venous or arterial blood. The clinical laboratory of the Blegdam Hospital had a pH meter that could be used for blood, but it required a large sample volume and could not be used for frequent or routine measurements. As stated earlier, the very high levels of carbon dioxide concentration were initially attributed to a mysterious alkalosis, and it was Ibsen who first recognized at the time of the epidemic that instead they signaled a severe respiratory acidosis although this had previously been suggested [25].

Astrup related that "this outright misinterpretation of a high CO_2 content as alkalosis in patients with respiratory insufficiency produced a very deep impression on me as a laboratory man" [3], p. 258). Astrup was able to persuade the Radiometer A/S in Copenhagen to provide him with a smaller pH meter that could be used for blood [28]. A quick measurement of blood pH at 38 °C soon proved Ibsen right, and this led to the 12-year-old girl being tracheotomized and given manual positive pressure ventilation that immediately caused the "alkalosis" to disappear. Astrup noted that the value of the carbon dioxide concentration of blood as an index of its alkalinity could be traced all the way back to 1877 when it was described by Friedrich Walter (born: 1850) and in its time was a very valuable contribution, but clearly here was a situation where the concept was misleading. In defense of the misconceptions of the clinicians in the early 1950s, it should be added that it was unusual to request laboratory data in patients with abnormalities of ventilation.

A top priority was to measure the PCO_2 in the blood, and this was done using the Henderson-Hasselbalch equation. The graphical depiction of this by Van Slyke and Sendroy [33] was well known to Astrup, and their original diagram is reproduced in Fig. 22.6. The vertical line on the extreme left shows the total carbon dioxide content of blood both in milliliters of CO_2 per deciliter of blood and in millimolar concentrations, whereas the central and right-hand lines show the plasma pH and PCO_2, respectively. A line has subsequently been added joining the normal pH of 7.4 and normal PCO_2 of 40 mmHg to show a total CO_2 content of about 56 ml CO_2 per deciliter of blood.

The carbon dioxide concentration was determined using the manometric method described by Van Slyke and O'Neill [32]. As mentioned earlier, the normal laboratory blood pH meter at the time required a large sample size. However, when the smaller pH meter became available from Radiometer, more than 700 pH determinations were made in the Blegdam Hospital over the next 4 months [4].

Nevertheless this method for determining the PCO_2 of blood was still cumbersome because the Van Slyke manometric method was so time consuming. Astrup later realized that if a sample of either plasma or whole blood was exposed to different CO_2 partial pressures, the resulting change in pH was linearly related to the logarithm of the PCO_2 within the clinical range [2]. First the pH of the sample of plasma or blood was measured. Then the sample was exposed to gas with high and

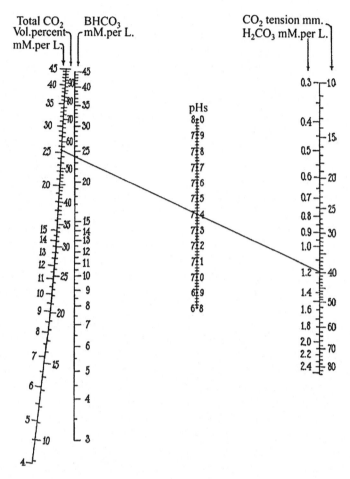

Fig. 22.6 Nomogram that was used for determining the blood PCO_2 from the total CO_2 content and plasma pH. (From Van Slyke and Sendroy [33])

low PCO_2 values (for example, about 80 and 15 mmHg), and the pH for each PCO_2 was measured. The actual PCO_2 was then obtained by interpolation. This rapid interpolation method was extensively used until the CO_2 electrode was eventually introduced several years later.

The early measurements of pH and PCO_2 on the patients from the Copenhagen epidemic who were manually ventilated sometimes showed dramatic changes within short periods of time. Table 22.1 shows an example from a 5-year-old boy who was almost moribund on admission and then was tracheotomized and manually ventilated [4]. Note that the pH rose from 6.99 to 7.65 over a 3.5 h period and the PCO_2 fell from 150 (in venous blood) to 14 mmHg (in arterial blood)!

In another patient, a woman aged 30 year, the pH and PCO_2 were monitored over a series of 13 days (Table 22.2). The fact that a number of the measurements were

Table 22.1 Blood pH and Pco_2 values in a 5-yr-old boy after the onset of manual ventilation

Hour	Blood	pH	Pco_2, mmHg	CO_2 conc, mmol	Bicarbonate, mmol
11:40 AM	venous	6.99	150	39.0	34.5
2:10 PM	arterial	7.52	32	24.4	24.5
3:05 PM	arterial	7.65	14	15.6	15.2

Modified from Astrup et al. [3]

made on venous rather than arterial blood complicates the interpretation somewhat, but it can be seen that the PCO_2 remained fairly stable in the low 30 s until September 10 when, in a venous sample, it had fallen to 17 mmHg! This was the result of a decision made at a conference on September 10 to increase the rate of manual ventilation in all patients from 20 to 30 breaths/min. However, because a number of patients subsequently showed very low PCO_2 values, it was then decided to reduce the ventilation frequency to 25 breaths/min. In the patient shown in Table 22.2, the PCO_2 then rose to 23 mmHg.

Note that all these PCO_2 values were abnormally low, and it is likely that most of the patients who were ventilated by the inexperienced medical students were in this situation. Of course, hyperventilation was better than hypoventilation under these conditions. A common observation in patients who are mechanically ventilated over long periods is that they complain of "air hunger" if the PCO_2 is allowed to rise to near the normal level of 40 mmHg.

In 1953, there was another poliomyelitis epidemic, this time in Stockholm, and with the experience obtained in Copenhagen in the previous year, the management

Table 22.2 Blood pH and Pco_2 values in a 5-yr-old woman during manual ventilation over 13 days

Date	Hour	Blood	pH	Pco_2, mmHg
Sept 1	1:00 PM	venous	7.49	32
	4:45 PM	"	7.47	31
2	9:15 AM	"	7.50	32
	10:30 AM	arterial	7.50	36
	10:35 AM	venous	7.47	34
	2:15 PM	"	7.48	36
4	9:10 AM	"	7.55	30
6	10:55 AM	arterial	7.55	30
8	12:35 PM	venous	7.56	31
10	10:10 AM	"	7.70	17
13	10:10 AM	"	7.56	23

Modified from Astrup et al. [3]

of patients with respiratory paralysis had improved. Nevertheless several pH values over 7.6 were reported in arterial blood [19].

A further fallout from these early measurements of pH, PCO_2, and bicarbonate in blood were other indexes that improved our understanding of the respiratory and metabolic components of acid-base disturbances particularly when complicated mixed situations occurred. One measurement was the "standard bicarbonate," which was the plasma bicarbonate concentration when the blood was exposed to a gas of normal PCO_2 of 40 mmHg. This was often obtained by having the laboratory technician exhale over it! In effect, this measurement removed the respiratory component of the acid-base disturbance and allowed any metabolic compensation to be recognized. A similar concept had been suggested earlier by Van Slyke and also by Hasselbalch.

However, the most useful fallout was the concept of "base excess." A simple way to represent this is the position of the "blood buffer line" on a diagram relating plasma bicarbonate concentration to pH, often known as the Davenport diagram [11]. The blood buffer line shows the relationship between bicarbonate concentration and pH as CO_2 (or carbonic acid) is added to or subtracted from a blood sample. The vertical position of the line is a measure of metabolic compensation for acidosis or alkalosis and is measured in milliequivalents per liter. These fundamental advances in our understanding of acid-base physiology by Astrup and collaborators were later amplified by Siggaard-Andersen and others [30].

22.2   The Renaissance in Clinical Respiratory Physiology During the 1950s

Respiratory physiology in textbooks around 1950. The 1952 Copenhagen poliomyelitis epidemic dramatically demonstrated the uncertainties in clinical respiratory physiology in the areas of mechanical ventilation and gas exchange and is remarkable in its own right. However, it was also symptomatic of the slowness in translating important advances that had been made in the 1940s into the clinical setting. One way to emphasize this is to examine textbooks of physiology for medical students published around 1950. A good example is the textbook *Physiological Basis of Medical Practice* by Charles Herbert Best (1899–1978) and Norman Burke Taylor (born:1885) which was first published in 1937 and met with great success. New editions appeared in 1939, 1943, and 1945, with a 5th edition in 1950. The book was reprinted 15 times between 1937 and 1950, and the 5th edition probably gives a good overview of what doctors learned at that time. Parenthetically, this was the textbook that I used during my medical course in Australia in the late 1940s.

Several sections were certainly adequate by the standards of the day. The basic gas laws were well summarized and there were good chapters on the carriage of oxygen and carbon dioxide by the blood, because these topics had been well worked out. However, the sections on pulmonary mechanics and pulmonary gas exchange can only be described as abysmal. There is a whole chapter titled "The mechanics of respiration," but this is mainly concerned with the respiratory muscles, the consequences of pneumothorax, and a discussion of bronchiectasis! The chapter includes two pages on "artificial respiration" with a diagram of the Drinker negative pressure "iron lung," but the rest of the discussion is about Schafer's method, which consists of pressing the lower ribs of a prone subject with the hands, and Eve's rocking method in which the patient is tilted from the head-up to head-down position and back again on a rocking bed, and the diaphragm is displaced by gravity. Positive pressure ventilation is not even mentioned. The only other place where the topic of the mechanics of breathing occurs is in short sections on asthma and "chronic emphysema" with vague references to the resistance of the airways and the elastic properties of the lung. There is no quantitative treatment of these important topics.

Pulmonary gas exchange also gets very short shrift mainly in a section titled "Anoxia." Again this is entirely qualitative. The importance of ventilation in eliminating carbon dioxide is not appreciated. Indeed one important factor in CO_2 retention is thought to be the slower rate of diffusion of CO_2 in the alveolar gas because of the large size of the molecule! There is no hint of the alveolar ventilation as determining the alveolar PCO_2 nor the alveolar gas equation that relates the alveolar PO_2 to the inspired PO_2 and alveolar PCO_2. It is hardly surprising in view of this state of knowledge that the clinicians faced with patients with respiratory paralysis were at a loss. It is extraordinary that the respiration section of the 5th edition of Best and Taylor published in 1950 is very similar to that in the 1st edition of 1937.

Another popular medical student textbook of physiology current at about the same time in 1948 had similar coverage of the respiratory system [36]. In particular, the section on "artificial respiration" included the same procedures and there was a remarkable footnote. "Schafer's method is, however, frequently not applicable in cases where patients stop breathing on the operating table. Here the prone position on the floor is rarely possible; the rocking method or compression of the chest or abdomen, may have to be employed." It is extraordinary that even in this situation the possibility of positive pressure ventilation was not envisaged. It was not until the 9th edition published in 1952 that positive pressure ventilation was mentioned and then only in one sentence in the setting of anesthesia.

Contributions of physiologists in the 1940s and early 1950s. This unhappy state of affairs contrasts greatly with the work that was being done in departments of physiology and in some departments of medicine in the 1940s. Just to take a few examples, Wallace Fenn, Hermann Rahn, and Arthur Otis at the University of Rochester, New York, had established many of the principles of pulmonary mechanics and pulmonary gas exchange during the war years with publications occurring during the mid- and late 1940s [14]. Much of this work did not appear initially in the open literature but was contained in classified reports that were later released in eight volumes, the first being Fenn et al. [15]. Subsequently some, although not all, of the papers were published in the *Journal of Applied Physiology* and elsewhere. Another productive group included Jere Mead, James Whittenberger, and their colleagues in the Harvard School of Public Health. Julius Comroe had an active research team in the Graduate School of Medicine at the University of Pennsylvania (see below). Richard Riley, Andre Cournand, and their coworkers were breaking new ground in New York on pulmonary gas exchange, and beginning in 1941 Andre Cournand and Dickinson Richards with their colleagues developed cardiac catheterization in humans [10]. Incidentally, this Nobel prize work is not mentioned in the 1948 Best and Taylor.

This rapid acceleration of research was not confined to departments of physiology or to the United States. In London, UK, there were active groups at St. Bartholomew's Hospital Medical School and the Post Graduate Medical School, and in South Wales important advances were made in pulmonary function in occupational lung disease. Other groups were active in Canada, France, and South Africa. Physiologists in Germany contributed developments of oximetry and other measuring devices, and a two-volume account of their work is available [16].

Stimulated in part by these advances, the *Journal of Applied Physiology* started publication in 1948. Part of the impetus for this was the desire to release to the open literature work that had been described in classified documents during the war. Other periodicals, such as the *Journal of Clinical Investigation*, published numerous articles in applied clinical physiology. Many of the advances of the 1940s were brought together by Comroe and colleagues in the book *Pulmonary Function Tests: Methods in Medical Research, Volume II*, which was published in 1950. Readers of this who compared it with the 5th edition of Best and Taylor, which was published in the same year might wonder if they were produced on the same planet.

The role of the book "The Lung..." by Comroe and others. But the most influential of the books in this renaissance of clinical pulmonary physiology was *The Lung: Clinical Physiology and Pulmonary Function Tests* by Julius Comroe, Robert Forster, Arthur DuBois, William Briscoe, and Elizabeth Carlsen, which was published in 1955. This book made an enormous impression. Comroe stated in the preface that it was based on the Beaumont lecture that he gave to the Wayne County Medical Society in Detroit in February 1954. However, the main basis was a course for pulmonary physicians that was given in March 1953 by Comroe and colleagues in the Department of Physiology and Pharmacology, Graduate School of Medicine, University of Pennsylvania. Robert Forster covered blood gases and diffusion, William Briscoe dealt with ventilation-perfusion inequality and other aspects of gas exchange, Arthur DuBois discussed the mechanics of breathing, and Comroe covered the control of ventilation. DuBois still has his notes of the course, including a memo from Comroe dated February 13, 1953, about the preparation of the "lantern slides." Fig. 22.7 shows two sketches made by DuBois for the course, and Fig. 22.8 depicts how they were eventually published in *The Lung*. The physics-based nature of these diagrams forms a striking contrast to the treatment of these topics in the 5th edition of Best and Taylor [5] or the paper by Rattenborg [27]. It is a remarkable coincidence that in early 1953 at the same time that the Danish physicians were struggling with the problems posed by the patients with respiratory paralysis, the group at the University of Pennsylvania were laying the foundations for the renaissance in clinical respiratory physiology.

The authors of The Lung had no illusions about what they were doing. The preface to the book began "Pulmonary physiologists understand pulmonary physiology reasonably well. Many doctors and medical students do not." It went on to add

Fig. 22.7 Sketches by Arthur Dubois that were prepared for the course on clinical respiratory physiology given at the University of Pennsylvania in March 1953

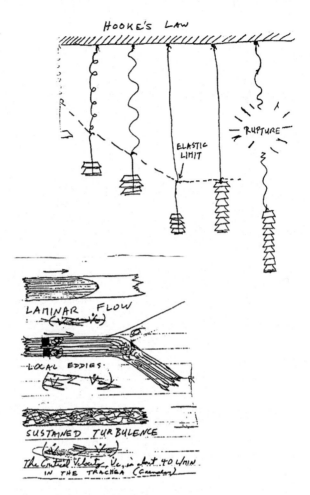

"This is *not* a book for pulmonary physiologists: it is written for doctors and medical students." This was the quintessential translation from physiology departments where the new work had been done in the 1940s to the medical students and others in the 1950s who desperately needed the new information.

One of the features of the book was the excellent clear illustrations that had been prepared by Carlsen (Fig. 22.8). The fact that the illustrator, who was a student at the time, was included in the list of authors emphasized the importance that Comroe gave to this feature. Interestingly, although the approach was rigorous and quantitative, the authors were reluctant to include equations in the main text because they felt that most medical students and physicians would be intimidated by these. For example, the alveolar ventilation equation relating the CO_2 production to the alveolar ventilation and fraction of CO_2 in alveolar gas was given as a footnote. However, all the important equations were summarized in an appendix at the end of the book.

Fig. 22.8 Illustrations from the book The Lung… by Comroe et al. [9] showing how the sketches of Fig. 22.7 appeared in the book

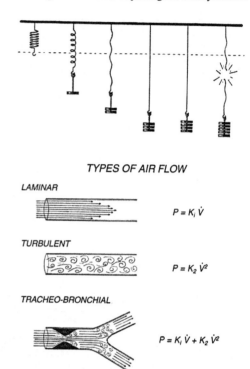

TYPES OF AIR FLOW

LAMINAR

$$P = K_1 \dot{V}$$

TURBULENT

$$P = K_2 \dot{V}^2$$

TRACHEO-BRONCHIAL

$$P = K_1 \dot{V} + K_2 \dot{V}^2$$

Other seminal publications appeared at about the same time, including the classical short monograph *A Graphical Analysis of the Respiratory Gas Exchange: The O_2-CO_2 Diagram* by Rahn and Fenn [26]. However, this and similar publications were targeted more at professional physiologists and physiology graduate students than medical students and pulmonary physicians. It was Comroe's book that really ushered in the renaissance in clinical respiratory physiology.

What was the essential difference between the treatment of pulmonary mechanics and gas exchange in Best and Taylor [5] compared with Comroe et al. [9]? The answer lies in the completely different approaches, one qualitative and the other rigorously quantitative. For example, in Best and Taylor there is not one equation in the sections on airway resistance, elastic properties of the lung, or pulmonary gas exchange. Airway resistance is largely dealt with in a section on asthma where there were statements such as "The expiratory muscles compress the chest and the abdominal muscles contract in the attempt to squeeze the air from the lungs. The intrapulmonary pressure is greatly elevated and the air escapes through the constricted tubes with a distinct wheezing sound." However, airway resistance is not defined in quantitative terms as the pressure difference divided by flow, and there is no mention of the different modes of air flow such as laminar and turbulent.

The elastic properties of the lung are dealt with in a similar nonquantitative way in the section on "Chronic Emphysema." For example, a typical statement is "Two factors are concerned in the production of emphysema (a) reduction in the elastic tissue of the lung and (b) increased distension of the alveolar spaces." There is no mention of the relationship between pressure and volume in an elastic structure nor is the term compliance mentioned.

There is a similar nonquantitative treatment of pulmonary gas exchange mainly in the section on "Chronic Emphysema." A typical paragraph begins "The cause of the impaired gaseous exchange is not altogether clear. Thickening of the alveolar and capillary walls and the obliteration of capillaries have been considered to be a factor. Yet if this were so one would not expect the retention of carbon dioxide which, owing to its greater solubility (30 times that of oxygen) has a much higher rate of diffusion through the pulmonary membrane, to be so much more pronounced that the anoxia." Interestingly this passage is followed by a brief statement that uneven ventilation "as a result of the loss of elasticity" could affect gas exchange in some way. However, that is about as far as the text goes on the subject of ventilation-perfusion inequality.

The treatment of these topics in *The Lung* by Comroe et al. [9] is completely different. The starting point is elementary physics and engineering, and lung elasticity for example is approached through Hooke's law. The principles of laminar and turbulent flow through tubes are discussed, both of these concepts being alluded to in Figs. 22.7 and 22.8. There is a full analysis of the changes in alveolar and intrapleural pressure during the respiratory cycle. Pulmonary gas exchange is dealt with in a quantitative way beginning with the alveolar ventilation equation leading to the alveolar gas equation. The differences between diffusion limitation of gases across the blood-gas barrier and ventilation-perfusion inequality as factors impairing gas exchange are clearly stated. Obviously, there was a revolution in the presentation of the principles of clinical respiratory physiology to medical students and physicians in the early 1950s, and we continue to benefit from this renaissance today.

In summary, the 1952 Copenhagen poliomyelitis epidemic provided momentous challenges in clinical respiratory physiology and is fascinating for that reason. In addition, it underscores the gap that had developed between the rapid advances in respiratory physiology that took place in the 1940s and their application to the clinical environment. This is highlighted by the backward state of respiratory physiology as exemplified in typical medical student textbooks around 1950. However, a remarkable change took place in that decade. The rapid advances made in departments of physiology were translated to the clinical setting with one of the most influential factors being the publication of The Lung by Comroe et al. in 1955. The result was a greatly improved understanding of applied respiratory physiology that continues to benefit patients today.

Appendix

Logistics of the long-term manual ventilation. One of the most remarkable features of the epidemic was the way medical students and others were organized to provide round-the-clock, long-term manual ventilation. Information about this is given in three contemporary articles in the Danish medical journal Ugeskrift for Laeger by Thomsen [31], Bjorneboe [6], and Hansen [17]. Thomsen was one of the medical student "ventilators," Bjorneboe was on the medical staff of the Blegdam Hospital, and Hansen was mayor of Copenhagen. Later Wackers [34] interviewed a number of the people involved.

The emotional demands on the patient and student were enormous [31]. A young patient would be admitted struggling to breathe, a tracheostomy was performed, and he (or she) would meet the 18–19 year-old students who were to keep him alive by squeezing the bag for an indefinite period. Bonding between the patient and student was strong, especially for young children. Four or five students were allocated to each patient because of the 24 h coverage, but the same students always ventilated the same patient. In the case of young children, the student read to them and played games. If a new student was added to the team for some reason, the patient initially reacted strongly. Communication between the patient and student was difficult because of the tracheostomy tube, but students learned to lip read, and some patients gave information by moving their eyes.

Very early in the epidemic, when it was clear that large numbers of students would be required, the medical student council was involved, and they accepted the responsibility for rostering the students, arranging for them to be excused from otherwise obligatory courses, and negotiating payment for their services. The students worked 6–8 h shifts, which was both emotionally and physically demanding. During an 8 h shift, there was a 10 min "smoke" break each hour, and a half-hour meal break in the middle, but otherwise the student was continually compressing the bag.

There were many technical problems connected with the manual ventilation [6]. The students first had instruction on the general principles from an anesthesiologist and then 3 or 4 h of practical instruction with the equipment. Equipment problems included unexpected emptying of the oxygen tank, kinking of the tube from the tank, and damage to the ventilating bag. The CO_2 absorber had to be changed periodically, and the tracheostomy tube could slip down and occlude a main bronchus. Periodic suctioning of the airways was necessary in some patients.

Observation of the patient was very important, and another student in the team or nurse would monitor the pulse rate and blood pressure from time to time. There was continual surveillance by doctors and nurses walking up and down the wards to give help where needed. Many students found the emotional and physical demands too much and gave up. Also there was a concern about developing polio, although this apparently never happened to any of the students. As the epidemic wore on, the second year students who had done most of the ventilating were partly replaced by first-year students, who, in the European system, would have come straight from high school and might be only 18 year old.

The economic impact of the epidemic was vast. Very quickly, the Blegdam hospital ran out of beds and three other hospitals were recruited. Large numbers of extra doctors, including anesthesiologists, nurses, and hospital staff were required as well as the medical students. The cost of the epidemic up to April 1953 was estimated to be 5–6 million Danish Kroner [17], that is about 30 million US$ at today's exchange rate.

Postscript

When this manuscript was being prepared, I wrote to several of the student "ventilators" but received no replies. However, after the paper was accepted for publication, I received a most interesting letter from Dr. Uffe Kirk who was 25 years old in 1952 at the time of the epidemic. He had just finished medical school and was asked to play a major role in organizing the medical student ventilators. Here are some extracts from his account.

"The difference between ordinary patients requiring ventilation and polio patients was characteristic: They were conscious! The students invented ways to communicate with their patients. Some patients holding a small stick in their mouths communicated by pointing at letters on a poster, laboriously spelling what they wanted to say. This went fairly well because the student learned to half-guess what the patient would say after only a few letters. The student would then say out loud what he or she thought the patient meant, and the patient would then wink in one way if the student had guessed right and in another way if not. If the student was in no way near the correct answer, the patient could point at the word "Idiot" written on the poster. This way the student always received a message from the patient if the ventilation required correcting. It was almost a safer way to correct ventilation than laboratory tests, blood pressure, and other medical controls.

The intimate relation made the students very concerned about the well being of their patients. They were exhilarated at every positive sign but were also very sad when things went downhill. And it did for many patients. Even though the students knew that death was a very real option, they were mentally strained when their patients died.

At worst, the patients died during the night. The light in the wards was dimmed in order not to disturb the patients in their sleep. But the faint light and the fact that the students were not able to tell anything from the ventilation made it impossible for the students to know that their patient had died. It was therefore a shock for the student when morning came and he/she realized that the patient had been dead for a while....

Not so long ago professor Bjørn Ibsen was lauded at a conference here in Denmark. He sat on a chair in the front row when a woman of~65 year quietly went up to him, kissed him on the cheek and said 'Thank you for my life!'

In 1952 she was the twelve-year old girl whom Bjørn Ibsen was permitted to try and save by means of tracheostomy and a tube through which he wanted to ventilate in replacement of her [paralysed] respiration. He succeeded and the woman was proof of that, and was the direct cause of 1500 medical and dental students ventilating polio patients for 165,000 h at the Blegdam Hospital in 1952 thereby saving ~100 people who would have been lost without this effort...."

The complete letter has been placed in an archive in the Mandeville Special Collections Library at the University of California, San Diego.

Acknowledgements I thank the following: Arthur DuBois for providing me with a copy of his notes of the course on pulmonary physiology held at the University of Pennsylvania in March 1953; DuBois and Robert Forster for reading the manuscript; Ger Wackers for sending me his doctoral dissertation; and Harrieth Wagner for translating several articles from Danish.

References

1. Andersen EW, Ibsen B. The anaesthetic management of patients with poliomyelitis and respiratory paralysis. Br Med J. 1954;1:786–8.
2. Astrup P. A simple electrometric technique for the determination of carbon dioxide tension in blood and plasma, total content of carbon dioxide in plasma, and bicarbonate content in separated plasma at a fixed carbon dioxide tension (40 mmHg). Scand J Clin Lab Invest. 1956;8:33–43.
3. Astrup P, Severinghaus J. The history of blood gases, acids and bases. Copenhagen: Munksgaard; 1986.
4. Astrup P, Gotzche H, Neukirch F. Laboratory investigations during treatment of patients with poliomyelitis and respiratory paralysis. Br Med J. 1954;4865:780–6.
5. Best CH, Taylor NB. The physiological basis of medical practice. Baltimore: Williams & Wilkins; 1950.
6. Bjorneboe M. Studenten og poliopatienten. Ugeskr Læger. 1953;115:469–71.
7. Bower AG, Bennett VR, Dillon JB, Axelrod B. Investigation on the care and treatment of poliomyelitis patients. Ann West Med Surg. 1950a;4:561–82.
8. Bower AG, Bennett VR, Dillon JB, Axelrod B. Investigation on the care and treatment of poliomyelitis patients; II. Physiological studies of various treatment procedures and mechanical equipment. Ann West Med Surg. 1950b;4:686–716.
9. Comroe JH, Forster RE, Dubois AB, Briscoe WA, Carlson RW. The lung: Clinical physiology and pulmonary function tests. Chicago: Year Book; 1955.
10. Cournand A, Ranges HA. Catherization of the right auricle in man. Proc Soc Exp Biol Med. 1941;46:462–466.
11. Davenport HW. The ABC of acid-base chemistry: the elements of physiological blood-gas chemistry for medical students and physicians. Chicago: University of Chicago Press; 1974.
12. Emerson CG. The clinical application of prolonged controlled ventilation. Acta Anaesthesiol Scand Suppl. 1963;13
13. Fenn WO. Mechanics of respiration. Am J Med. 1951;10:77–90.
14. Fenn WO, Rahn H, Otis AB. A theoretical study of the composition of the alveolar air at altitude. Am J Physiol. 1946;146:637–53.
15. Fenn WO, Otis AB, Rahn H. Studies in respiratory physiology. AF Technical Report No. 6528. 1951.
16. German Aviation Medicine: World War II. Washington DC: US Air Force; 1950.
17. Hansen J. Den økonomiske baggrund for poliobekæmpelsen. Ugeskr Læger. 1953;471–3.
18. Ibsen B. The anaesthetist's viewpoint on the treatment of respiratory complications in poliomyelitis during the epidemic in Copenhagen, 1952. Proc Roy Soc Med. 1954;47:72–4.
19. Jungner I, Laurent B. The poliomyelitis epidemic in Stockholm 1953. Biochemical laboratory investigations. Acta Med Scand Suppl. 1956;316:71–9.
20. Lassen HCA. A preliminary report on the 1952 epidemic of poliomyelitis in Copenhagen with special reference to the treatment of acute respiratory insufficiency. Lancet. 1953;261:37–41.
21. Lassen HCA. The management of respiratory and bulbar paralysis in poliomyelitis. In: Poliomyelitis. No editor stated Geneva: World Health Organization; 1955. p. 157.
22. Lassen HCA. Management of Life-Threatening Poliomyelitis, Copenhagen, 1952–1956, With a Survey of Autopsy-Findings in 115 Cases [translated from the Danish by Hans Andersen and others]. Livingstone, Edinburgh; 1956.
23. Mead J, Whittenberger JL. Physical properties of human lungs measured during spontaneous respiration. J Appl Physiol. 1953;5:779–96.
24. Medicinske Studenterrådet. En orientering om poliomyelitis. Især med henblik på epidemien 1952. Udarbejdet af de medicinske studenterråd ved Københavns Universitet (på basis af referat fra et møde om poliomyelitis på med. anotomisk institut den 5 november 1952) N. Olaf Møller; København, 1952.

25. Nielsen HK. Om repiratorbehndling af respirations-pareser ved poliomyelitis anterior acuta. Ugeskr Læger. 1946;108:1341–8.

26. Rahn H, Fenn WO. A graphical analysis of the respiratory gas exchange. Am Physiol Soc.; Washington, DC: 1955.

27. Rattenborg C, Lassen HCA, editor. Basic mechanics of artificial ventilation. In: Lassen HCA, editor. Management of life-threatening poliomyelitis. Copenhagen, 1952–1956, With a Survey of Autopsy-Findings in 115 Cases [translat from the Danish by Hans Andersen and others]. Edinburgh: Livingstone, 1956.

28. Severinghaus JW, Astrup PB. History of blood gas analysis; II. pH and acid-base balance measurements. J Clin Monit. 1985;1:259–77.

29. Severinghaus JW, Astrup P, Murray JF. Blood gas analysis and critical care medicine. Am J Respir Crit Care Med. 1998;157:114–22.

30. Siggaard-Andersen O. The acid-base status of the blood. Copenhagen: Munksgaard; 1974.

31. Thomsen VF. Kort redegorelse for den midlertidige studenterhlelp pa blegdamshospitalet. Ugeskr Læger. 1953;115:468–9.

32. Van Slyke DD, O'Neill JM. The determination of blood gases in blood and other solutions by vacuum extraction and manometric measurement. J Biol Chem. 1924;61:523–73.

33. Van Slyke DD, Sendroy J Jr. Studies of gas and electrolyte equilibria in blood; XV. Line charts for graphic calculation by the Henderson–Hasselbalch equation, and for calculating plasma carbon dioxide content from whole blood content. J Biol Chem. 1928;79:781–798.

34. Wackers GL. Constructivist medicine (Dissertation) Maastricht. The Netherlands: University of Maastricht; 1994a.

35. Wackers GL. Modern anaesthesiological principles for bulbar polio: manual IPPR in the 1952 polio-epidemic in Copenhagen. Acta Anaesthesiol Scand. 1994b;38:420–431.

36. Wright S. Applied physiology. London: Oxford University Press; 1948.

37. Zorab J. The resuscitation greats: Bjørn Ibsen. Resuscitation. 2003;57:3–9.

Chapter 23
Historical Aspects of the Early Soviet/Russian Manned Space Program

Abstract Human spaceflight was one of the great physiological and engineering triumphs of the twentieth century. Although the history of the United States manned space program is well known, the Soviet program was shrouded in secrecy until recently. Konstantin Edvardovich Tsiolkovsky (1857–1935) was an extraordinary Russian visionary who made remarkable predictions about space travel in the late nineteenth century. Sergei Pavlovich Korolev (1907–1966) was the brilliant "Chief Designer" who was responsible for many of the Soviet firsts, including the first artificial satellite and the first human being in space. The dramatic flight of Sputnik 1 was followed within a month by the launch of the dog Laika, the first living creature in space. Remarkably, the engineering work for this payload was all done in less than 4 week. Korolev's greatest triumph was the flight of Yuri Alekseyevich Gagarin (1934–1968) on April 12, 1961. Another extraordinary feat was the first extravehicular activity by Aleksei Arkhipovich Leonov (1934–) using a flexible airlock that emphasized the entrepreneurial attitude of the Soviet engineers. By the mid-1960s, the Soviet program was overtaken by the United States program and attempts to launch a manned mission to the Moon failed. However, the early Soviet manned space program has a preeminent place in the history of space physiology.

Human spaceflight was one of the great physiological and engineering triumphs of the twentieth century. The history of the early United States manned space program is well known, beginning with the suborbital flight of Alan Shepard on May 5, 1961, and climaxing with Neil Armstrong's "giant leap for mankind" on the Moon on July 20, 1969. By contrast, the Soviet/Russian manned space program has been shrouded in secrecy until recently. However, much new information has now become available, and the story has many remarkable features. Although much of the information necessarily comes from secondary sources, a recent scholarly book [5] and three excellent television documentaries [11–13] contain interviews with many of the principal people, and a reasonably authoritative account has emerged.

© American Physiological Society 2015
J. B. West, *Essays on the History of Respiratory Physiology,*
Perspectives in Physiology, DOI 10.1007/978-1-4939-2362-5_23

23.1   Tsiolkovsky, An Early Russian Space Visionary

Konstantin Edvardovich Tsiolkovsky (1857–1935) was a remarkable visionary who can be considered the father of human space travel not only by Russia but the rest of the world as well. Tsiolkovsky (Fig. 23.1, *left*) was a young, almost deaf, mathematics teacher in a small Russian provincial town when he sketched a spacecraft design as early as 1883 (Fig. 23.1, *right*). He published his first article on space travel in 1895 [7]. There is a splendid memorial to the Russian space program in Moscow showing the upward-sweeping trajectory of a spacecraft after liftoff, and a statue of Tsiolkovsky has a prominent place. He derived the basic equations of rocket dynamics that relate the speed of rocket flight, jet exhaust velocity, propellant mass, and the mass of the rocket vehicle. He also recognized the importance of the "orbital velocity" of ~7900 m/s, which would allow a spacecraft to orbit the Earth, and also the "escape velocity" of ~11,200 m/s, which is the speed required for a spacecraft to escape the gravitational attraction of the Earth. He also realized that spaceflight would require liquid propellants because of their greater efficiency compared with solid propellants. However, as was the case with many early rocket pioneers, he received little acknowledgement in his lifetime.

Tsiolkovsky's extraordinary vision and imagination are illustrated in his description of the liftoff of an imaginary spacecraft (Ref. [7], p. 99). Recall that this was written at the end of the nineteenth century and beginning of the twentieth century, when the Wright brothers were experimenting with the first aircraft.

 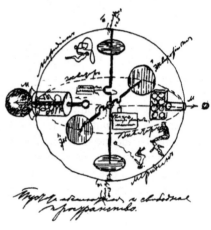

Fig. 23.1 *Left*: Konstantin Edvardovich Tsiolkovsky (1857–1935), an almost deaf mathematics teacher in provincial Russia but a remarkable space visionary (from Tass-Sovfoto). *Right*: sketch of a spacecraft made by Tsiolkovsky in 1883. (From Ref. [14])

The signal is given; the explosion, attended by a deafening noise, starts setting off. The rocket shakes and takes off. We have the sensation of terrible heaviness. My weight has increased tenfold. I am knocked down to the floor, severely injured and perhaps have even been killed—can there be any talk of observations? There are ways of standing up to this terrible weight, but only in a, so to say, compact form or being submerged in a liquid. [Volunteers have now tolerated 30 G during water immersion [4]]

Even when submerged in liquid we will hardly be inclined to observe anything outside. Be all that as it may, the gravity in the rocket has apparently increased tenfold since takeoff. We would be informed of this by a spring balance or a load gauge by the accelerated swinging of a pendulum (some three times faster) [the period of a pendulum is inversely proportional to the square root of G], by the faster fall of bodies, by the diminished size of droplets (their diameter decreasing tenfold), by all things carried aboard the rocket becoming heavier, and many other phenomena.

Tsiolkovsky gave an equally imaginative account of the sensation of weightlessness that he predicted would occur when the spacecraft reached orbit and the engines were turned off (Ref. [7], p. 100).

The awful gravity that we experience will last until the explosion and the noise come to an end. Then, as dead silence sets in, the gravity pull will diminish instantaneously, just as it appeared. We are now out beyond the limits of the atmosphere at an altitude of 575 km. The gravity pull did not only diminish in force but vanished completely without a trace; we no longer even experience the terrestrial gravitation that we are accustomed to just as we accustomed to the air, though it is not at all so necessary as the latter. The altitude of 575 km is very little, it is almost at the surface of the Earth, and the gravity should have diminished ever so slightly. And that actually is the case. But we are dealing with relative phenomena, and for them there is no gravity.

The force of terrestrial gravitation exerts its influence on both the rocket and bodies in it in the same way. For this reason there is no difference in the motion of the rocket and the bodies in it. They are carried along by the same stream, the same force, and, as far as the rocket is concerned, there is no gravity.

There are many things that convince us of this. All objects in the rocket, that were not attached, have left their places and are hanging in the rocket's air, out of contact with anything; and if they touch something, they do not exert any pressure on each other or on the support. We ourselves do not touch the floor and can have any position and be in any direction: we can stand on the floor, on the ceiling or on the wall; we can stand perpendicularly or have an inclined attitude; we float in the middle of the rocket like fish but without any effort whatsoever, and we do not come in contact with anything; no object exerts pressure on any other if they are not pressed together.

Water does not pour from a decanter, a pendulum does not swing and hangs to the side. An enormous mass hung from the hook of a spring balance does not make the spring taut—it always indicates zero. Lever scales are also useless: the balance beam takes up any position, quite irrespective of and indifferent to the equality or inequality of the weights in the pans. Gold cannot be sold by weighing its mass. Conventional ways of measuring mass cannot be employed here.

Subsequent to Tsiolkovsky's theoretical studies, several rocket pioneers were active in the 1920s and 1930s. In 1926, the American Robert Hutchings Goddard (1882–1945) launched the first liquid-propelled rocket at his Aunt Effie's farm in Auburn, Massachusetts, having previously published his classical treatise *A Method of Reaching Extreme Altitudes* in 1919 [3]. However, many of his ideas were ridiculed in the New York Times and other quarters, and he largely retired from the public eye to work on his ideas in relative seclusion. Although he offered his ideas to the military in World War II, there was little interest and he worked on jet-assisted take-off devices for aircraft.

After his death, the importance of his work was acknowledged, a million dollar settlement was made for the use of his patents, and his name is perpetuated in the Goddard Spaceflight Center in Greenbelt, Maryland.

In Germany, the potential value of rockets was more clearly seen, and Herman Julius Oberth (1894–1989) formulated many of the technical problems of spaceflight, but even there many of his ideas were dismissed as fantasies. Later, he moved to Peenemünde to work with Wernher von Braun and eventually came to the United States.

23.2   Wernher von Braun and Peenemünde

Any historical survey of manned spaceflight must include a section on Wernher von Braun (1912–1977) because of the critical developments in rocketry that took place in Peenemünde from 1936 to 1945. Although von Braun (Fig. 23.2) did not directly influence the Soviet space program, remnants of rockets were examined by Soviet engineers, and, as described below, one of the first rockets in the Soviet space program was an identical copy of the main Peenemünde rocket.

Von Braun assisted Oberth in work on liquid-propelled rockets in a German rocket society in 1930; however, a few years later, the work was taken over by

Fig. 23.2 Wernher von Braun (1912–1977) who developed the A4 rocket in Peenemünde, Germany during World War II. This had a great influence on both the Soviet and U.S. space programs. (From Ref. [14])

the army under Gen. Walter Dornberger. The group worked first near Berlin but later moved to Peenemünde in northwest Germany on the Baltic Sea where there was ample space. The most important development was the A4 rocket (Fig. 23.3), the first rocket to demonstrate the potential of suborbital flight, partly because of its enormous thrust and also the development of a sophisticated guidance system. In the television documentary *Spaceflight* [12], there are dramatic interviews with Wernher von Braun and Krafft Ehricke describing the first successful launch of the A4. The first toppled over to be followed by an enormous explosion, the second had

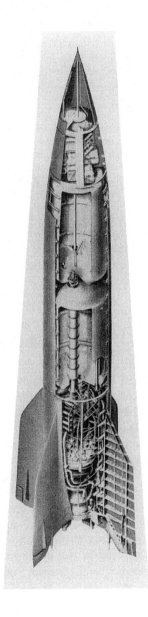

Fig. 23.3 A4 rocket developed by Wernher von Braun and his coworkers. This was the first successful suborbital rocket and played an important role in the development of the Soviet manned space program. (From Ref. [17])

better luck but went out of control, and the third streaked skyward until it was out of sight. Everybody immediately recognized that this was the beginning of a new age when humans would enter space. Ehricke vividly describes the exhilaration of the launch team following the first successful flight.

At the end of World War II, von Braun and several others from the Peenemünde facility were transferred to the United States, where they played crucial roles in the development of the American space program. All that remained of the A4 rockets in Peenemünde was systematically destroyed, but some of the remnants were studied by Soviet rocket experts. Here is a graphic description of what Boris Chertok, one of the engineers, found (Ref. [5], p. 65)

> So I come into this hall. Several hours before me our engine man, Alexei Mikhailovich Isaev—one of the future stars of our rocket technology—was let in. I see the lower part of his body and his legs sticking out of the rocket engine nozzle, while his head is somewhere inside…. I approach Bolkhovitinov.
>
> "What is this?"
>
> "This is what cannot be," he replies …. Understand, one of our most talented aircraft designers simply did not believe that in wartime conditions it would be possible to develop such a huge and powerful rocket engine. We had at the time liquid engines of our experimental rocket planes with thrusts of hundreds of kilograms. One and one half tons was the limit of our dreams. Yet here we quickly calculated, based on the nozzle dimensions, that the engine thrust was at least 20 t.

In fact, the A4 rocket had been redesignated as the V-2 (for Vengeance-2), and many were launched against London with terrifying results. The V-2 had a gross weight of nearly 13 t and could carry a 1-t payload 200 km. It was the first missile to incorporate a cryogenic turbo pump and the first to use a sophisticated guidance system.

23.3   Sergei Korolev, the Principal Architect of the Soviet/ Russian Manned Space Program

Sergei Pavlovich Korolev (1907–1966, Fig. 23.4) was the genius who achieved preeminence of the Soviet/Russian manned space program from its beginning to the mid-1960s. He was born in Zhitomir in Ukraine near Kiev, and, while in his teens, he built and flew gliders. In 1926, he enrolled in the Moscow Technical High School where one of his lecturers was Tupolev, the famous Russian aircraft designer. Korolev became interested in rocketry and in 1933 was responsible for the first flight of a liquid-fuelled rocket in the Soviet Union. During 1936–1938, Korolev and his talented colleague, Valentin Glushko, used rocket engines to propel gliders.

However, in 1938, disaster befell Korolev, and the subsequent story is one that could only have occurred in the Soviet Union. He was arrested by Stalin's regime on a trumped-up charge as an "enemy of the people" and sentenced to 10 years of hard labor. First, he was incarcerated in one of the most dreaded prisons, Kolmya, in far eastern Siberia. He spent 5 months in the winter, digging in a surface gold mine, and

Fig. 23.4 Sergei Pavlovich
Korolev (1907–1966) who
was the Chief Designer of the
early Soviet manned space
program. (From Ref. [14])

many of his fellow prisoners died. Later, Korolev was moved to Moscow where he was held under house arrest in a prison for scientists known as a "sharaga," where he was allowed to do some engineering design work. This was the type of institution described by Aleksandr Solzhenitsyn in *The Gulag Archipelago* [16]. One of his fellow prisoners was Tupolev. Korolev was in prison for a total of about 7 years but was eventually released after World War II when the Soviets recognized the importance of developing a missile program and realized that he was one of their most talented rocket engineers. In 1944 Korolev was discharged from prison and his convictions were expunged, and in 1945 he was commissioned as a colonel in the Red Army. Amazingly, he then continued to work until his death 21 years later at breakneck speed and appeared to harbor no resentment toward the regime. On the contrary, according to a sequence in one of the television documentaries [11], he told Gagarin and Leonov shortly before his death of his complete loyalty and devotion to the Soviet Union.

Initially, Korolev was asked by Stalin to build a replica of the German A4 rocket that had been developed in Peenemünde (Fig. 23.3). Korolev apparently argued that he could come up with a better design, but Stalin was suspicious of technical innovation and wanted to start with something that would certainly work. Korolev complied and produced a replica of the A4 within 2 years. He then went on to mastermind the Soviet rocket and space systems, and his successes were spectacular (Table 23.1). By 1957, he had developed the R-7 booster (Fig. 23.5), which propelled a 5-t dummy warhead 6400 km to Kamchatka, thus making it the world's

Table 23.1 "Firsts" in the soviet space program attributable to Sergei Korolev

1957 (October 4)	First satellite in space: Sputnik 1
1957 (November 3)	First living being in space: dog Laika
1959	First spacecraft to reach another celestial body: Luna 2 lands on the moon
1959	First photographs of the dark side of the moon
1961 (April 12)	First human being in space: Yuri Gagarin
1962–1964	First spacecraft to reach Mars and Venus
1963	First woman in space: Valentina Tereshkova
1964	First 3-man crew in space
1965	First Extravehicular activity: Aleksei Leonov

first intercontinental ballistic missile (ICBM). This was followed by the launch of Sputnik 1, the world's first artificial satellite; the launch of the dog Laika, the first living being in space; and his most spectacular success, the launch of Yuri Gagarin, the first human being in space. Subsequent successes included the first woman in space, Valentina Tereshkova; the first three-person crew in space; and the first extravehicular activity (EVA). This was an amazing catalog of firsts.

A feature of the early Sputnik launches was their great mass (for example, that of Sputnik 3 was 1300 kg or 1.3 t), and this meant that the boosters had an enormous thrust, a realization that caused considerable alarm in the United States at the time. The R-7 booster with its five engines, each of which had four thrust chambers (Fig. 23.5, *right*), generated a thrust of 500 t. By contrast, the Atlas booster, which was the most powerful rocket in the United States at the time, had a thrust of only 200 t. Ironically, the large size of the Soviet boosters was the result of their less sophisticated nuclear technology. The ICBM was developed to propel an H bomb, which at some 5 t was substantially heavier than the comparable weapon developed in the United States.

The military of the Soviet Union placed enormous importance on the development of an ICBM that could carry a nuclear warhead. Vassily Mishin explains in one of the documentaries [13] how the Soviet Union felt tremendously threatened because it was ringed by NATO airbases from which strategic bombers could deliver a nuclear bomb to anywhere in the Soviet Union. By contrast, the Soviet military were unable to reach the United States with a nuclear weapon, and so they saw the development of an ICBM as the only way to establish parity.

An extraordinary feature of Korolev's program was that throughout his career he was never referred to by name because of security reasons but only as the "Chief Designer." In fact, some of the cosmonauts who worked directly under him were apparently not aware of his last name. A dramatic reminder of his anonymity was when Gagarin was welcomed after his historic flight by Khrushchev in an enormous ceremony on Red Square. Korolev was nowhere to be seen because it was thought that his safety could be threatened. Only when he died in 1966 was his identity officially acknowledged, and there was subsequently a tremendous outpouring of affection by the Russian people.

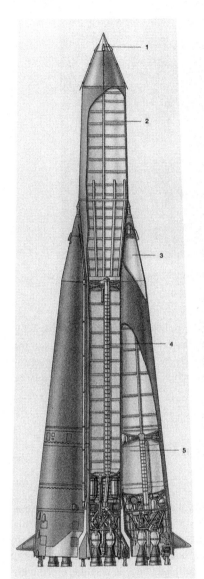

Fig. 23.5 Two views of the R-7 booster designed by Korolev and others that put Sputnik 1 and many other firsts into Earth orbit. This had five clustered engines each with four thrust chambers. Modifications of this design are still used. (From Refs. [1] and [9])

One of Korolev's greatest achievements was the launch of Sputnik 1 on October 4, 1957 (Fig. 23.6). This used an R-7 booster, although the payload was only about 80 kg. Of course, Sputnik 1 created a sensation everywhere, not least in the United States, because it could be seen crossing the sky at dusk, and, by tuning into its frequency on a small radio, anyone could hear the beep-beep. It is interesting that the

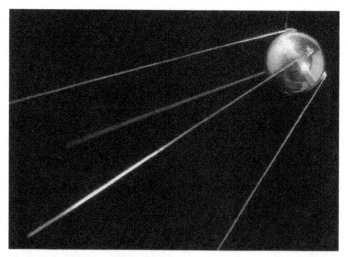

Fig. 23.6 Sputnik 1, the first artificial satellite. Its total weight was just over 80 kg. (From http://www.fht-esslingen.de/telehistory/sputnik.html)

Soviet authorities completely underestimated the political fallout of Sputnik 1 (as related by Sergei Khrushchev) because the October 5 issue of Pravda relegated the launch of Sputnik to a small section on the front page. It was only when the sensation that the event occasioned in the West was recognized that the October 6 issue of Pravda carried the full story.

23.4   Flight of the Dog Laika, the First Living Creature in Space

The events leading up to the launch of the dog Laika provide a graphic demonstration of the differences between the Soviet and United States space programs at this stage. As can be imagined, there were many delays before the successful launch of Sputnik 1, and, according to Vassily Mishin, one of the key Soviet rocketeers, the living conditions at the launch site were appallingly bad. It was therefore with tremendous relief that everybody connected with Sputnik 1 looked forward to a period with their families in a more comfortable environment after the successful launch. However, this was not to be. To the chagrin of the launch team, they were immediately ordered back to work. Cosmonaut Grechko relates the events as follows (Ref. [5], p. 132)

> I heard this from Korolev himself with my own ears. After *Sputnik I* Korolev went to the Kremlin and Khrushchev said to him, "We never thought that you would launch a sputnik before the Americans. But you did it. Now please launch something new in space for the next anniversary of our [October] revolution."

The anniversary would be in one month! I'll bet that even with today's computers nobody would launch something into space in one month. It was, I think, the happiest month of his [Korolev's] life. He told his staff, and his workers, that there would be no special drawings, no quality check, everyone would have to be guided by his own conscience. And we launched on November 3, 1957, in time for the celebration of the Revolution [because of a calendar change, the October revolution is now celebrated on November 6].

The payload of Sputnik 2 included the mongrel dog Laika, who was placed in an environmentally controlled chamber immediately below a replica of the Sputnik 1 sphere (Fig. 23.7). The physiological variables that were monitored and telemetered back to Earth included electrocardiogram (chest lead), blood pressure, respiration rate, and motor activity [15]. Food was automatically delivered twice a day. The heart rate was 103 beats/min before launch and increased to 240 beats/min during the early acceleration. However, after 3 h of weightlessness, it was back to 102 beats/min. Because the orbit of the spacecraft was very elliptical, the solar irradiation was higher than planned and the environmental chamber overheated. Apparently, no engineering work had been done on the payload of Sputnik 2 until after Sputnik 1 went up a month before. It is almost unbelievable that such a complicated payload could have been put together in such a short time, and the triumph emphasizes the resourcefulness and the fly-by-the-seat-of-the-pants attitude of the Soviet

Fig. 23.7 The payload of Sputnik 2, which included the dog Laika, the first living being in space. The compartment for the dog is below a replica of Sputnik 1. (From Ref. [1])

designers in contrast to the much more cautious attitude of NASA, which came into being about 10 months after the launch of Sputnik 1.

The flight of Laika was followed by the launch of Sputnik 3, which contained a large load of scientific instruments to measure upper atmosphere phenomena. Actually, Sputnik 3 required two attempts because the first launch on April 27, 1958, failed as a result of problems with a rocket engine. The successful launch took place on May 15, 1958, and, although its payload was devoted to atmospheric physics, there was an interesting sidelight that is relevant to manned spaceflight. Part of its instrumentation was designed to map the radiation belts in the atmosphere, but the experiment was unsuccessful because of the failure of a tape recorder. In fact, Sputnik 3 detected high levels of radiation, but it was not possible to say whether these were local or distributed in a belt around the Earth. It was left to the United States satellite, Explorer 1, with only one-sixth the payload weight of Sputnik 2 to discover the Van Allen radiation belt as it is now called. This intense region of radiation is of considerable importance to astronauts and cosmonauts: it is avoided so that radiation doses can be limited to acceptable levels and it partly determines the altitude of orbiting manned spacecraft.

23.5   Flight of Yuri Gagarin, the First Human Being in Space

The large amount of information now available about Sergei Korolev makes it clear that, from the outset, his main objective was to launch a human being into space and, if possible, have him or her reach the Moon or even Mars. Korolev was a remarkable visionary! However, in practice, the Soviet space program was dominated by the military who recognized its importance for delivering nuclear weapons. Sergei Khrushchev (son of Nikita Khrushchev) tells the story that Korolev met Premier Khrushchev and told him that he wanted to launch the first artificial satellite (Sputnik 1). "What's that?" asked the Premier, and "will it interfere with the military program?" On hearing Korolev say that it would not, he gave the assent. However, the documentaries make it clear that there was considerable tension between Korolev and the military when he successfully pursued his manned space program.

Actually, it was obvious to outsiders that when Korolev launched Laika and then Sputnik 3 with its 1.3-t payload, he was leading up to the launching of a human being. Alexei Leonov in one of the documentaries describes how Korolev met with the cosmonauts in training and chose Gagarin (Fig. 23.8) apparently more because of his personality rather than objective data. All of the 20 or so cosmonauts in training were pilots of fighter aircraft, as was the case with the 7 astronauts who made up the U.S. Mercury team.

Yuri Alekseyevich Gagarin (1934–1968) was launched into low Earth orbit by means of an R-7 booster on April 12, 1961. According to the documentaries, some people in the launch team had serious misgivings about the proposed flight because there had been a number of rocket failures. For example, there was an explosion

Fig. 23.8 Yuri Alekseyevich
Gagarin (1934–1968), who
was the first human being to
enter space

of a ballistic missile at the launch site only 6 months before that killed 165 people, including a number of leaders in the space program. Before Gagarin's flight, much thought had been given to the effects of spaceflight on the human body and extensive research on the effects of weightlessness on various animals had been carried out by the Institute for Biomedical Problems in Moscow under the directorship of Oleg Georgievich Gazenko (1918–2007) using free-fall from high-altitude balloons. This institute was responsible for various aspects of human environmental physiology, including space, high altitude, and diving [2]. In fact, there was some crossover between these disciplines in part because of the Russian belief in cross-adaptation. The thrust of this was that adaptation to one environmental extreme improved the ability of the body to withstand another environmental stress. For example, it was believed that acclimatization to high altitude improved human tolerance to very high accelerations. The Soviets were just as macho as anybody else in this area; in one of the television documentaries, Alexei Leonov proudly states that he has tolerated 14 G. Of course Gagarin's flight produced a sensation throughout the world, but many of the details were unknown until recently. As an example, it was assumed for a long time that Gagarin had landed in his space capsule, and one of the television documentaries implies this. However, it is now known that Gagarin ejected from the spacecraft at an altitude of 7000 m and came down by parachute.

Excerpts from Gagarin's personal account of his flight are now available (Ref. [5], pp. 170–175), and they make fascinating reading. The acceleration during the launch exceeded 5 Gx (eyeballs in) but did not prevent Gagarin from communicating with the ground. Once in orbit he described the appearance of the Earth and the unexpected complete blackness of the sky as has now been done many times since. He had no problem with eating or drinking. During the flight (duration of 1 h and 48 min), a number of physiological variables were monitored, including electrocardiogram and chest movements by pneumography. A television camera also recorded the cosmonaut's activities.

The main physiological variables that were monitored were the electrocardiogram and respiration rate [17]. Four hours before launch, the heart rate was 65 beats/min and the respiration rate was 12 breaths/min. Interestingly, 5 min before launch, the heart rate had risen to 108 beats/min and the respiration rate to 25 breaths/min, presumably because of anxiety. One minute after the start of the launch, the heart rate exceeded 150 beats/min. At this time, Gagarin was exposed to increased acceleration forces, vibration, and noise from the rocket engines, but he reported that these sensations were easily tolerated. As the spacecraft ascended and then entered orbit, the heart rate steadily fell to about 100 beats/min. After 20 min of weightlessness, the rate had fallen slightly farther with a mean of 97 and range of 85–113 beats/min. However, the respiration rate tended to remain high so that during the period of 10–15 min after the beginning of weightlessness it ranged between 24 and 37 breaths/min. During descent after the braking rockets were fired, the pulse rate rose to 112 beats/min and the respiration rate was in the range of 25–30 breaths/min.

However, the reentry and landing did not go according to plan and here are excerpts from Gagarin's account (Ref. [5], pp. 173–174)

> At precisely the appointed time the third [reentry] command was issued…. I felt the braking rocket kick in…. The braking rocket operated for exactly 40 s…. As soon as [it] shut off there was a sharp jolt, and the craft began to rotate around its axis at a very high velocity…. about 30 degrees per second at least…. I waited for the separation. There wasn't any. I knew that, according to plan, that was to occur 10–12 s after the braking rocket switched on. I decided that something was wrong…. I estimated that all the same I would land normally…. The Soviet Union was 8000 km long, which meant I would land somewhere in the far east…. I reasoned that it was not an emergency situation. I transmitted the all-normal signal with a key.

> The craft's rotation was beginning to slow, but it was about all 3 axes…. Suddenly, a bright crimson light appeared along the edges of the shade…. I felt the oscillations of the craft and the burning of the coating…. It was audibly crackling. Then the g-load began to steadily increase. It felt as if the g-load was 10 g. There was a moment for about 2–3 s when the indicators on the instruments began to become fuzzy. Everything seemed to go gray.

> I'm awaiting the ejection…. At an altitude of approximately 7000 m hatch number 1 was shot off…. I'm sitting there thinking, that wasn't me that was ejected, was it? Then I calmly turned my head upward, and at that moment the firing occurred and I was ejected…. without a hitch. I didn't hit anything…. I flew out in the seat.

> I immediately saw a large river. I thought that's the Volga…. When I was parachute training, we had jumped many times over this very site. I see that I am going to land in a plowed field…. The landing was very soft…. I was alive and well.

> I went up on a knoll and saw a woman and a little girl coming toward me…. I saw the woman slow her pace, and the little girl broke away…. I began to wave my arms and yell "I am one of yours, a Soviet, don't be afraid, don't be scared, come here…." I went up to her and said I was a Soviet and that I had come from space.

There is no doubt that Yuri Alekseyevich Gagarin had the right stuff!

23.6 Alexei Leonov Performs the First Extravehicular Activity

Aleksei Arkhipovich Leonov (1934–) comes over in the television documentaries, especially in *Russian Right Stuff* [11], as one of the most engaging personalities in the early Soviet manned space program. He recounts how Korolev told him in 1962, "Any sailor on a ship has to know how to swim, and so each cosmonaut has to know how to swim and do construction work outside his vehicle in space. Now orlyonok [eaglet], put on a space suit and go through the procedures for the engineers." In fact, Leonov became the first human being to perform EVA or space walking on March 18, 1965; again, this is a remarkable example of Soviet seat-of-the-pants resourcefulness.

The spacecraft called Voskhod that was used was a larger version of Vostok in which Gagarin had been launched 4 years earlier. For the second flight of Voskhod, it was fitted with a collapsible cylindrical airlock about 1 m in diameter and 2.5 m long with hatches at both ends. The airlock was stove-piped into the side of the spacecraft and constructed of a double-thickness rubber-like material covered with protective fabric. It was folded down like an accordion in the shroud of the space-craft during liftoff and was subsequently expanded. This arrangement allowed the cosmonaut to enter the airlock, close a hatch in the spacecraft wall, decompress to the hard vacuum of space, and then open another hatch and emerge at the other end. The pressure suit itself was developed in only 9 months, and it was necessary for Leonov to prebreathe oxygen for some time to washout the body nitrogen and so avoid decompression sickness (bends). The atmosphere of all the Soviet spacecraft was air at 760 Torr pressure. At the end of the EVA period when the cosmonaut was safely back in the capsule, the airlock was unfastened and allowed to drift away. It is difficult to imagine relying on such a makeshift device, but it worked. There is an unforgettable television sequence showing Leonov pushing off from the airlock and floating out into space waving his right hand while tethered by an umbilical (Fig. 23.9). His account in his own words is dramatic [8].

Fig. 23.9 Aleksei Arkhipovich Leonov (1934–) performing the first extravehicular activity or space walk. (From Ref. [10])

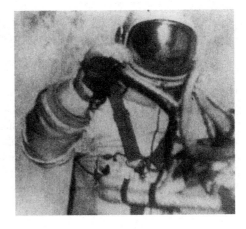

The photographs of Leonov floating alone in space provided another sensation throughout the world. However, what was not known at the time was that a crisis developed when he tried to reenter the airlock. As he describes in the documentary, his spacesuit had expanded so much that it took him 12 min of struggle to get back into the airlock so that he could close the outer hatch. His oxygen supply was nearly exhausted; only by depressurizing the suit to some extent with a hand-operated valve was he able to reduce its volume and eventually make his way back into the safety of the capsule itself. The stress that Leonov was subjected to can be realized by the respiration rate and heart rate, which were both monitored. The respiration rate reached 26–30 breaths/min (vs. 10–15 under ordinary conditions on Earth), and the heart rate was 152–162 beats/min [6]. Again, we have to marvel at the bravery and enterprise of these early cosmonauts.

Some people at the time thought that the EVA was something of a circus act with little practical value in space exploration. Of course, we now know that EVA plays a critical role in maintaining a space station such as Mir; in addition, the construction of the International Space Station, which is in progress at the present time, relies heavily on EVA. In fact, there will be more EVAs in the next 4 years than in all previous years of the manned space program, although they carry an appreciable physiological risk, for example, from decompression sickness.

The resourcefulness of the Soviet space engineers in developing the collapsible airlock for the first EVA is further highlighted by a particularly colorful anecdote told to Harford by Arkady Ostashov, an engineer (Ref. [5], p. 118)

In 1961 we had the first test of the R-9 ICBM. The test pad had two parts—a movable part attached to the missile, and a fixed part on the pad itself. The two parts were mated on the pad. When the engine was ready to start we detected a small leak of liquid oxygen between the two parts. You could see a small cloud of oxygen vapor. Voskresensky said, "Let's go to the rocket."

He and I and another guy approached to make sure that the leak was small. Fortunately nobody was watching. Voskresensky pissed on the leaky joint, the liquid froze, and the joint held until ignition.

Perhaps this anecdote is too good to be true.

23.7 Salyut and Mir, the First Permanently Manned Space Station

The Soviet/Russian manned space program continued its successes with the launch of Salyut 1 in April 1971, which became the world's first space station. Even more impressive was the launch of the first components of the Mir space station in February 1986. This was the first permanently inhabited space station and remained in orbit for some 15 years, far beyond its design lifetime of about 5 years. However, these projects will not be considered in detail here because they do not belong to the early Soviet/Russian manned space program. In fact, by the mid 1960s, the Soviet program was running into difficulties for various reasons. Korolev died in 1966, and, according to the documentaries, the loss of this charismatic leader resulted in political infighting within the Soviet space program. The Soviet program to put a human on the Moon, which was strongly supported by Korolev, foundered in part because of the enormous cost of developing a suitable booster. This was the N-1 rocket, which was approximately the same size as the Saturn V that allowed the American astronauts to reach the Moon, but its construction was plagued with difficulties, and at least one blew up on the launch pad. Although the Soviets marshaled enormous resources toward their Moon program, they could not compete with the great range of private aerospace companies in the United States. Nonetheless, the extraordinary enterprise of Korolev, in particular, and many others in the early Soviet manned space program will never be forgotten.

Appendix

Sources of Material

Of course, a historical account like this must be derivative. However, the three television documentaries contain interviews with many of the principal people concerned, including Wernher von Braun, Krafft Ehricke, Lt. General Victor Favorsky, Konstantin Feoktistov, Lt. General Kevim Kerimov, Sergei Khrushchev, Natalya Korolev (daughter of Sergei Korolev), Sergei Kryukov, Alexei Leonov, and Vassily Mishin. The three documentaries are (1) *Red Files: Race to the Moon*, including interviews with Leonov, Mishin, Khrushchev, Feoktistov, Natalya Korolev, etc. [13]; there is a Web page with additional information at http://www.pbs.org/redfiles/; (2) *Russian Right Stuff* in the NOVA series [11] introduced by Alexei Leonov; and (3) *Spaceflight* by PBS [12], including interviews with Wernher von Braun and Krafft Ehricke.

In addition, the *Korolev* book by James Harford [5] is well-researched and contains many verbatim quotations from some of the most important Soviet people involved.

References

1. Clark P. The Soviet manned space program: an illustrated history of the men, the missions, and the spacecraft. New York: Orion Books; 1988.
2. Gazenko OG. Milestones of space medicine development in Russia (establishment and evolution of the Institute of Biomedical Problems). J Grav Physiol. 1997;43:1–4.
3. Goddard RH. A method of reaching extreme altitudes. Washington, DC: Smithsonian Institution; 1919.
4. Gray RF, Webb MG. High G protection. Aerospace Med. 1961;32:425–430.

5. Harford J. Korolev: how one man masterminded the soviet drive to beat America to the moon. New York: Wiley; 1997.
6. Kas'yan II, Makarov GF. External respiration, gas exchange and energy expenditures of man in weightlessness [in Russian]. Kosm Biol Aviakosm Med. 1984;18:4–9.
7. Kosmodemiansky A. Konstantin Tsiolkovsky, 1857–1935. Moscow: General Editorial Board for Foreign Publications, Nauka; 1985.
8. Leonov A. My first steps in space. UNESCO Courier. 1965;18:4–11.
9. McHenry R. editor. Exploration. In: The new encyclopedia Britannica. Chicago: Encyclopedia Britannica; 1993. pp. 44–58.
10. Nicogossian AE, Pool SL, Uri JJ. Historical perspectives. In: Nicogossian AE, Huntoon CL, Pool SL, editors. Space physiology and medicine. Philadelphia: Lea & Febiger; 1994. pp. 3–49.
11. NOVA. Russian Right Stuff. Alexandria, VA: PBS, 1991, videocassette.
12. PBS. Spaceflight. Alexandria, VA: PBS, 1985, videocassette.
13. PBS. Red Files: Secrets of the Russian Archives Revealed. Part 3: Secret Soviet Moon Missions. Alexandria, VA: PBS, 1999, videocassette.
14. Rauschenbach BV, Sokolskiy VN, Gurjian AA. Historical aspects of space exploration. In: Nicogossian AE, Mohler SR, Gazenko OG, Grigoryev AI, editors. Space biology and medicine. Vol. 1. Washington, DC: American Institute of Aeronautics and Astronautics; 1993. pp. 1–50.
15. Sisakian NM editor. Problems of space biology. Digest of the papers. Moscow: URSS Academy of Sciences; 1962. pp. 359–70.
16. Solzhenitsyn AI. The Gulag Archipelago, 1918–1956 (translated from the Russian by Whitney TP). New York: Harper & Row; 1974.
17. Volynkin YM, Yazdovskiy VI, et al. First manned space flights. Wright-Patterson Air Force Base, OH: Foreign Technology Division, Air Force Systems Command, FTD-TT-62–16196, Translation of Pervyye kosmicheskiye polety chelovek, Moskva, Mediko-biologicheskiye Issledovaniya; 1962.
18. von Braun W. Multi-stage rockets and artificial satellites. In: Marbarger JP, editor. Space medicine: the human factor in flights beyond the earth. Urbana: University of Illinois; 1951. pp. 14–30.

CPSIA information can be obtained at www.ICGtesting.com
Printed in the USA
LVOW02*1121310515

440588LV00001B/65/P